Carbon Nanotube–Polymer Composites

RSC Nanoscience & Nanotechnology

Series Editors:
Professor Paul O'Brien, *University of Manchester, UK*
Professor Sir Harry Kroto FRS, *University of Sussex, UK*
Professor Ralph Nuzzo, *University of Illinois at Urbana-Champaign, USA*

Titles in the Series:
1: Nanotubes and Nanowires
2: Fullerenes: Principles and Applications
3: Nanocharacterisation
4: Atom Resolved Surface Reactions: Nanocatalysis
5: Biomimetic Nanoceramics in Clinical Use: From Materials to Applications
6: Nanofluidics: Nanoscience and Nanotechnology
7: Bionanodesign: Following Nature's Touch
8: Nano-Society: Pushing the Boundaries of Technology
9: Polymer-based Nanostructures: Medical Applications
10: Metallic and Molecular Interactions in Nanometer Layers, Pores and Particles: New Findings at the Yoctolitre Level
11: Nanocasting: A Versatile Strategy for Creating Nanostructured Porous Materials
12: Titanate and Titania Nanotubes: Synthesis, Properties and Applications
13: Raman Spectroscopy, Fullerenes and Nanotechnology
14: Nanotechnologies in Food
15: Unravelling Single Cell Genomics: Micro and Nanotools
16: Polymer Nanocomposites by Emulsion and Suspension
17: Phage Nanobiotechnology
18: Nanotubes and Nanowires: 2nd Edition
19: Nanostructured Catalysts: Transition Metal Oxides
20: Fullerenes: Principles and Applications, 2nd Edition
21: Biological Interactions with Surface Charge Biomaterials
22: Nanoporous Gold: From an Ancient Technology to a High-Tech Material
23: Nanoparticles in Anti-Microbial Materials: Use and Characterisation
24: Manipulation of Nanoscale Materials: An Introduction to Nanoarchitectonics
25: Towards Efficient Designing of Safe Nanomaterials: Innovative Merge of Computational Approaches and Experimental Techniques
26: Polymer–Graphene Nanocomposites
27: Carbon Nanotube–Polymer Composites

How to obtain future titles on publication:
A standing order plan is available for this series. A standing order will bring delivery of each new volume immediately on publication.

For further information please contact:
Book Sales Department, Royal Society of Chemistry, Thomas Graham House, Science Park, Milton Road, Cambridge, CB4 0WF, UK
Telephone: +44 (0)1223 420066, Fax: +44 (0)1223 420247
Email: booksales@rsc.org
Visit our website at www.rsc.org/books

Carbon Nanotube–Polymer Composites

Edited by

Dimitrios Tasis
Department of Materials Science, University of Patras, Greece
Email: dtassis@upatras.gr

RSCPublishing

RSC Nanoscience & Nanotechnology No. 27

ISBN: 978-1-84973-568-1
ISSN: 1757-7136

A catalogue record for this book is available from the British Library

© The Royal Society of Chemistry 2013

All rights reserved

Apart from fair dealing for the purposes of research for non-commercial purposes or for private study, criticism or review, as permitted under the Copyright, Designs and Patents Act 1988 and the Copyright and Related Rights Regulations 2003, this publication may not be reproduced, stored or transmitted, in any form or by any means, without the prior permission in writing of The Royal Society of Chemistry or the copyright owner, or in the case of reproduction in accordance with the terms of licences issued by the Copyright Licensing Agency in the UK, or in accordance with the terms of the licences issued by the appropriate Reproduction Rights Organization outside the UK. Enquiries concerning reproduction outside the terms stated here should be sent to The Royal Society of Chemistry at the address printed on this page.

The RSC is not responsible for individual opinions expressed in this work.

Published by The Royal Society of Chemistry,
Thomas Graham House, Science Park, Milton Road,
Cambridge CB4 0WF, UK

Registered Charity Number 207890

For further information see our web site at www.rsc.org

Printed in the United Kingdom by Henry Ling Limited, Dorchester, DT1 1HD, UK

Preface

The combination of carbon nanotube special properties (structural and conductive) makes them the ideal filler material in various polymer matrices for a wide range of applications. With the aim of mimicking the conditions existing in metal rod-reinforced concrete, Ajayan and co-workers reported the first ever fabrication of a carbon nanotube/polymer composite in 1994. Since then, there have been some thousands of works published, revealing the potential of polymers reinforced with one-dimensional graphitic nanostructures. In the early years of related research, as-prepared carbon nanotube material was actually a mixture of different metallicities, diameters and lengths, not to mention the presence of impurities and sidewall defects. Furthermore, self-aggregation phenomena due to van der Waals interactions and a lack of chemical affinity between the filler and the matrix have been found to hamper the homogeneous distribution of the graphitic nanostructures within a polymer matrix. Finally, due to their relatively short length and the presence of sidewall defects, carbon nanotubes are normally curled, and therefore individual tubes embedded in a polymer only exhibit a fraction of their potential. Thus, the superb properties of the graphitic nanomaterial have been partially translated into high stiffness and conductivity polymer composites. Only in recent years have been great advances towards the development of multifunctional carbon nanotube/polymer assemblies.

The purpose of this book is to summarize the basic principles of carbon nanotube chemistry in relation to the fabrication of polymer composites, but also to highlight some of the most remarkable advances that have occurred in the topic during the last recent years. Indeed, the rapid advances in the chemical functionalization of carbon nanotubes have paved the way towards the fabrication of hybrid polymer-based assemblies with enhanced potential in a wide range of applications. Recent studies have shown that such carbon nanotube/polymer composites exhibit interesting activities, as supercapacitors, battery electrodes, organic light-emitting diodes, photovoltaic cells, actuators and infrared sensors, as well as conductive coatings. All these aspects of carbon nanotube/polymer composite science are summarized in the different chapters collected in this book. Two additional chapters are devoted to a related and very important field, namely the development of characterization tools for carbon nanotube/polymer samples.

The different chapters of the present book have been delivered by prominent experts and I would like to sincerely thank all of them for their effort and their enthusiasm to participate to this journey. My gratitude is extended to the Royal Society of Chemistry (RSC) for its high standard of support in this adventure and to all of the RSC members who contributed to the process regarding the efficient handling of all the chapters. Especially, I would like to acknowledge the help of Mrs Rosalind Searle and Dr Merlin Fox during the whole editing process. Further, I would like to thank the RSC Design team for the design of the front cover image.

Contents

Chapter 1	**Conducting Polymer-based Carbon Nanotube Composites: Preparation and Applications** *Sang-Ha Hwang, Jeong-Min Seo, In-Yup Jeon, Young-Bin Park and Jong-Beom Baek*	**1**
	1.1 Discovery of Conducting Polymers	1
	1.2 Synthesis of Conducting Polymers	2
	1.3 Conductivity and Doping of Conducting Polymers	3
	1.4 Conducting Polymers as Carbon Nanotube (CNT) Composite Matrices	6
	1.5 Applications of CNT/Conducting Polymer Composites	9
	1.5.1 Supercapacitors	9
	1.5.2 Rechargeable Lithium-ion Battery Electrodes	12
	1.5.3 Photovoltaic Devices	14
	1.5.4 Organic Light-emitting Diodes (OLEDs)	16
	1.6 Conclusions	17
	Acknowledgements	17
	References	18
Chapter 2	**Actuators and Infrared Sensors Based on Carbon Nanotube–Polymer Composites** *Jian Chen*	**22**
	2.1 Introduction	22
	2.2 Shape-Memory CNT–Polymer Composites	23
	2.2.1 IR Heating of CNT–SMP composites	25
	2.2.2 Inductive Heating of CNT–SMP composites	28
	2.2.3 Resistive Heating of CNT–SMP composites	28

RSC Nanoscience & Nanotechnology No. 27
Carbon Nanotube-Polymer Composites
Edited by Dimitrios Tasis
© The Royal Society of Chemistry 2013
Published by the Royal Society of Chemistry, www.rsc.org

	2.3	Shape-Changing CNT–Polymer Composites	31
		2.3.1 Light-driven Shape-changing CNT–Polymer Composites	32
		2.3.2 Electroactive Shape-changing CNT–Polymer Composites	36
	2.4	CNT–Polymer Composite IR Sensors	42
	2.5	Conclusions	44
	Acknowledgements		45
	References		45

Chapter 3 Photoelectrical Responses of Carbon Nanotube–Polymer Composites 51
Yumeng Shi and Lain-Jong Li

3.1	Introduction	51
3.2	Band Structure and Chirality Dependence	52
3.3	Band-to-band Transition of SWNTs	53
3.4	Wrapping SWNTs with Polymers	56
3.5	Energy Transfer from Photosensitive Polymers to SWNTs	58
3.6	Photoelectric Responses from the SWNTs Coated with Photosensitive Polymers	58
	3.6.1 SWNT Optoelectronic Devices Based on Photosensitive Polymers	59
	3.6.2 Electrostatic Force Microscopy (EFM) Measurement of SWNT–Polymer	63
	3.6.3 The Ability for Hole and Electron Discrimination in SWNTs	64
3.7	Conclusions	69
References		69

Chapter 4 Chemical Functionalisation of Carbon Nanotubes for Polymer Reinforcement 72
Yurii K. Gun'ko

4.1	Introduction	72
4.2	Non-covalent Functionalisation of CNTs	73
4.3	Covalent Functionalisation	77
	4.3.1 "Grafting From" Approach	79
	4.3.2 "Grafting To" Approach	81
4.4	Combination of Non-covalent and Covalent Approaches	83
4.5	Main Techniques for Fabrication of CNT–Polymer Composites	85
	4.5.1 Solution Processing of Composites	86

		4.5.2	Melt Processing	86

       4.5.2 Melt Processing   86
       4.5.3 *In Situ* Polymerisation Processing   87
       4.5.4 Processing of Composites Based on Thermosets   89
       4.5.5 Coagulation Spinning and Electrospinning   90
       4.5.6 Buckypaper-based Approaches   91
       4.5.7 Layer-by-layer (LBL) Technique   92
       4.5.8 Swelling Under Ultrasound Technique   93
 4.6   Influence of Nanotube Functionalisation on
       Mechanical Properties of CNT–Polymer Composites   93
 4.7   Role of Fabrication and Processing Techniques in
       Reinforcement of Polymers by CNTs   100
       4.7.1 Mechanical Properties of Solution-processed
              Composites   100
       4.7.2 Mechanical Properties of Melt-processed
              Composites   101
       4.7.3 Mechanical Properties of Composites Based on
              Thermosetting Polymers   103
       4.7.4 Mechanical Properties of Composites Prepared
              by *In Situ* Polymerisation   104
       4.7.5 Mechanical Properties of Composites Fibres
              Prepared by Spinning   105
       4.7.6 Mechanical Properties of Composites Prepared
              Using Buckypaper   106
       4.7.7 Mechanical Properties of Composites Prepared
              Using LBL Approach   107
 4.8   Conclusions and Future Outlook   108
 References   110

Chapter 5 Polymer-grafted Carbon Nanotubes *via* "Grafting From"
           Approach   120
           Chao Gao, Zheng Liu, Liang Kou and Xiaoli Zhao

 5.1   Linear Polymer-functionalized Carbon Nanotubes
       (CNTs)   120
       5.1.1 Atom Transfer Radical Polymerization (ATRP)
              Approach to Polymer-grafted CNTs   120
       5.1.2 Reversible-addition Fragmentation Chain-
              transfer (RAFT) Polymerization Approach to
              Polymer-grafted CNTs   142
       5.1.3 Nitroxide-mediated Radical Polymerization
              (NMRP) Approach to Polymer-grafted CNTs   146
       5.1.4 Ring-opening Polymerization (ROP) Approach
              to Polymer-grafted CNTs   148
       5.1.5 Ring-opening Metathesis Polymerization
              (ROMP) Approach to Polymer-grafted CNTs   155

	5.1.6	Anionic Polymerization Approach to Polymer-grafted CNTs	157
	5.1.7	Other "Grafting From" Methods to Polymer-grafted CNTs	158
	5.1.8	Binary-grafting Approach to Polymer-grafted CNTs	160
5.2	Dendritic Polymer-functionalized CNTs		162
	5.2.1	Self-condensing Vinyl (Co)Polymerization (SCVP/SCVCP)	165
	5.2.2	ROP Approach	166
	5.2.3	Polycondensation Approach	169
5.3	Concluding Remarks		173
Acknowledgements			173
References			173

Chapter 6 Metallic Single-walled Carbon Nanotubes for Electrically Conductive Materials and Devices 182
Ankoma Anderson, Fushen Lu, Mohammed J. Meziani and Ya-Ping Sun

6.1	Introduction	182
6.2	Harvesting Metallic SWNTs	184
6.3	Electrically Conductive Nanocomposites	191
	6.3.1 Composites with Non-enriched SWNTs	191
	6.3.2 Composites with Separated Metallic SWNTs	195
6.4	Transparent Conductive Coatings and Films	197
6.5	Perspective	203
Acknowledgements		204
References		204

Chapter 7 Characterization of Dispersability of Industrial Nanotube Materials and their Length Distribution Before and After Melt Processing 212
B. Krause, M. Mende, G. Petzold, R. Boldt and P. Pötschke

7.1	Introduction	212
7.2	Experimental	213
	7.2.1 Materials	213
	7.2.2 Centrifugal Separation Analysis (CSA)	214
	7.2.3 Melt Processing	215
	7.2.4 Morphological Characterization	216
7.3	Results and Discussion	216
	7.3.1 Characterization of the Dispersability of CNT Materials	216

	7.3.2	Length Analysis of CNTs Before and After Melt Processing	221

        7.3.2 Length Analysis of CNTs Before and After Melt Processing   221
        7.3.3 Relation Between SME, CNT Dispersion, CNT Length, and Electrical Properties of Melt Mixed Composites   224
  7.4 Summary and Conclusion   230
Acknowledgements   231
References   232

Chapter 8 Methods for Improving the Integration of Functionalized Carbon Nanotubes in Polymers   234
L. Valentini, D. Puglia and J. M. Kenny

  8.1 Introduction   234
  8.2 *In situ* Polymerization Methods   235
        8.2.1 Poly(methyl methacrylate) (PMMA)-based Nanocomposites   235
        8.2.2 Hybrid Conducting Polymers   237
  8.3 Melt Blending and Solvent Dispersion   238
        8.3.1 Thermoplastic Polymers   238
        8.3.2 Confinement of CNTs in Block Copolymer Matrix   239
        8.3.3 Solvent Dispersion   241
  8.4 Chemical and Physical Methods of CNT Dispersion   242
        8.4.1 Epoxy Nanocomposites   242
        8.4.2 Assembly of CNTs   245
  8.5 Conclusions   247
References   248

Chapter 9 Raman Spectroscopy of Carbon Nanotube–Polymer Hybrid Materials   253
Konstantinos Papagelis

  9.1 Introduction   253
  9.2 Chemical Modification of CNTs with Polymers   254
  9.3 Background of Raman Spectroscopy of CNTs   255
        9.3.1 Electronic Structure of CNTs   255
        9.3.2 Raman Spectrum of CNTs   257
  9.4 Raman Characterization of CNT–Polymer Hybrid Materials   260
  9.5 Conclusions   266
Acknowledgements   267
References   267

Subject Index   270

CHAPTER 1

Conducting Polymer-based Carbon Nanotube Composites: Preparation and Applications

SANG-HA HWANG[a], JEONG-MIN SEO[b], IN-YUP JEON[b], YOUNG-BIN PARK*[a] AND JONG-BEOM BAEK*[b]

[a] Ulsan National Institute of Science and Technology, School of Mechanical and Advanced Materials Engineering/Low Dimensional Carbon Materials Center, 100, Banyeon, Ulsan 689-798, South Korea; [b] Ulsan National Institute of Science and Technology, Interdisciplinary School of Green Energy/Low Dimensional Carbon Materials Center, 100, Banyeon, Ulsan 689-798, South Korea
*E-mail: jbbaek@unist.ac.kr or ypark@unist.ac.kr

1.1 Discovery of Conducting Polymers

Polymers have traditionally been considered to be good electrical insulators, and a variety of applications have relied on this insulating property. However, in 1958, Natta *et al.*[1] succeeded in synthesizing polyacetylene (PA), a semi-conducting conjugated polymer, which paved the way for the upsurge in conjugate polymer research that followed in the decades to come. Alan Heeger, Alan G. MacDiarmid and Hideki Shirakawa made the revolutionary discovery that plastic can be electrically conducting in the 1970s.[2] As a result, they were jointly awarded the Nobel Prize in Chemistry in 2000. For a polymer to be electrically conducting, it must "imitate" a metal—that is, the electrons must be freely mobile and not bound to the atoms. One way to achieve this, is to have the polymer backbone consisting of alternating single and double bonds, called "conjugated double bonds", between carbon atoms. It must also be "doped", which means that electrons are removed (through oxidation) or introduced (through reduction). The resulting holes or extra electrons can move along the macromolecule, which would, in turn, make the polymer electrically conducting.[3,4]

The chemical origins of such a remarkable difference in the material properties between various types of polymers can be readily rationalized. Traditional polymers, such as polyethylene or polypropylene, are made up of essentially σ-bonds; hence, a charge once created on any given atom on the polymer chain is not mobile (static charge). The presence of an extended π-conjugation in polymers, however, confers the required mobility to charges that are created on the polymer backbone (by the process of doping) and makes them electrically conducting. One problem is that, due to the presence of this extended conjugation along the polymer backbone, the chains are rigid and possess strong inter-chain interactions, resulting in insoluble and infusible materials.[5,6] These conjugated polymers, hence, lack one of the most important and useful properties of polymers, namely the ease of processability. More recently, however, it was demonstrated that when lateral substituents were introduced even conjugated polymers can be made soluble (hence, processable) without significant decrease in their electrical conductivity.[7,8] Another problem in technological application is the inherent chemical instability, especially, in the doped form, to ambient conditions.[9] Today, conducting polymers that are stable even in the doped form have been developed. We shall highlight some specific examples of such systems and discuss some of their research progress and potential applications.[10–12]

PA, in view of possessing the simplest molecular framework, has attracted the most attention, especially of physicists, with an emphasis on understanding the mechanism of conduction. However, its insolubility, infusibility and poor environmental stability had rendered it rather unattractive for technological applications.[13] The technologically relevant front runners belong to essentially four families: polyaniline (PANI), polypyrrole (PPy), polythiophene (PT) and poly(phenylene vinylene) (PPV). PANI is rather unique as it is the only polymer that can be doped by a protonic acid and can exist in different forms depending upon the pH of the medium.[14] A few of the more important conducting polymers and their molecular structures are shown in Figure 1.1.

1.2 Synthesis of Conducting Polymers

Conducting polymers can be synthesized either chemically or electrochemically, and each of their advantages and disadvantages are summarized in Table 1.1.[15] Through the chemical polymerization approach, the conjugated monomers react with an excess amount of an oxidant in a suitable solvent, such as acid. The polymerization takes place spontaneously and requires constant stirring as the reaction progresses.[16] The second method is *via* electrochemical polymerization, which involves placing both the counter and

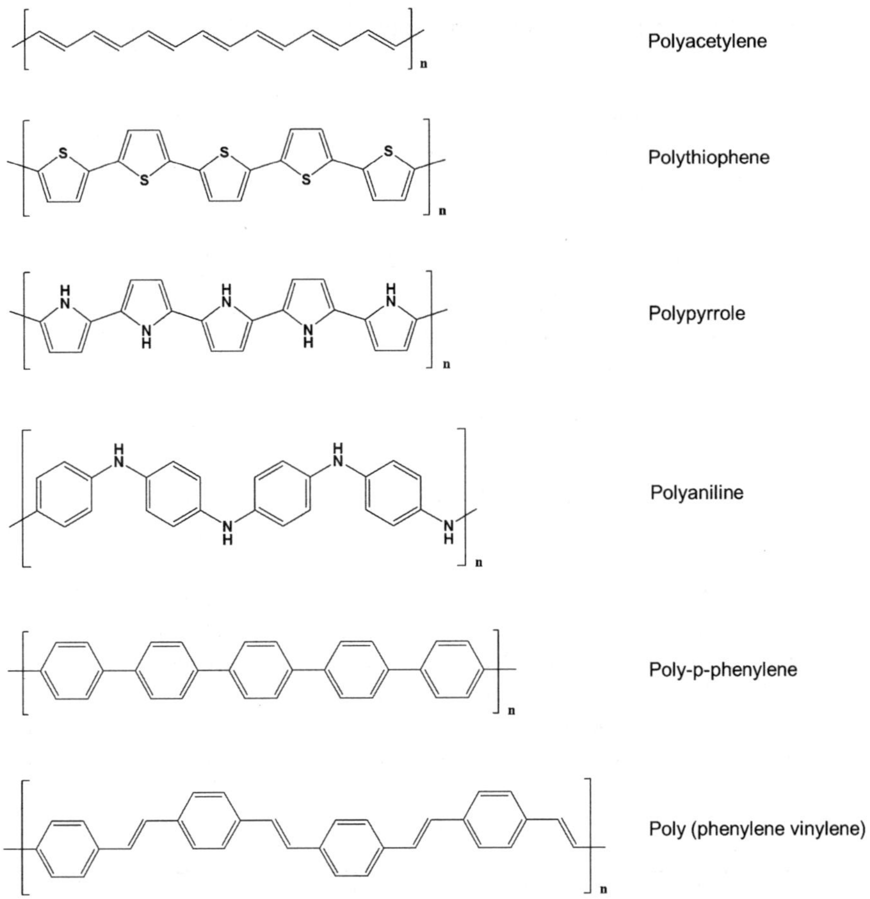

Polyacetylene

Polythiophene

Polypyrrole

Polyaniline

Poly-p-phenylene

Poly (phenylene vinylene)

Figure 1.1 Chemical structures of some undoped conducting polymers.

reference electrodes (such as platinum) into the solution containing the diluted monomer and electrolyte (the dopant) in a solvent. After applying a proper voltage, the polymer film immediately begins to form on the working electrode. A major advantage of chemical polymerization is associated with the mass reproducibility or scalability at a reasonable cost, which is a problem associated with electrochemical methods. On the other hand, an important feature of the electropolymerization technique is the direct formation of conducting polymer films that are highly conductive for use especially in electronic devices.[17]

1.3 Conductivity and Doping of Conducting Polymers

In general, materials with conductivities less than 10^{-8} S cm^{-1} are considered as insulators, materials with conductivities between 10^{-8} and 10^{3} S cm^{-1} are

Table 1.1 Comparison of conducting polymer polymerization methods.[15]

Polymerization method	Advantages	Disadvantages
Chemical polymerization	Possibility of large scale production Post-covalent modification of bulk polymer is possible More options to modify polymer backbone covalently	Cannot make thin film Relatively complicated
Electrochemical polymerization	Easy to synthesize Directly applicable to thin film synthesis Simultaneous doping	Detachment from electrode surface is difficult Post-covalent modification of bulk polymer is difficult

considered as semiconductors, and materials with conductivities greater than 10^3 S cm^{-1} are considered as conductors. Conducting polymers in their pristine (undoped) states are usually considered as semiconductors or insulators, having a band gap energy excessively high for the thermal excitation of a significant number of charge carriers. Therefore, undoped conducting polymers, such as PA and PT, show electrical conductivities of only 10^{-10}–10^{-8} S cm^{-1} (Figure 1.2).[18] Upon the doping of conducting polymers, there is a dramatic increase in the electrical conductivity by several orders of magnitude up to values of approximately 10^{-1} S cm^{-1}, even at very

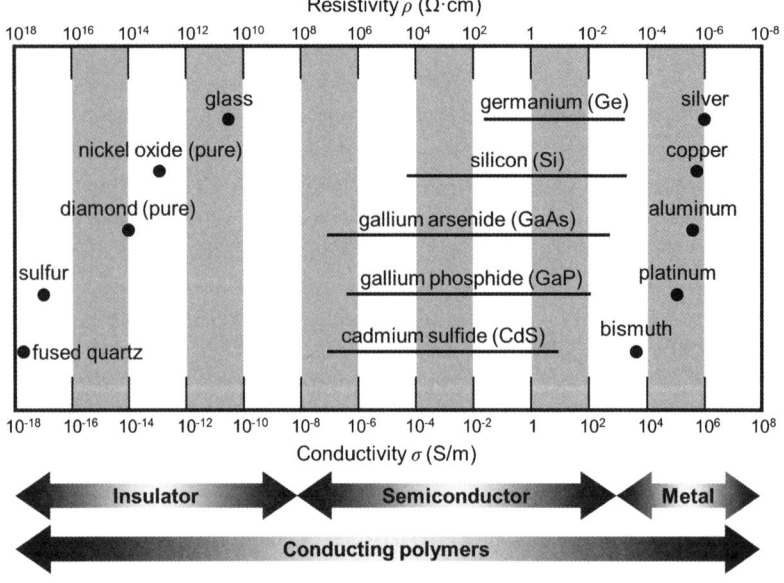

Figure 1.2 Full range of conductivity from insulators to metals.[18]

low level of doping, for example, less than 1%. Subsequent doping to higher levels results in the saturation of the conductivity at values in the range 10^2–10^5 S cm^{-1}, as shown in Figure 1.3.[1] Heavily doped, stretch-oriented PA has the highest reported electrical conductivity among the conducting polymers, with the confirmed value of 80 000 S cm^{-1}.[19]

In the case of conventional inorganic crystalline semiconductors, such as silicon and germanium, doping occurs through the substitutional insertion of atoms (dopant) with a higher or lower valence band than the host semiconductor for n-type or p-type doping, respectively. For example, phosphorous (5 valence electrons) providing an extra electron to the silicon (4 valence electrons) lattice semiconductor band structure leads to n-type doping. However, doping in a conducting polymer is the process of oxidizing (p-doping) or reducing (n-doping) a neutral polymer and providing a counter anion or cation (*i.e.*, dopant), respectively. Upon doping, a conducting polymer with a net charge of zero is produced due to the close association of the counter ions with the charged polymer backbone. This process introduces

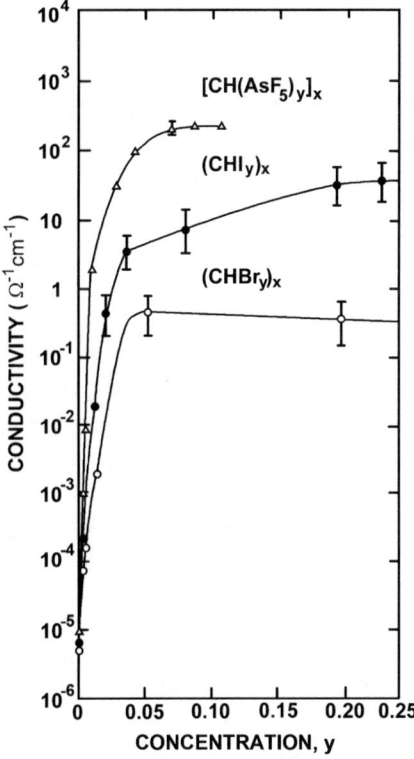

Figure 1.3 Electrical conductivity (room temperature) of PA as a function of the dopant concentration. Stretch-oriented samples exhibit much higher conductivity values.[2]

charge carriers, in the form of charged polarons (*i.e.*, radical ions) or bipolarons (*i.e.*, dications or dianions), into the polymer (Figure 1.4).[20]

As summarized in Figure 1.5, charge injection by doping can be achieved in several ways.[21] Each of the methods of doping illustrated in Figure 1.5 leads to unique and important phenomena. For chemical and electrochemical doping, the induced electrical conductivity is permanent until the carriers are either chemically compensated or until the carriers are purposely removed by "dedoping". In the case of photoexcitation, the photoconductivity is transient and lasts only until the excitations are either trapped or decay back to ground state. For charge injection at an interface, electrons reside in the π^*-band, and holes reside in the π-band only as long as the bias voltage is applied. Doping also leads to a reversible shift in the electrochemical potential, thereby making possible polymer batteries and electrochemically active polymer electrodes, electrochromic phenomenon and polymer p–n junctions (light-emitting electrochemical cells). In the case of photodoping, the redistribution of oscillator interband (π^*–π) transition provides a route to a non-linear optical (NLO) response. Real occupation of low energy excited states and virtual occupation of higher energy excited states (perturbation theory) lead to resonant and non-resonant NLO responses.

1.4 Conducting Polymers as Carbon Nanotube (CNT) Composite Matrices

The increasing demand for efficient products has driven a trend towards downsizing and lesser power consumption but greater performance. The progression relies upon searching for new desirable materials and the ability to

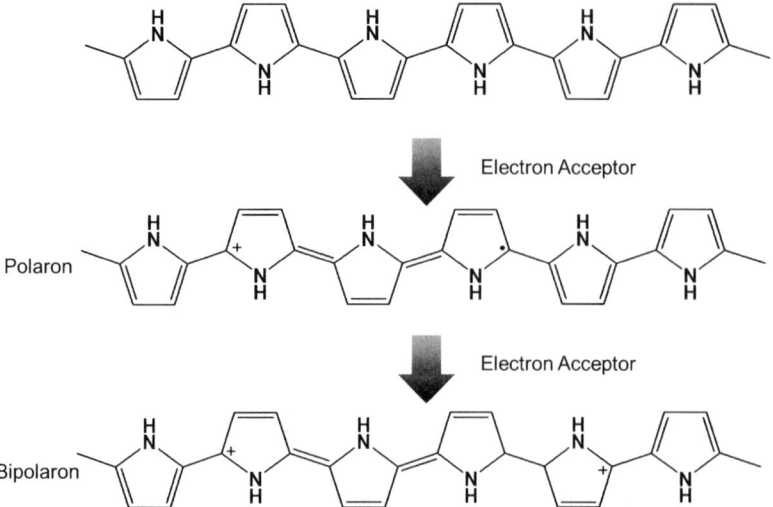

Figure 1.4 Polaron and bipolaron formation on the conjugated backbone of PPy.[20]

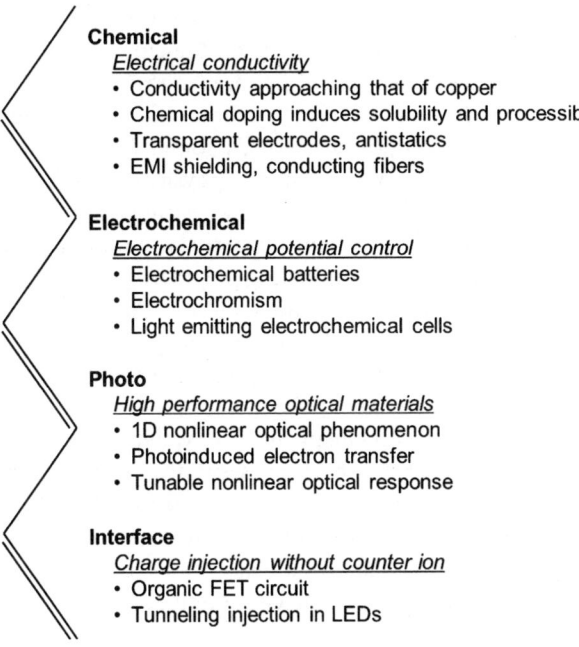

Figure 1.5 Doping methods of conducting polymers.[21]

make micro/nanoscale structures with high accuracy and precision. However, it is not so easy to satisfy these seemingly contradicting demands. For instance, current silicon-based semiconductor devices are reliable, and their performances have been constantly enhanced. However, they are reaching a barrier of modern quantum physics, as they approach the nanoscale. The search for new alternative materials and fabrication methods, therefore, is an urgent task for future developments.

CNTs and conducting polymers have both shown exceptional properties and characteristics. Coupling the two materials has furthermore revealed synergistic effects, which offers an attractive route to create a new breed of multifunctional materials with greater application potentials. Envisioned applications from CNT/conducting polymer composite systems involve mechanical, thermal, electrical, electrochemical features and their applications in various areas such as supercapacitors, sensors, organic light emitting diodes, solar cells, electromagnetic absorbers and advanced electronic devices.

Conducting polymers are conjugated polymers in doped states, which consist of alternating single and double bonds along its linear chains (sp^2 hybridized structure). The conductivity of conducting polymers relies on these double bonds, which are sensitive to physical or chemical interactions.[21,22] Similarly, CNTs also have sp^2 hybridized bonds over the structure. CNTs possess unique structures and exhibit extraordinary electrical, optical, chemical and mechanical properties, which are somewhat complementary to those of conducting polymers.[23–25] For instance, CNTs have a very long mean free path, ultra-high carrier mobility and can be either very good conductors or narrow-bandgap semiconductors depending on the chirality and diameter.

When mixed together, the two materials show a strong interfacial coupling *via* donor–acceptor binding and π–π interaction.[26,27] Combining CNTs and conducting polymers in a composite has been found to affect their chemical and electronic structures. Beyond a simple physical combination of their properties, some synergistic effects and new features appear and can be developed into applications.[26–34] From a chemistry standpoint, two possible impacts may take place in a CNT/conducting polymer system: either the CNTs are functionalized by the conducting polymers or the conducting polymers are modified (doped) with CNTs. Therefore, morphological modifications, electronic interactions, charge transfers or a combination of these effects may occur between the two constituents in the system.[35–39]

Due to the nanoscale confinement in the system, the interaction *via* interfacial bonding is considered to play an essential role in the impacts.[37,40–42] Morphologically, the interfacial interaction sites on the CNT surface are: (a) defect sites at the tube ends and side-walls; (b) covalent side-wall bindings; (c) non-covalent exohedral side-wall bindings; and (d) endohedral filling (Figure 1.6).[43] Three routes have been commonly used for preparation of

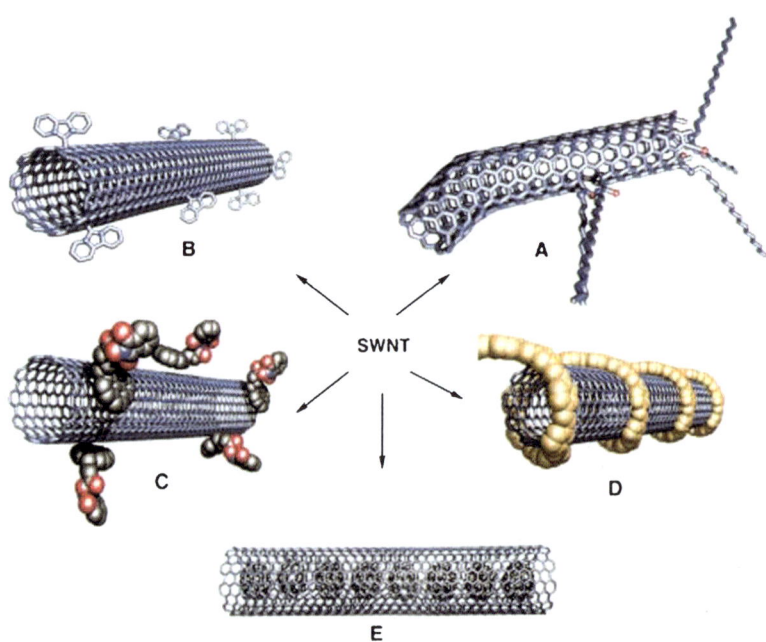

Figure 1.6 Several possible functionalization mechanisms for SWCNTs.[43] (A) defect-group functionalization, (B) covalent sidewall functionalization, (C) noncovalent exohedral functionalization with surfactants, (D) noncovalent exohedral functionalization with polymers, (E) endohedral functionalization with fullerene (C_{60}).

CNT/conducting polymer composites: (i) direct mixing; (ii) chemical polymerization; and (iii) electrochemical synthesis.

With regards to the development, studies on CNT/conducting polymer composites started in the early 1990s of the last century. Since 1992, Heeger and co-workers[44] observed the photovoltaic effects arising from the photo-induced charge transfer at the interface between conducting polymers as donors and a C_{60} film as acceptor, although the conversion efficiency was extremely low (less than 1%). However, the conversion efficiency of conducting polymer-based solar cell have been significantly improved ($\sim$6%) by further use of CNT derivatives acting as the electron acceptors in the CNT/conducting polymer composite matrices.[45] The conversion efficiency is still moderate in comparison with that of inorganic systems, but the simple processing and low cost enable CNT/conducting polymer systems to be a promising choice for photovoltaic applications.

On the other hand, the introduction of CNTs into a polymer matrix improves the electrical conductivity, while possibly providing an additional active means for capacitive energy storage and secondary batteries.[46,47] Gas sensors fabricated with single-walled carbon nanotube (SWCNT)/PPy nanocomposites have shown a higher sensitivity than that of PPy. The improvement stems from the effects of the increase of specific surface area and anion doping in the PPy matrix. The gas sensing capability in SWCNT/PANI composites has also been similarly improved. For biosensing applications, it has been demonstrated that the CNT/conducting polymer nanocomposites are very attractive as transducers, because they provide the best electron transfer and assure a faster ion mass transfer.[48,49] In some host polymers, the CNT additives have served as a hole-blocking material, causing a shift of the recombination emission.[50] The interaction between host polymer and CNT additive is considered to be the main reason accounting for all of these modifications.

Studies on CNT/conducting polymer systems will further contribute to the fundamental understanding of the nucleating capability of CNTs, epitaxial interaction and templated crystallization of the polymer at the CNT–polymer interface, and may ultimately lead to more efficient production of bulk nanocomposites. The combination of a strong CNT–polymer interaction, the nucleation ability of CNTs, the templating of polymer orientation of CNTs and crystallinity are all features that can be built on to develop high-performance composites. With that goal, the aim of this Chapter is to provide an outlook on the development and trend in future research on CNT/conducting polymer systems.

1.5 Applications of CNT/Conducting Polymer Composites

1.5.1 Supercapacitors

As shown in Figure 1.7, conducting polymers are one of three groups of candidate materials for supercapacitors (Figure 1.8)—along with carbon

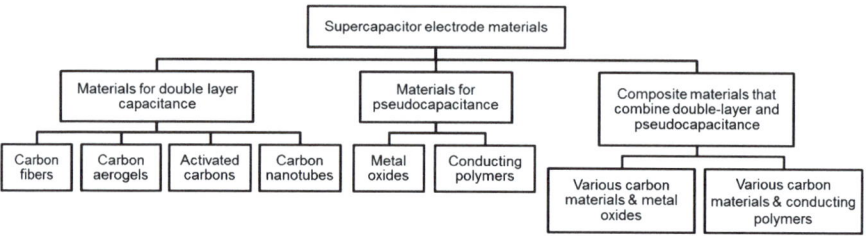

Figure 1.7 Taxonomy of the supercapacitor materials.[47]

materials and metal oxides—due to their good electrical conductivity, large pseudo-capacitance and relatively low cost. The most commonly used conducting polymers include PANI, PPy and poly(3,4-ethylenedioxy thiophene) (PEDOT). The electrochemical capacitance and charge storage properties of conducting polymers have been studied by cyclic voltammetry, electrochemical impedance spectroscopy and chronopotentiometry. Conducting polymers have a very large specific capacitance that is close to ruthenium oxides, *e.g.* 775 F g^{-1} for PANI,[51] 480 F g^{-1} for PPy[52] and 210 F g^{-1} for PEDOT.[53] However, conducting polymers commonly have poor mechanical stability due to repeated intercalation and depletion of ions during charging and discharging.

As one of the conducting and porous carbons, CNTs possess high mechanical strength, good electrical properties, high specific area and high dimensional ratios. Their application in electrochemical double-layer capacitors has been studied in detail. Aside from the three categories of pure materials, there is a new tendency to synthesize composite materials combining two or more pure materials for supercapacitors. The great promise is to combine CNTs with either metal oxides or conducting polymers. Composites of CNTs combined with RuO$_2$,[54] NiO[55] and MnO[56] have been prepared and have exhibited good potential for supercapacitor applications. However, composites consisting of CNTs and conducting polymers are even more

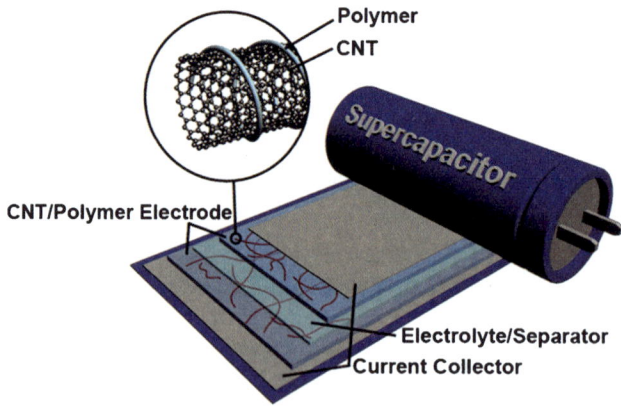

Figure 1.8 Schematic presentation of the working principle for a supercapacitor.

interesting and promising, as they can combine two relatively affordable materials to gain the large pseudocapacitance of the conducting polymers, coupled with the conductivity and mechanical strength of CNTs.

The composites can be obtained both chemically and electrochemically. Using electropolymerization, CNT/conducting polymer nanocomposites can be readily deposited from a monomer-containing solution on to a CNT preform or from a CNT/monomer-containing solution on to a traditional conductive substrate. The first attempt to electrochemically deposited conducting polymers with CNTs was made in 1999,[57] in which multi-walled carbon nanotube (MWCNT) electrodes were used for deposition of PANI films, and higher current density and more effective polymerization were found compared with those deposited on Pt electrode. In 2000, CNT/PANI co-axial nanowires were prepared electrochemically on aligned CNTs.[58] The thickness of the PANI layer was estimated to be 40–50 nm by transmission electron microscopy (TEM) images, with the CNT framework expected to offer a higher mechanical strength as compared with the bare co-axial nanowires. Thin and uniform PPy films, which were also coated on individual CNTs of well-aligned CNT arrays, were produced via potentiodynamic polymerization in aqueous solution. The faradaic current for PPy deposition on CNTs is much higher than that on Ti or Pt substrate. The ion diffusion and migration pathways are shortened due to the unique structure of the electrode material. PPy-coated CNT array electrodes show significantly improved capacitance as compared with PPy coated on Ti or Pt substrates. In 2002, PPy was deposited on CNTs through galvanostatic oxidation of monomers in sulfuric acid in order to find an alternative and relatively cheap method to enhance the capacitance of CNTs.[59] The PPy-modified electrodes had an elevated specific capacitance of 180 F g^{-1} as compared with 50 F g^{-1} for the unmodified CNT electrode. The charge loss of the PPy-modified CNT electrode was less than 20% after 2000 galvanostatic charging-discharging cycles.

Following these studies, much work has been done to investigate the properties and applications of conducting polymers on CNT electrodes. Among others, PPy and PANI have been most successful in forming a coating on CNTs via electrochemical polymerization. Both SWCNTs and MWCNTs were studied as electrodes. Electrochemically grown conducting polymers decreased the contact resistance between CNTs, producing composites with improved electrical conductivity. Raman and Fourier transform infrared (FTIR) spectroscopies showed evidence of possible interaction between CNTs and conducting polymers. As electrode materials for supercapacitors, the composites also exhibited much higher specific capacitance, specific energy and power than CNTs or conducting polymers alone. A specific capacitance of 463 F g^{-1} was achieved for 73 wt% PANI on SWCNTs.[60] However, as conducting polymers were deposited on CNT preforms rather than individual tubes, it is impossible to obtain homogeneous composites of CNTs and conducting polymers. Scanning electron microscopy (SEM) images showed that as the polymerization charge increased, the conducting polymer coating grew thicker and eventually formed a fouling, which prevented conducting polymer from growing on inner CNTs, *i.e.* the electro-deposition only took place on CNTs that have good contact with the monomer solution, whereas the CNTs inside had little or no polymer coatings, resulting in a heterogeneous structure.[60,61] Therefore, as the polymer grows thicker, the material will behave like pure conducting polymers, and the effects of the CNTs will be diminished. Electro-deposition of conducting polymers on aligned CNT arrays is likely to give a more homogeneous structure and a higher utility of CNT surface for deposition.[62] However, due to the inefficient charge transfer in the interfacial region, partial polymer coverage on CNT with a bias on the tips was observed, *i.e.*, only a predetermined portion of CNT length was covered by polymer.[58]

Conducting polymers have also been deposited on porous carbon electrochemically. The combination of double-layer capacitance and pseudo-capacitance led to a greatly improved specific capacitance of 180 F g^{-1} as compared with 92 F g^{-1} for bare carbon electrodes.[63] Recently, a high specific capacitance of 1600 F g^{-1} and a high current density of 45 mA cm^{-2} were achieved for PANI-coated porous carbon. Conducting polymers deposited on CNTs or other porous carbon preforms both utilize the good conductivity and high surface area of the substrate carbon materials.[64] Intrinsically, there is little difference in the charge storage mechanisms between these methods.

The CNT/conducting polymer nanocomposites can also be prepared chemically by an oxidant. Frackowiak *et al.*[59] deposited PPy on MWCNTs by chemical polymerization with $(NH_4)_2S_2O_8$ as an oxidant in an acidic solution (0.1 M HCl). The electrode was prepared by blending the resultant PPy-coated MWCNTs with acetylene black and poly(vinylidene fluoride) (PVDF) binder. The capacitance obtained from the MWCNTs coated with PPy reached 170 F g^{-1}, approximately twice that given by either pure MWCNTs (*ca.* 80 F g^{-1}) or pure PPy (*ca.* 90 F g^{-1}). The open entangled network of the nanocomposite seems to favor a better efficiency for the formation of the electrical layer in PPy. An *et al.*[65] used $FeCl_3$ as an oxidant to polymerize PPy and deposit it on SWCNTs. The electrode was then prepared by mixing the SWCNT/PPy composite with acetylene black and PVDF binder. Due to the uniform PPy coating on the porous and conductive support of SWCNTs, the SWCNT/PPy nanocomposite electrode showed a much higher capacitance (265 F g^{-1}) than pure PPy and pure SWCNT electrodes as well.

1.5.2 Rechargeable Lithium-ion Battery Electrodes

Despite having all of the good properties, including high conductivity, thermal and chemical stability, processability, ease of synthesis and low material cost, conducting polymers suffer from poor cyclability when they are used as an electrode material in rechargeable lithium-ion batteries (Figure 1.9).[66,67] During the cycles of charge (doping) and discharge (dedoping), polymer chains undergo shrinkage and swelling. The volume change during continuous charge–discharge cycling causes degradation of the polymer, resulting in the

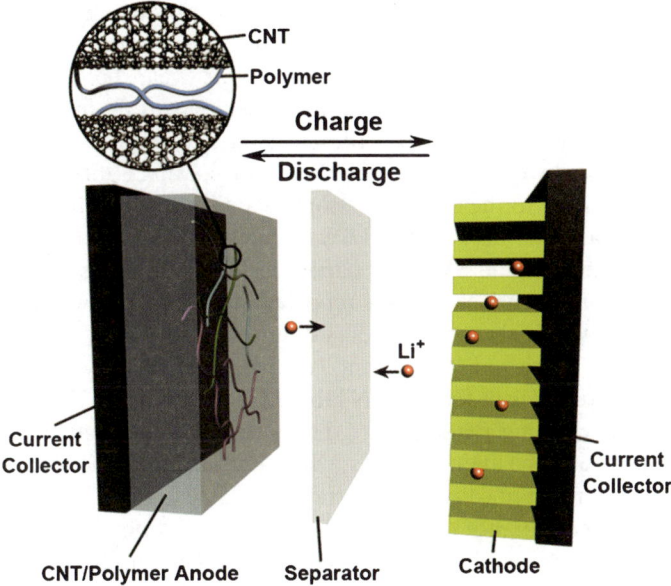

Figure 1.9 Schematic presentation of the working principle for a lithium-ion battery.

capacity fading to an undesirable level. As the poor cyclability of conducting polymers has been realized by many researchers, a lot of studies have been devoted to mitigate the degradation problem. One of the strategies that have been adopted is to employ the CNT/conducting polymer composites as an electrode.[68–70] In the application area of supercapacitors, the composite electrodes have been tested with the intention to prolong the cycle life and to enhance the charge storage ability.[71,72] Also, it has been expected to enhance the cycling performances of rechargeable lithium-ion batteries.

The first study concerning the use of the CNT/conducting polymer composites as active materials in rechargeable lithium-ion batteries was in 2006.[73] Thus, using an electrolytic solution containing $LiPF_6$, the current density was 10 mA g^{-1} and the potential range was between 1 and 3.5 V during the first 20 charge–discharge cycles. High specific discharge capacities of approximately 45 and 115 mA h^{-1} g^{-1} were reported for a positive electrode based on SWCNTs and MWCNTs, respectively, functionalized with poly(N-vinylcarbazole) (PVK). Baibarac et al.[73] reported a striking difference between SWCNT/PVK and MWCNT/PVK composites in the range of 1.5–2.5 V. In the case of MWCNT/PVK composites, they invoke a two-stage interaction process, which increases intensity with the number of cycles. These results indicate that the electrochemical performance of the PVK battery is much improved when a CNT functionalized with PVK is used as the active material because of the apparent synergy between the host polymeric matrix and guest nanomaterials.[72]

One year later, Chen et al.[46] reported that a membrane electrode based on "free-standing" aligned CNTs, PEDOT and PVDF, which is lightweight, flexible, highly conductive and mechanically robust, can be easily fabricated into a rechargeable battery without using a metal substrate or binder. In this lithium-ion battery, the weight of the electrode is reduced significantly as compared with a conventional electrode made by coating a mixture containing an active material on to the metal substrate. The reported result shows that the capacity of the aligned CNT/PEDOT/PVDF electrode is 50% higher than that observed for free-standing SWCNT paper. In fact, a highly stable dischargeable capacity of 265 mA h^{-1} g^{-1} was reported after 50 cycles when the aligned CNT/PEDOT/PVDF electrode in a lithium-ion cell was tested under a constant current density of 0.1 mA cm^{-2}. This is significantly higher than the value obtained previously for SWCNT-based buckypaper (173 mA h^{-1} g^{-1}) under identical working conditions.[74] This result was attributed to the high accessible surface of the aligned CNTs, which coupled with the robust polymer layer provides a mechanically stable array.

In 2009, the performance of MWCNT/PANI composites as the cathode for rechargeable lithium-ion batteries was tested.[75] The discharge capacity of MWCNT/PANI composites was as high as 122.8 mA h^{-1} g^{-1} compared with 98.9 mA h^{-1} g^{-1} for PANI. The composites showed a stable discharge behavior, whereas the discharge capacity of pure PANI exhibited a decreasing tendency during 50 cycles. The main reason may be attributed to the fact that the addition of MWCNTs to PANI makes the composites similar to an interwoven fibrous structure, which improves the conductivity and facilitates access of the electrolyte. A particular result of these studies is that the specific capacity and Coulombic efficiency values are much higher than other rechargeable lithium-PANI cells assembled with a gel polymer electrolyte.[76,77] The promising results reported until now in the lithium rechargeable batteries field make this application of primary importance.

1.5.3 Photovoltaic Devices

Photovoltaic devices are one of the main applications of conducting polymer-based devices (Figure 1.10). Efficient photovoltaic action requires optimization of some important factors: (i) generation of excitons; (ii) breaking of excitons to individual charge carriers; and (iii) transportation of these carriers to their respective electrodes. Photon absorption in the conjugated polymer produces bound-state excitons, and dissociation of these charge pairs is accomplished *via* electron-accepting impurities. CNTs with their high surface area[78] and high electron affinity[79] not only provide a tremendous opportunity for exciton dissociation, but also provide a percolation network to carry electrons to the electrode, while the holes are transported through polymer chains. This process is known as "photo-induced charge transfer".

At low doping levels, percolation pathways are established, providing the means for high carrier mobility and efficient charge transfer. This has been a

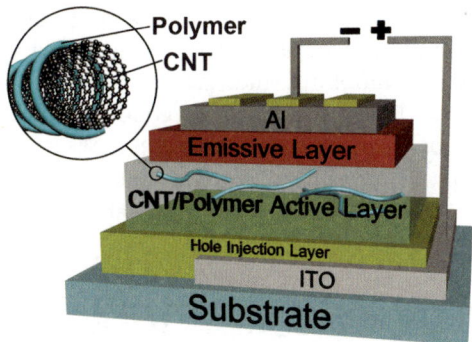

Figure 1.10 Schematic presentation of the working principle for a photovoltaic cell.

problem in the majority of polymer-based solar cells developed to date, even with the advent of semiconductor nanorods. Since the diffusion distances for excitons in conducting polymers, for example, PPV,[80] have been reported at less than 10 nm, the requirement for a sufficient percolation network of electron-accepting dopants in the polymer composite is substantiated. Electrical conductivity data have validated that CNT-doped polymer composites demonstrated this extremely low percolation threshold.[81]

Other beneficial properties of SWCNTs relevant to polymeric photovoltaic development include composite reinforcement and thermal management. SWCNTs have shown promises in the development of CNT/conducting polymer composites with enhanced mechanical strength by load transfer from the polymer matrix to the dopant.[82] The tensile strength of SWCNT has been estimated to be 20 GPa,[83] and the Young's modulus measured by atomic force microscopy was 1 TPa.[84] In addition, thermal conductivity of isolated SWCNTs has been theoretically predicted to be as high as 6600 W m^{-1} K^{-1}.[85] These high mechanical properties and thermal conductivity could provide improved mechanical stability and more design options for thermal management to large-area thin-film arrays.

SWCNT/poly(3-octylthiophene) (P3OT) composites have been used for the fabrication of new photovoltaic devices.[45] P3OT, acting as the photo-excited electron donor, is blended with SWCNTs, which act as the electron acceptors. In such devices the transferred electrons are transported by percolation paths provided by the addition of SWCNTs. Diodes [aluminum/CNT-conducting polymer composite/indium tin oxide (ITO)] with a low CNT concentration (<1%) have shown a photovoltaic behavior with an open circuit voltage of 0.7–0.9 V. The short circuit current was increased by two orders of magnitude compared with the pristine conducting polymer-based diodes and the fill factor also increased from 0.3 to 0.4 for the CNT/conducting polymer cells. As the main reason for this increase, the authors proposed a photo-induced electron transfer at the CNT/conducting polymer interface. It was shown that the internal CNT/conducting polymer junctions act as dissociation centers, which are able to split up the excitons and also create a continuous pathway for the electrons to be efficiently transported to the negative electrode. This results in an increase in electron mobility, and, hence, balances the charge carrier transport to the electrodes. In addition, the conductivity of the composite is increased by a factor of 10, indicating percolation paths within the materials. A conclusion of these results is that the SWCNT/conducting polymer composites represent an alternative class of hybrid organic semiconducting materials that are promising for organic photovoltaic cells with improved performance.

1.5.4 Organic Light-emitting Diodes (OLEDs)

OLEDs are devices that operate inversely to photovoltaic devices (Figure 1.11), transforming radiation from electricity into the optical range. They have been of significant interest in recent years for their potential applications in the lighting and display industries. The simplest version of an OLED comprises a layer of an electroluminescent organic material sandwiched between two electrodes. One of the electrodes must be transparent to transmit light emitted during the electroluminescent process. The luminescent emission of OLEDs is due to the radiative recombination of excitons, a process somewhat opposite to that of a solar cell. Fabrication of high-efficiency OLEDs depends not only on the electronic and the optical properties of the pure organic materials, but also on the control of charge transport, holes or electrons through the electroluminescent layers and on the enhancement of charge migration by doping the emissive materials.[86,87] A proper layer combination in OLEDs can also balance the injected charges in an emissive layer, thus increasing the external efficiency. The buffer layer leads to a reduction of the charge injection barrier and an even charge distribution with a large contact area at the electrode–organic interface.

A recent work showed that the dispersion of SWCNTs in a host polymer poly[(*p*-phenylenevinylene)-*co*-(2,5-dicotoxy-*m*-phenylenevinylene)] (PmPV)

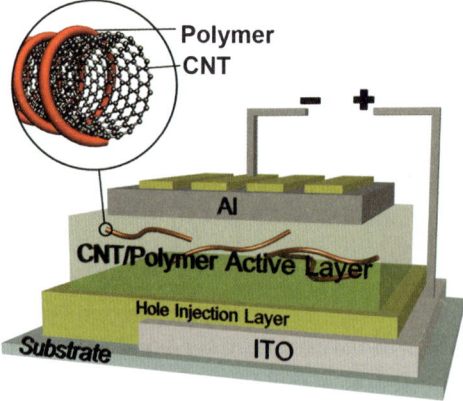

Figure 1.11 Schematic presentation of the working principle for an OLED.

traps the holes in a double-emitting OLED.[50] The device fabricated without SWCNTs dispersed in the PmPV showed a dominant emission near red at 600 nm, which is in the range of the characteristic emission of Nile Red-doped Alq3, while the addition of a small amount of SWCNTs enhanced a green emission. In addition, the devices fabricated with the SWCNT/conducting polymer composites have shown an increase in the oscillator strength of the green emission with a dominant emission peak near 500 nm, which is the characteristic emission of PmPV. The shift in the emission indicates that the SWCNTs in the PmPV matrix act as a hole-blocking material that causes shifting of the recombination region from the Nile Red-doped Alq3 layer to the PmPV composite layer. The addition of CNTs in a conducting polymer has also found to modify the electronic properties of the polymer composite. For example, OLEDs fabricated with hole-conducting polymers, such as SWCNT dispersed in PEDOT or PVK, show a change in electroluminescence, photoluminescence and I–V data. The modification in electronic structure of the composite originates from the hole-trapping nature of SWCNTs and SWCNT–conducting polymer interactions.

1.6. Conclusions

Owing to their light weight, corrosion resistance, low toxicity and ability to impart colors, conducting polymers have opened up a new field of materials with the capability to replace conventional inorganic materials such as metals and ceramics. The addition of CNTs to conducting polymers has not only enhanced the performance of the polymers, but also given rise to new value-added functionalities enabled by nanostructure control, formation of percolated network, *etc.* The synergies between CNTs and conducting polymers have allowed the application of their composites to energy conversion and storage, including rechargeable lithium-ion batteries, super-capacitors, photovoltaic devices, organic electronics, including transistors and printable conductors, and gas- and bio-sensors. With continued advancements in CNT and conducting polymer research, the demands and markets for CNT/conducting polymer composites are expected to grow. In particular, owing to the organic and biocompatible nature of conducting polymers, the application of CNT/conducting polymer composites is expected to be pronounced in the biomedical sector, *e.g.*, nerve regeneration. However, technological issues associated with large-scale manufacturability, affordability and applicability of these materials, as well as the environmental issues associated with carbon nanomaterials, will need to be addressed and eventually be resolved.

Acknowledgements

This work was supported by the Basic Science Research, WCU, US-Korea NBIT and BRL programs through the National Research Foundation (NRF) of Korea.

References

1. G. Natta, G. Mazzanti, P. Corradini, *Rend. Accad. Naz. Lincei*, 1958, **25**, 3–12.
2. H. Shirakawa, E. J. Louis, A. G. MacDiarmid, C. K. Chiang and A. J. Heeger, *J. Chem. Soc., Chem. Commun.*, 1977, 578–580.
3. T. C. Clarke, R. H. Geiss, J. F. Kwak and G. B. Street, *J. Chem. Soc., Chem. Commun.*, 1978, 489–490.
4. J. L. Brédas, R. R. Chance and R. Silbey, *Phys. Rev. Sect. B*, 1982, **26**, 5843–5854.
5. P. S. Rao, S. Subrahmanya and D. N. Sathyanarayana, *Synth. Met.*, 2003, **139**, 397–404.
6. S. Y. Park, M. S. Cho and H. J. Choi, *Curr. Appl. Phys.*, 2004, **4**, 581–583.
7. P. J. Kinlen, J. Liu, Y. Ding, C. R. Graham and E. E. Remsen, *Macromolecules*, 1998, **31**, 1735–1744.
8. W. Liu, A. Anagnostopoulos, F. F. Bruno, K. Senecal, J. Kumar, S. Tripathy and L. Samuelson, *Synth. Met.*, 1999, **101**, 738–741.
9. S. P. S. Yen, R. Somoano, S. K. Khanna and A. Rembaum, *Solid State Commun.*, 1980, **36**, 339–343.
10. X. Chen, J. P. Issi, J. Devaux and D. Billaud, *J. Mater. Sci.*, 1997, **32**, 1515–1518.
11. L. Brožová, P. Holler, J. Kovářová, J. Stejskal and M. Trchová, *Polym. Degrad. Stab.*, 2008, **93**, 592–600.
12. W. F. da Cunha, P. H. de Oliveira Neto, R. Gargano and G. Magela e Silva, *Int. J. Quantum Chem.*, 2008, **108**, 2448–2453.
13. C. B. Gorman, E. J. Ginsburg and R. H. Grubbs, *J. Am. Chem. Soc.*, 1993, **115**, 1397–1409.
14. H. Kiess, W. Meyer, D. Baeriswyl and G. Harbeke, *J. Electron. Mater.*, 1980, **9**, 763–781.
15. N. K. Guimard, N. Gomez and C. E. Schmidt, *Prog. Polym. Sci.*, 2007, **32**, 876–921.
16. P. Soudan, P. Lucas, H. A. Ho, D. Jobin, L. Breau and D. Belanger, *J. Mater. Chem.*, 2001, **11**, 773–782.
17. H. Goto, H. Yoneyama, F. Togashi, R. Ohta, A. Tsujimoto, E. Kita, K.-i. Ohshima and R. Daniel, *J. Chem. Educ.*, 2008, **85**, 1067.
18. http://www.britannica.com/EBchecked/media/139/Typical-range-of-conductivities-for-insulators-semiconductors-and-conductors (Accessed May 2012).
19. Y. W. Park, C. O. Yoon, C. H. Lee, H. Shirakawa, Y. Suezaki, K. Akagi, *Synth. Met.*, 1989, **28**, D27–D34.
20. P. A. Kilmartin and G. A. Wright, *Electrochim. Acta*, 2001, **46**, 2787–2794.
21. A. J. Heeger, N. S. Sariciftci and E. B. Namdas, *Semiconducting and Metallic Polymers*, Oxford University Press, Oxford, 2010.
22. T. A. Skotheim, *Handbook of Conducting Polymers*, M. Dekker, New York, 1986.
23. S. Iijima, *Nature*, 1991, **354**, 56–58.

24. T. W. Ebbesen and P. M. Ajayan, *Nature*, 1992, **358**, 220–222.
25. E. T. Thostenson, Z. Ren and T.-W. Chou, *Compos. Sci. Technol.*, 2001, **61**, 1899–1912.
26. H. Zengin, W. Zhou, J. Jin, R. Czerw, D. W. Smith, L. Echegoyen, D. L. Carroll, S. H. Foulger and J. Ballato, *Adv. Mater.*, 2002, **14**, 1480–1483.
27. J.-E. Huang, X.-H. Li, J.-C. Xu and H.-L. Li, *Carbon*, 2003, **41**, 2731–2736.
28. J. Wang, J. Dai and T. Yarlagadda, *Langmuir*, 2004, **21**, 9–12.
29. T.-M. Wu, Y.-W. Lin and C.-S. Liao, *Carbon*, 2005, **43**, 734–740.
30. H. G. Chae, J. Liu and S. Kumar, in *Carbon Nanotubes*, CRC Press, London, 2006, pp. 213–274.
31. W. Wang, F. Lu, L. M. Veca, M. J. Meziani, X. Wang, L. Cao, L. Gu and Y.-P. Sun, in *Encyclopedia of Inorganic Chemistry*, John Wiley & Sons, Hoboken, NJ, 2006.
32. Rajesh, T. Ahuja and D. Kumar, *Sens. Actuators B*, 2009, **136**, 275–286.
33. M. Baibarac, I. Baltog and S. Lefrant, in *Nanostructured Conductive Polymers*, John Wiley & Sons, Hoboken, NJ, 2010, pp. 209–260.
34. Y. Suckeveriene Ran, E. Zelikman, G. Mechrez and M. Narkis, in *Reviews in Chemical Engineering*, 2011, vol. 27, p. 15.
35. A. G. MacDiarmid and A. J. Epstein, *Synth. Met.*, 1994, **65**, 103–116.
36. K. Yoshino, H. Kajii, H. Araki, T. Sonoda, H. Take and S. Lee, *Fullerene Sci. Technol.*, 1999, **7**, 695–711.
37. Z. Wei, M. Wan, T. Lin and L. Dai, *Adv. Mater.*, 2003, **15**, 136–139.
38. C. Wang, Z.-X. Guo, S. Fu, W. Wu and D. Zhu, *Prog. Polym. Sci.*, 2004, **29**, 1079–1141.
39. T.-M. Wu and Y.-W. Lin, *Polymer*, 2006, **47**, 3576–3582.
40. A. Choudhury and P. Kar, *Composites Part B: Engineering*, 2011, **42**, 1641–1647.
41. A. Ubul, R. Jamal, A. Rahman, T. Awut, I. Nurulla and T. Abdiryim, *Synth. Met.*, 2011, **161**, 2097–2102.
42. T. H. Ting, Y. N. Jau and R. P. Yu, *Appl. Surf. Sci.*, 2012, **258**, 3184–3190.
43. A. Hirsch, *Angew. Chem. Int. Ed.*, 2002, **41**, 1853–1859.
44. N. S. Sariftci, L. Smilowitz, A. J. Heeger and F. Wudl, *Science*, 1992, **258**, 1474–1476.
45. E. Kymakis and G. A. J. Amaratunga, *Appl. Phys. Lett.*, 2002, **80**, 112–114.
46. J. Chen, Y. Liu, A. I. Minett, C. Lynam, J. Wang and G. G. Wallace, *Chem. Mater.*, 2007, **19**, 3595–3597.
47. C. Peng, S. Zhang, D. Jewell and G. Z. Chen, *Prog. Nat. Sci.*, 2008, **18**, 777–788.
48. Z. Ting, M. Syed, V. M. Nosang and A. D. Marc, *Nanotechnology*, 2008, **19**, 332001.
49. J.-Y. Park and S.-M. Park, *Sensors*, 2009, **9**, 9513–9532.
50. H. S. Woo, R. Czerw, S. Webster, D. L. Carroll, J. Ballato, A. E. Strevens, D. O'Brien and W. J. Blau, *Appl. Phys. Lett.*, 2000, **77**, 1393–1395.

51. V. Gupta and N. Miura, *Mater. Lett.*, 2006, **60**, 1466–1469.
52. L.-Z. Fan and J. Maier, *Electrochem. Commun.*, 2006, **8**, 937–940.
53. Y. Xu, J. Wang, W. Sun and S. Wang, *J. Power Sources*, 2006, **159**, 370–373.
54. X. Qin, S. Durbach and G. T. Wu, *Carbon*, 2004, **42**, 451–453.
55. J. Y. Lee, K. Liang, K. H. An and Y. H. Lee, *Synth. Met.*, 2005, **150**, 153–157.
56. M. Wu, G. A. Snook, G. Z. Chen and D. J. Fray, *Electrochem. Commun.*, 2004, **6**, 499–504.
57. C. Downs, J. Nugent, P. M. Ajayan, D. J. Duquette and K. S. V. Santhanam, *Adv. Mater.*, 1999, **11**, 1028–1031.
58. M. Gao, S. Huang, L. Dai, G. Wallace, R. Gao and Z. Wang, *Angew. Chem.*, 2000, **112**, 3810–3813.
59. E. Frackowiak, K. Jurewicz, K. Szostak, S. Delpeux and F. Béguin, *Fuel Process. Technol.*, 2002, **77–78**, 213–219.
60. V. Gupta and N. Miura, *J. Power Sources*, 2006, **157**, 616–620.
61. V. Gupta and N. Miura, *Electrochim. Acta*, 2006, **52**, 1721–1726.
62. J. H. Chen, Z. P. Huang, D. Z. Wang, S. X. Yang, W. Z. Li, J. G. Wen and Z. F. Ren, *Synth. Met.*, 2001, **125**, 289–294.
63. W.-C. Chen, T.-C. Wen and H. Teng, *Electrochim. Acta*, 2003, **48**, 641–649.
64. S. K. Mondal, K. Barai and N. Munichandraiah, *Electrochim. Acta*, 2007, **52**, 3258–3264.
65. K. H. An, K. K. Jeon, J. K. Heo, S. C. Lim, D. J. Bae and Y. H. Lee, *J. Electrochem. Soc.*, 2002, **149**, A1058–A1062.
66. P. Novák, K. Müller, K. S. V. Santhanam and O. Haas, *Chem. Rev.*, 1997, **97**, 207–282.
67. K. S. Ryu, Y.-S. Hong, Y. J. Park, X. Wu, K. M. Kim, Y.-G. Lee, S. H. Chang and S. J. Lee, *Solid State Ionics*, 2004, **175**, 759–763.
68. P. C. Ramamurthy, W. R. Harrell, R. V. Gregory, B. Sadanadan and A. M. Rao, *J. Electrochem. Soc.*, 2004, **151**, G502–G506.
69. V. Mottaghitalab, G. M. Spinks and G. G. Wallace, *Synth. Met.*, 2005, **152**, 77–80.
70. N. Grossiord, J. Loos, O. Regev and C. E. Koning, *Chem. Mat.*, 2006, **18**, 1089–1099.
71. Y.-k. Zhou, B.-l. He, W.-j. Zhou, J. Huang, X.-h. Li, B. Wu and H.-l. Li, *Electrochim. Acta*, 2004, **49**, 257–262.
72. V. Khomenko, E. Frackowiak and F. Béguin, *Electrochim. Acta*, 2005, **50**, 2499–2506.
73. M. Baibarac, M. Lira-Cantú, J. Oró-Solé, N. Casañ-Pastor and P. Gomez-Romero, *Small*, 2006, **2**, 1075–1082.
74. S. H. Ng, J. Wang, Z. P. Guo, J. Chen, G. X. Wang and H. K. Liu, *Electrochim.a Acta*, 2005, **51**, 23–28.
75. B.-L. He, B. Dong, W. Wang and H.-L. Li, *Mater. Chem. Phys.*, 2009, **114**, 371–375.
76. F. Cheng, W. Tang, C. Li, J. Chen, H. Liu, P. Shen and S. Dou, *Chem. – Eur. J.*, 2006, **12**, 3082–3088.

77. S. R. Sivakkumar, W. J. Kim, J.-A. Choi, D. R. MacFarlane, M. Forsyth and D.-W. Kim, *J. Power Sources*, 2007, **171**, 1062–1068.
78. M. Cinke, J. Li, B. Chen, A. Cassell, L. Delzeit, J. Han and M. Meyyappan, *Chem. Phys. Lett.*, 2002, **365**, 69–74.
79. M. Bansal, R. Srivastava, C. Lal, M. N. Kamalasanan and L. S. Tanwar, *J. Exp. Nanosci.*, 2010, **5**, 412–426.
80. J. J. M. Halls, K. Pichler, R. H. Friend, S. C. Moratti and A. B. Holmes, *Appl. Phys. Lett.*, 1996, **68**, 3120–3122.
81. M. J. Biercuk, M. C. Llaguno, M. Radosavljevic, J. K. Hyun, A. T. Johnson and J. E. Fischer, *Appl. Phys. Lett.*, 2002, **80**, 2767–2769.
82. L. S. Schadler, S. C. Giannaris and P. M. Ajayan, *Appl. Phys. Lett.*, 1998, **73**, 3842–3844.
83. F. Li, H. M. Cheng, S. Bai, G. Su and M. S. Dresselhaus, *Appl. Phys. Lett.*, 2000, **77**, 3161–3163.
84. J.-P. Salvetat, G. A. D. Briggs, J.-M. Bonard, R. R. Bacsa, A. J. Kulik, T. Stöckli, N. A. Burnham and L. Forró, *Phys. Rev. Lett.*, 1999, **82**, 944–947.
85. S. Berber, Y.-K. Kwon and D. Tománek, *Phys. Rev. Lett.*, 2000, **84**, 4613–4616.
86. T. M. Brown, J. S. Kim, R. H. Friend, F. Cacialli, R. Daik and W. J. Feast, *Appl. Phys. Lett.*, 1999, **75**, 1679–1681.
87. M. Matsumura, A. Ito and Y. Miyamae, *Appl. Phys. Lett.*, 1999, **75**, 1042–1044.

CHAPTER 2

Actuators and Infrared Sensors Based on Carbon Nanotube–Polymer Composites

JIAN CHEN*

Department of Chemistry and Biochemistry, University of Wisconsin-Milwaukee, Milwaukee, WI 53211, USA
E-mail: jianchen@uwm.edu

2.1 Introduction

Carbon nanotubes (CNTs), which have one-dimensional (1D) hollow cylinder nanostructures, are recognized as the ultimate carbon fibers for high-performance multifunctional composites owing to their high aspect ratio, small diameter, lightweight, high mechanical strength, high electrical and thermal conductivities, and unique optical and optoelectronic properties. Such properties can result in the broad application of CNTs in aerospace vehicles, military uniforms, biomedical devices, sporting goods, artificial bones and muscles.[1–5] However, the remarkable properties of pristine CNTs proved to be difficult to harness simultaneously in polymer composites.[6–23] This is largely due to the smooth CNT surface (*i.e.*, sidewall) that is incompatible with most solvents and polymers, which leads to poor dispersion of CNTs in polymer matrices and weak CNT–polymer interface adhesion. Chemical functionalization of CNTs is therefore critical to fully realize the diverse potentials of nanotubes.[18,24–32]

The two most common commercially available CNT materials are single-walled CNTs (SWNTs) and multi-walled CNTs (MWNTs). Unlike MWNTs and graphene, which have featureless visible/near-infrared (IR) absorption, semiconducting SWNTs show strong and discrete absorptions in the visible/near-IR region owing to optical transitions that are inversely proportional to the nanotube diameters.[33,34] Therefore semiconducting SWNTs can absorb near-IR light better than other types of nanocarbons in the specific near-IR region.[35–37] Although as-produced SWNTs are normally a mixture of metallic and semiconducting tubes with different diameters and lengths, rapid advances in nanotube separation make it possible to obtain single chirality SWNTs with tunable, sharp optical absorptions in the near-IR region (Figure 2.1).[38–40]

While most efforts in the field of CNT–polymer composites have been focused on passive material properties such as mechanical, electrical, and thermal properties,[20–23] there is growing interest, inspired by nature, in harnessing active material functions such as actuation, sensing, healing, and power generation in the design of CNT–polymer composites.[41] The synergy between CNTs and the polymer matrix has been judiciously exploited to create highly desirable active material functions. In this Chapter, recent progress in active CNT–polymer composites is highlighted with a focus on shape-memory and shape-changing actuators, and IR sensors.

2.2 Shape-Memory CNT–Polymer Composites

Shape-memory polymers (SMPs) are polymeric smart materials that can memorize a temporary shape and are able to return from the temporary shape to their permanent shape upon exposure to an external stimulus such as heat (Figure 2.2), light, or magnetic field.[42–57] Compared with shape-memory alloys and ceramics, SMPs offer a number of distinctive advantages, which include high recoverable strain (up to 400%), low density, ease of processing, low cost, programmable and controllable recovery behavior, and tunable recovery temperature. Such advantages could enable a broad spectrum of applications, including smart textiles, microelectromechanical systems, information storage, reversible adhesives, biomedical devices, drug delivery, and deployable aerospace structures such as hinges, mirrors, and antenna for satellites.[42–57]

Although SMPs show promising shape-memory effects, there are several limitations affecting their performance and desirability as follows. 1) *Slow recovery speed.* The rate of shape recovery process is mostly determined by the thermal conductivity of a thermo-responsive SMP. Typical SMPs have very poor thermal conductivity ($\leq$0.3 W m^{-1} K^{-1}).[44,45] 2) *Low recovery stress.* The relatively low recovery stresses ($\leq$10 MPa) exhibited by typical SMPs over their alloy, ceramic, and glass counterparts are mainly due to their intrinsically lower modulus, in the order of 0.1–1 GPa below the thermal transition temperature.[44,45,48] 3) *Lack of remote control for thermo-responsive SMP.*

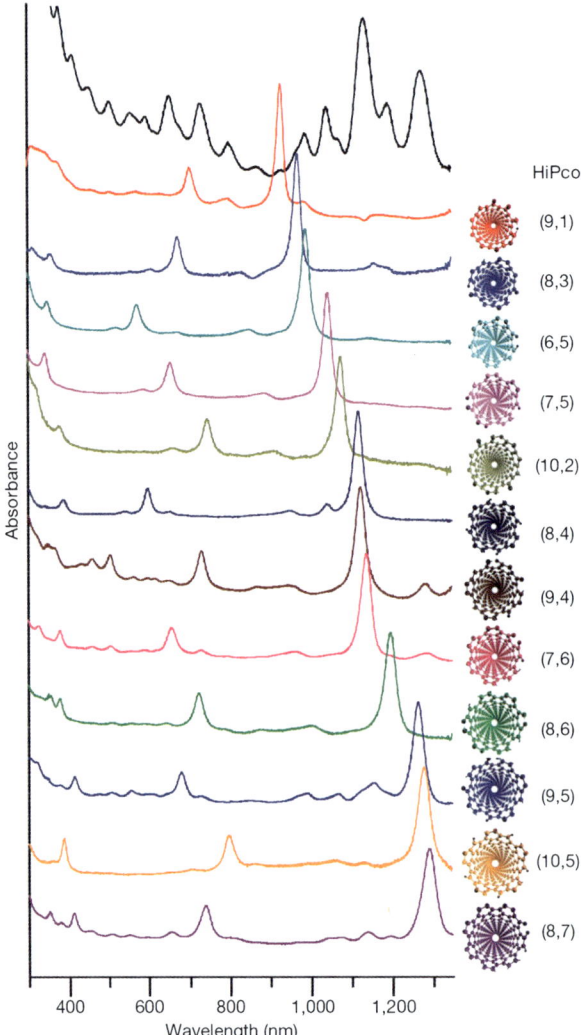

Figure 2.1 Ultraviolet (UV)-visible (Vis)-near-IR absorption spectra of 12 purified semiconducting SWNTs and the starting HiPco-produced SWNT mixture (SWNTs$_{HiPco}$). The structure of each purified SWNT species (viewed along the tube axis) and its (n, m) notation are given at the right side of the corresponding spectrum. The baseline level of each spectrum was offset to facilitate visual comparison. Reprinted by permission from Macmillan Publishers Ltd.[40]

Thermal programming of SMPs using a direct heating source (*e.g.*, oven or heating chamber), although widely used in research laboratories, is impractical for many potential applications. To address these issues, various types of fillers have been added to the SMP matrix to form SMP composites with enhanced multifunctional properties.[45,47–49,51,57] Herein we focus on major developments

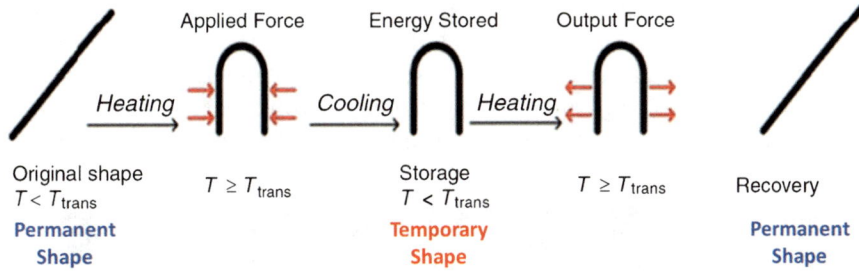

Figure 2.2 Shape-memory cycle of a thermo-responsive SMP. The typical shape-memory cycle of a thermo-responsive SMP consists of the following steps: 1) start with a SMP in its original shape (permanent shape); 2) heat the SMP above its thermal transition temperature (T_{trans}) and deform the SMP by applying an external force, cool well below T_{trans}, and remove the constraint to obtain the temporary shape with energy stored; and 3) heat the pre-deformed SMP above T_{trans}, at which point the SMP releases the stored energy and recovers the permanent shape (shape recovery). Reprinted and adapted by permission from Cambridge University Press.[48]

in thermo-responsive CNT–SMP composites that can be controlled by IR light, magnetic or electric field.

2.2.1 IR Heating of CNT–SMP composites

The ability to control mechanical devices remotely by IR light with high speed and spatial precision offers many intriguing possibilities.[58,59] The low recovery speed (up to several minutes) of thermo-responsive SMPs originates from their intrinsically low thermal conductivity (<0.3 W m^{-1} K^{-1}).[44,45] Therefore the heat transfer from an external heating source to the core of a SMP sample will take considerable amount of time. In CNT–polymer composites, CNTs can efficiently absorb and transform IR light into thermal energy, thereby serving as numerous nanoscale heaters being uniformly embedded in the polymer matrix. This IR heating approach does NOT hinge on dramatically improved thermal conductivity of a composite as a whole. Recent studies have shown that although CNTs impart great electrical conductivity to polymers, a similar degree of thermal conductivity enhancement is NOT achieved due to phonon scattering.[22,23,45] In fact, the lack of dramatic thermal conductivity enhancement in CNT–polymer composites will be essential to remote and local programming of CNT–polymer composites by a focused IR beam at both the macro- and microscale.

CNTs have been successfully used to aid in the remote recovery of stored energy in a shape-memory thermoplastic polyurethane (TPU) matrix (Figure 2.3).[60] Pristine TPU exhibits relatively modest shape-memory properties. However, by adding MWNTs, the shape-memory properties are greatly enhanced. Non-radiative decay of IR photons absorbed by MWNTs raises the internal temperature of the MWNT–TPU composites. The temperature increase melts the strain-induced soft-segment crystallites, which serve as physical cross-links, and releases the stored entropic energy of deformed amorphous chains, enabling the system to recover to its original shape. The 1

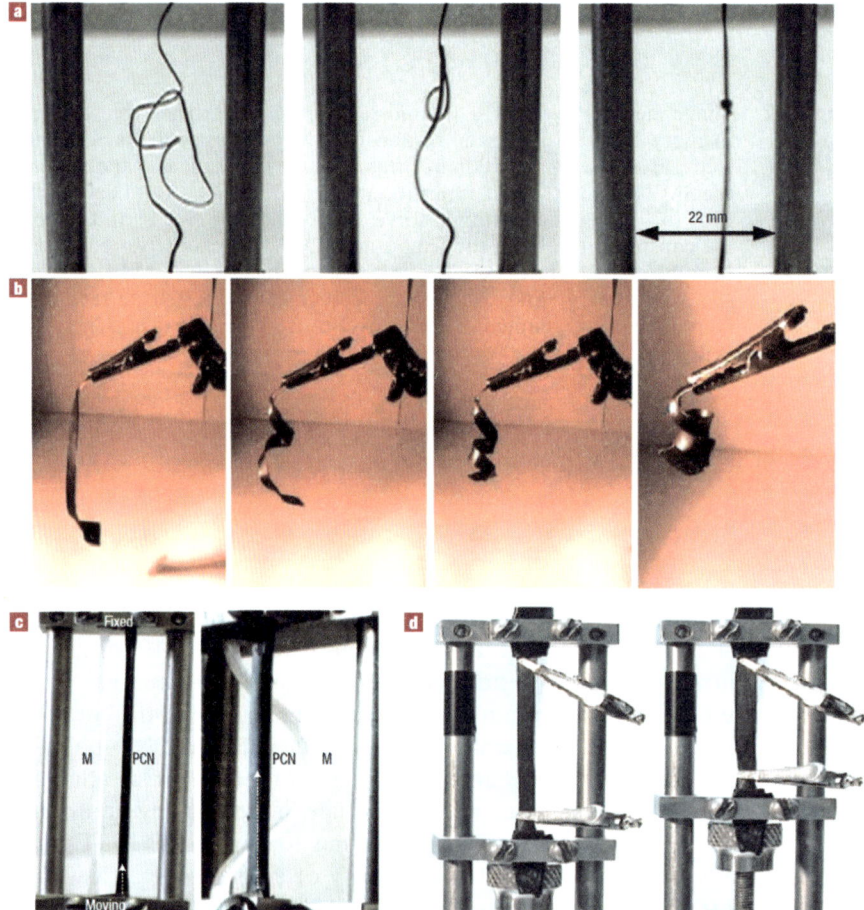

Figure 2.3 Shape-memory properties of MWNT-TPU composites. (a) Stretched (800%) 1 wt% MWNT–TPU composite ribbon, tied into a loose knot and heated at 55 °C. The knot closes on strain recovery. (b) Strain recovery and curling of the 1 wt% MWNT-TPU composite ribbon upon IR irradiation within 5 s. (c) Comparison of the stress recovery before (left) and after (right) remote actuation by IR irradiation. Neat TPU (M) bends and does not recover. In contrast, the 1 wt% MWNT–TPU composite (PCN) contracts on exposure to IR irradiation (arrow indicates moving direction). (d) Electrically stimulated stress recovery of a 16.7 wt% MWNT–TPU composite. Reprinted by permission from Macmillan Publishers Ltd.[60]

wt% MWNT–TPU composite displays a rapid shape recovery within 5 s upon exposure to an IR light (Figure 2.3b). Additionally, on exposure to the IR irradiation, the 1 wt% MWNT–TPU sample deformed to 300% exerts $\sim$19 J to lift 60 g of weight by more than 3 cm with $\sim$588 N of force (Figure 2.3c). Comparable effects occur for electrically induced actuation associated with resistive heating of the matrix when a current is passed through the conductive percolative network of MWNTs in a 16.7 wt% MWNT–TPU composite (Figure 2.3d).

The low recovery stress (<10 MPa) of thermo-responsive SMPs originates from their intrinsically low modulus, in the order of 0.1–1 GPa below the thermal transition temperature.[44,45,48] CNTs have proven to enhance the mechanical properties, particularly modulus, of various polymers.[22,23,45–48] For example, Koerner et al.[60] have found that 8.5 wt% MWNTs could increase the room temperature rubbery modulus of TPU by a factor of 5. In fact, incorporation of 5 wt% of MWNTs into the TPU matrix results in an increase of the recovery stress by $\sim$130%. Similarly, Ni et al.[61] have reported that adding 3.3 wt% MWNTs into the TPU matrix leads to an increase in the recovery stress by $\sim$100%. The recent study on mechanical properties of CNT–TPU at various nanotube loadings suggests that the reinforcement mechanism is largely due to the immobilization of TPU soft segments by adsorption on to the nanotube surface, which suppresses the glass transition and increases the stress at low strains.[62]

Chemical functionalization of CNT surfaces could improve their dispersion in the polymer matrix and enhance the nanotube–polymer interfacial interaction and the mechanical load transfer. The effects of nanotube functionalization on the properties of CNT–TPU composites have been investigated in details.[63,64] Xia and Song have synthesized polycaprolactone polyurethane (PU)-grafted SWNTs (PU-g-SWNTs) and corresponding PU-g-SWNT–PU composites by in-situ polymerization.[63] The results show that PU-g-SWNTs improve the dispersion of SWNTs in the PU matrix and strengthen the interfacial interaction between the PU and SWNTs. Compared with neat PU and pristine SWNT–PU composites, PU-g-SWNT–PU composites demonstrate remarkable enhancement on Young's modulus. The Young's modulus of a 0.7 wt% PU-g-SWNT–PU composite increases by $\sim$178% over the blank PU and $\sim$88% over the 0.7 wt% pristine SWNT–PU composite, respectively.

Khan et al.[64] have prepared various composites from a TPU reinforced with two types of chemically functionalized SWNTs: water-soluble tubes functionalized with poly(ethylene glycol) (PEG–SWNTs) or poly(amino benzene sulfonic acid) (PABS–SWNTs) and tetrahydrafuran-soluble tubes functionalized with octadecylamine (ODA–SWNTs). SWNT–TPU composites prepared with water- or tetrahydrafuran-soluble tubes display markedly different properties. The addition of water-soluble tubes tends to result in crystallization of the PU soft segments, whereas addition of the tetrahydrafuran-soluble tubes promotes crystallization of the PU hard segments. This observation suggests that the ODA–SWNTs selectively interact with the PU hard segments, whereas the PEG– and PABS–SWNTs selectively interact with the PU soft segments. This interpretation is supported by the differences in mechanical properties of the composites. The water-based composites tend to be stiffer and display higher plateau stress, consistent with reinforcement of the soft segments. However, the tetrahydrafuran cast composites tend to maintain their strength and ductility at higher nanotube loading levels, whereas the water-based composites become weak and brittle above ~10 vol% nanotubes. This is consistent with the water-based nanotubes impeding the extension and motion of the soft segments, resulting in loss of ductility. In contrast, the tetrahydrafuran-soluble nanotubes become segregated in the hard segments and so do not negatively impact on the mechanical properties at higher nanotube content.

2.2.2 Inductive Heating of CNT–SMP composites

An alternative approach to remotely actuate SMP composites involves inductive heating in an alternating magnetic field.[65] In general, the incorporated magnetic particles consist of iron/iron oxide, nickel, or cobalt compounds could generate heat in an alternating magnetic field by hysteresis losses, eddy current losses, and/or other relaxational losses depending on the nature and size of the particle.[49] He et al.[66] have developed remote-controlled multishape polymer nanocomposites with selective radiofrequency (RF) actuations. Their approach is based on the principle that Fe_3O_4 nanoparticles and CNTs can selectively induce heat at two very different RF ranges (296 kHz and 13.56 MHz frequency, respectively). By embedding the two nanofillers into different regions of a shape-memory epoxy composite, independent heating (thus recovery) of each region are enabled when exposed to the two selected RF fields (Figure 2.4a). The target multi-composite SMP consists of three regions: the Fe_3O_4-SMP (right region), the neat SMP (center region), and the CNT-SMP (left region). The permanent straight shape is first deformed into the temporary shape #1, in which strains are introduced into all the three regions. For temporary shape #1, the recovery of the Fe_3O_4–SMP or CNT–SMP region can be remotely actuated by the selected RF without enacting the recovery of other regions. As such, its recovery to the permanent shape can follow five different numbered recovery routes, with a total of five different shapes involved.[66] Figure 2.4(b) shows experimental demonstration of the multicomposite shape recovery route 3.

2.2.3 Resistive Heating of CNT–SMP composites

CNTs are recognized as the ultimate carbon fibers for high-performance, multifunctional polymer composites, where the addition of only a small amount of nanotubes, if engineered appropriately, could lead to simultaneously enhanced

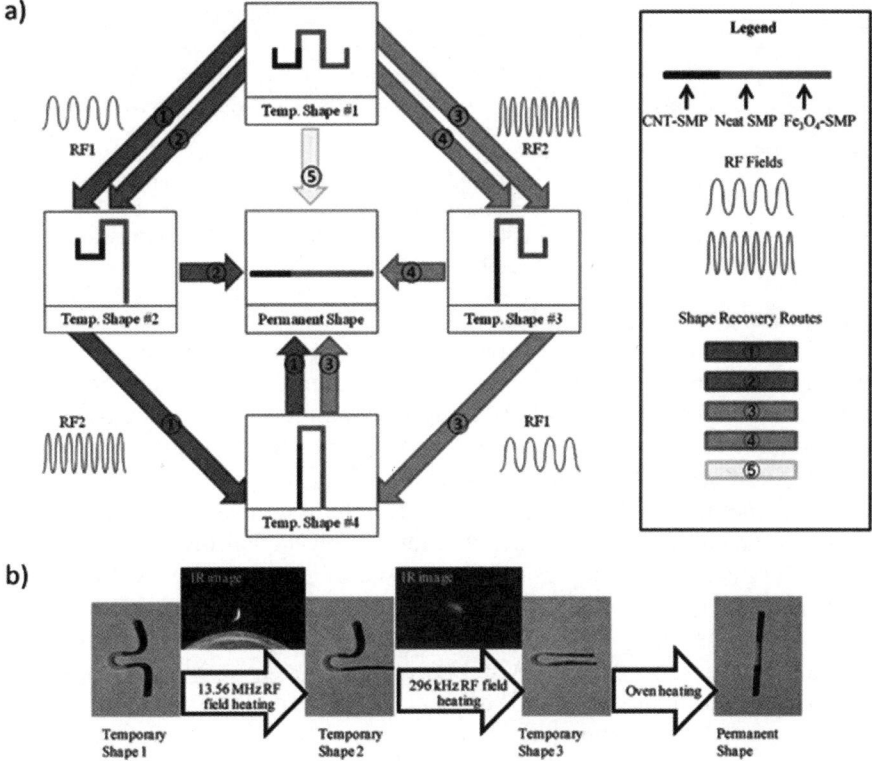

Figure 2.4 Selective RF actuation of a multicomposite SMP. (a) Conceptual illustration of various shape recovery routes from temporary shape #1 to the permanent shape in a multicomposite SMP, featuring remote and selective triggering of the recovery of multiple temporary shapes. Five recovery routes are possible depending on the actuation sequences. RF1 and RF2 represent low and high radio frequencies, respectively. In all of recovery routes, the last recovery step to the permanent shape is always achieved by direct heating, as the central neat SMP region does not contain any filler. (b) Experimental demonstration of the multicomposite shape recovery route 3. The sample was subjected to RF fields of 13.56 MHz and 296 kHz sequentially. IR images show selective heating of the corresponding regions. Reprinted by permission from Wiley-VCH Verlag GmbH & Co. KGaA.[66]

mechanical strength and electrical conductivity.[16–23] CNT–SMP composites can be electrically actuated *via* resistive heating when the loading level of CNTs is above its electrical conductivity percolation threshold.[48,49,60,67–73]

Koerner *et al.*[60,67] have reported that the addition of relatively small amounts (1–16.7 wt%) of MWNTs to the TPU matrix produces polymer nanocomposites with high electrical conductivity (1–10 S cm^{-1}), low electrical percolation, and enhancement of mechanical properties including increased modulus and yield stress without loss of the ability to stretch the elastomer

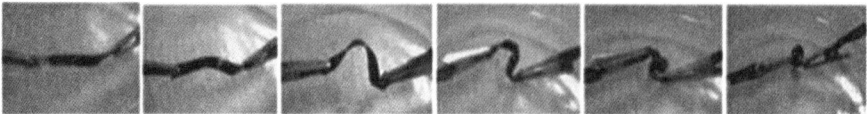

Figure 2.5 Electroactive shape-recovery behavior of a 5 wt% MWNT–TPU composite. The sample undergoes the transition from temporary shape (linear, left) to permanent shape (helix, right) within 10 s when a constant voltage of 40 V is applied. Reprinted by permission from Wiley-VCH Verlag GmbH & Co. KGaA.[68]

above 1000% before rupture. Electrically stimulated stress recovery of a 16.7 wt% MWNT–TPU nanocomposite (20 mm × 4 mm × 0.4 mm, deformed 100%) exerts ∼6 J to lift a 60 g mass by 1 cm (Figure 2.3d).[60] Cho et al.[68] have shown that a 5 wt% MWNT–TPU composite sample undergoes the transition from a linear temporary shape to a permanent shape (helix) within 10 s when a constant voltage of 40 V is applied (Figure 2.5).

By incorporating ∼20 wt% of SWNTs, with the assistance of sodium dodecyl sulfate surfactant and probe sonication, into a polyvinyl alcohol (PVA) matrix and deforming the resulting nanocomposite fibers at a low deformation temperature, Miaudet et al.[69] have been able to achieve remarkable recovery stress as high as nearly 150 MPa, which is approximately up to two orders of magnitude higher than typical SMP. However, their shape recovery (R_r), which characterizes the ability of a SMP to memorize its permanent shape,[45] is lower than 60%.[56,69] The low R_r implies that the long-range molecular movements are only partially suppressed by the presumably OH–π interactions between PVA and nanotube surfaces.[56] As SWNT–PVA fibers are electrically conductive, the thermal shape-memory effect can be triggered by resistive heating when an electrical current is passed through the fiber.[69,71]

Luo and Mather[72] have developed a unique shape-memory nanocomposite with a high-speed electrical actuation capability by incorporating ∼9.2 wt% of continuous, non-woven carbon nanofibers (CNFs) into an epoxy-based SMP matrix (Figure 2.6). With a DC voltage of 20 V, the shape recovery of 9.2 wt% epoxy–CNF nanocomposite is complete within 2 s, which is significantly faster than all previously reported conductive SMPs, which typically recover in 10 to 120 s with applied voltages ranging from 20 to 40 V.[72] Besides the simple processing and good electrical conductivity (0.3 S cm^{-1}), this non-woven CNF-based filler system simultaneously enhances the recovery stress (by raising the rubbery modulus above T_g) and the thermal conductivity of the SMP.[72]

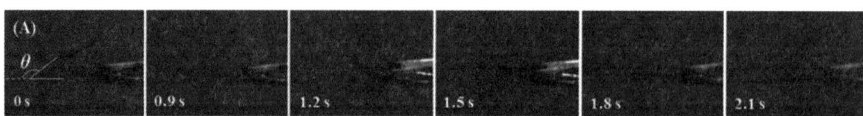

Figure 2.6 With a DC voltage of 20 V, the shape recovery of 9.2 wt% epoxy–CNF nanocomposite is complete within 2 s. Reprinted by permission from the Royal Society of Chemistry, Thomas Graham House.[72]

MWNTs can be electrically induced into aligned chains in a styrene-based SMP/carbon black (CB) composite and serve as long-distance conductive channels to bridge the CB aggregates in the SMP matrix.[73] Compared with the samples without nanotubes [*e.g.*, CB (15 wt%)/SMP composite] or with randomly distributed nanotubes [*e.g.*, MWNT (random, 1 wt%)/CB (15 wt%)/SMP composite], the electrical resistivity is reduced by over 100-fold in composites with chained nanotubes [*e.g.*, MWNT (chained, 1 wt%)/CB (15 wt%)/SMP composite]. With a DC voltage of 25 V, the MWNT (chained, 1 wt%)/CB (15 wt%)/SMP composite demonstrates much faster shape recovery than either the CB (15 wt%)/SMP composite or the MWNT (random, 1 wt%)/CB (15 wt%)/SMP composite (Figure 2.7).[73]

2.3 Shape-Changing CNT–Polymer Composites

SMPs generally exhibit one-way shape-memory effect, i.e., the shape change only follows the arrow direction in Figure 2.2 and is not reversible.[50,56] It is important to note, however, that any one-way shape-memory cycles can be repeatedly run, but going from the original (or a recovered) shape to the previous temporary shape always requires the application of an external mechanical manipulation (*i.e.*, programming).[56] In contrast to SMPs, shape-changing polymers (SCPs) are polymers that deform in response to an external stimulus such as heat, light, or electric field, and they normally recover their original shapes spontaneously once the stimulus is terminated.[50] In other words, SCPs can reversibly switch shapes without the need for external mechanical manipulation. The potential

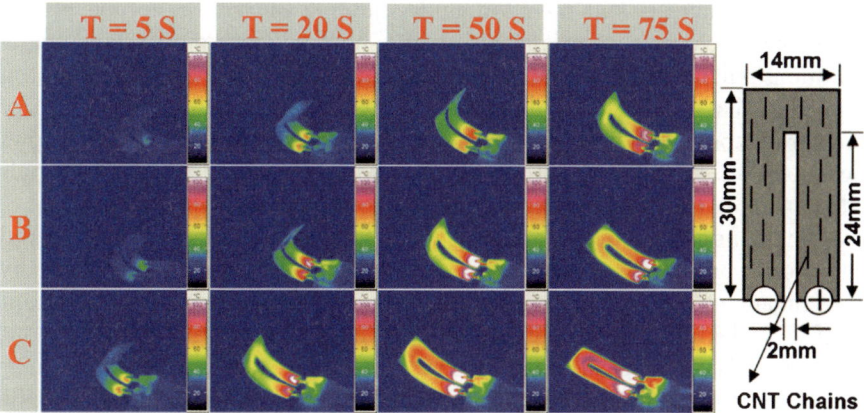

Figure 2.7 With a DC voltage of 25 V, snap shots of shape recovery and temperature distributions. Sample A: CB (15 wt%)/SMP composite; sample B: MWNT (random, 1 wt%)/CB (15 wt%)/SMP composite; sample C: MWNT (chained, 1 wt%)/CB (15 wt%)/SMP composite. Right panel shows the sample dimension and experimental setup for sample C. Reprinted by permission from the American Institute of Physics.[73]

applications of SCPs include artificial muscles, mini- and microrobots, "smart skins", pumps and valves in microfluidic systems for drug delivery, ventricular assist devices for failing hearts, and sensors for mechanical strain, humidity, and gases.[50,74,75] Researchers have recently demonstrated that the synergistic combination of CNTs and a suitable polymer matrix (including a non-SCP matrix) could lead to promising shape-changing CNT–polymer composites with either new or significantly enhanced actuation properties. In this section, we highlight some interesting shape-changing CNT–polymer composites that are responsive to either light or electric field.

2.3.1 Light-driven Shape-changing CNT–Polymer Composites

Ahir *et al.* have reported a novel phenomenon of photo-induced mechanical actuation observed in elastomeric MWNT–polydimethylsiloxane (PDMS) composites when exposed to IR irradiation.[76–78] That is, the direction of the actuation (expansion or contraction) depends on the extent to which the MWNT–PDMS nanocomposite is strained (Figure 2.8). If the material is being slightly pulled, it will expand when exposed to IR light. Moreover, this expansion is two orders of magnitude greater than the pristine unfilled PDMS elastomer. Conversely, if the material is pulled strongly (a strain higher than 10%), it will contract under an identical exposure to IR light. This process is completely reversible and persists after numerous cycles. Furthermore, a similar phenomenon has been observed in other polymer matrices, suggesting that the MWNTs are the origin of the observed intriguing actuation response.[78] There is still no good understanding of such nanotube photo-mechanical behavior when embedded in a host polymer matrix, and many questions remain unclear. One possible explanation considers CNTs as photon absorbers that locally redistribute the energy as heat, causing contraction of anisotropic polymer chains aligned near the nanotube surfaces.[78] This shows how nanotubes could impart new properties to otherwise passive materials such as the PDMS elastomer.

Nematic liquid crystalline elastomers (LCEs) have fascinated scientists and engineers continually since 1981 when they were first synthesized by Finkelmann *et al.*[79] LCEs bring together, as no other polymer, three important features: orientational order exhibited by the mesogenic units in amorphous soft materials; topological constraints *via* cross-links; and responsive molecular shape due to the strong coupling between the orientational order and the mechanical strain.[80] Acting together, they create many new physical phenomena, which can lead to a promising new generation of actuators.[80–85] Nematic LCEs can dramatically and reversibly elongate or contract in response to temperature changes (Figure 2.9); however, this type of LCE actuator tends to have a slow response time due to the low thermal conductivity of LCEs and cannot be actuated remotely, thereby limiting its potential applications.

Actuators and Infrared Sensors Based on Carbon Nanotube–Polymer Composites 33

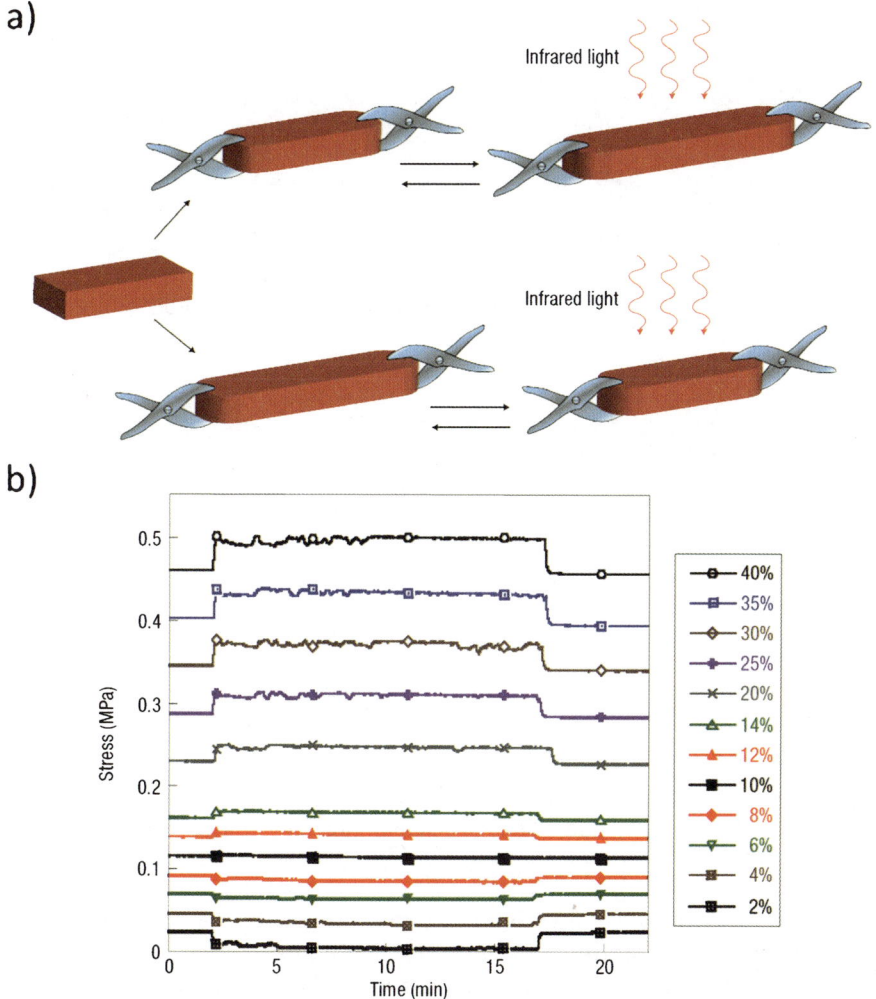

Figure 2.8 Photomechanical actuation of MWNT–elastomer composites. a) Bimodal and reversible actuation of a MWNT–elastomer nanocomposite induced by IR irradiation. Reversible expansion occurs at small pre-strains (top) and reversible contraction at large pre-strains (bottom). b) Response to IR irradiation at different values of pre-strain. Raw data on stress measured at fixed sample length of a 1 wt% MWNT–PDMS nanocomposite (different pre-strain curves labeled on the plot). Reprinted by permission from Macmillan Publishers Ltd.[76,77]

As mentioned earlier,[78] Ahir et al.[86] have studied the photomechanical behavior of three types of MWNT–elastomer nanocomposites irradiated with near-IR light under isostrain conditions, including PDMS, styrene-isoprene-styrene (SIS), and a nematic LCE with a polysiloxane backbone.

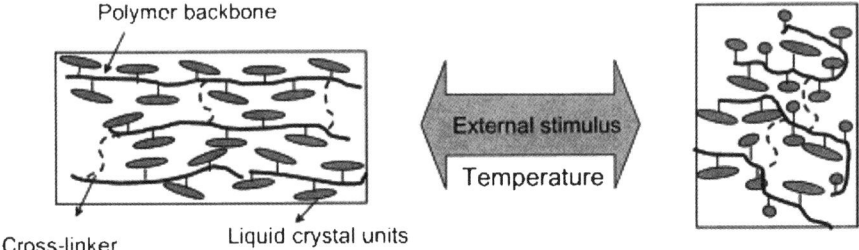

Figure 2.9 Conceptual illustration of a thermo-responsive LCE with side-on mesogenic units and its shape-changing behavior. Reprinted by permission from the American Chemical Society.[84,85]

They have shown that these composite materials have the novel ability to change their actuation direction upon IR irradiation.

Unlike MWNTs, which have featureless visible/near-IR absorption and are relatively easier to be dispersed in solutions, pristine SWNTs show strong and specific absorptions in the visible/near-IR region but can only be dispersed poorly in most solvents and polymer matrices. Although SWNTs, due to rapid progress in nanotube separation,[38–40] may ultimately offer unique opportunities for the wavelength-selective IR actuation of LCE nanocomposites, the effective use of SWNTs in polymer-composite applications strongly hinges on the ability to disperse them uniformly in a polymer matrix without damaging or destroying their integrity. For example, in one study, the maximum pristine SWNT loading level in a nematic LCE with a polysiloxane backbone was only 0.02 wt%, because of the difficulty to disperse SWNTs uniformly in the LCE matrix at higher concentrations.[87] Similar issue arises in another study, although SWNT–LCE nanocomposites display reversible photoactuation upon IR or white light irradiation, the inhomogeneous dispersion of pristine SWNTs in the LCE matrix with a polysiloxane backbone is noticeable at macroscale.[88]

Chen et al.[31] have developed a versatile, non-damaging chemistry platform that enables them to modify specific CNT surface properties, while preserving the intrinsic properties of the CNT. Rigid, conjugated macromolecules, such as poly(*p*-phenylene ethynylene) (PPE), can be used to non-covalently functionalize and solubilize CNTs and disperse CNTs homogeneously in polymer matrices.[16,18,19,31,89–94] Fast, reversible IR-actuation behavior of a new type of nematic SWNT–LCE nanocomposite has been reported using PPE-functionalized SWNTs (PPE–SWNTs) as a filler in a LCE matrix with a polyacrylate backbone and side-on mesogenic units.[95] SWNT–LCE nanocomposites have been prepared from a mixture of PPE–SWNTs and two monomers (**1a** and **1b**, Figure 2.9) through a two-stage photopolymerization process coupled with a hot-drawing technique, which enables the fabrication of relatively thick films (~160–260 μm). The excellent dispersion of PPE–SWNTs in the LCE matrix enables a significant and reversible IR-induced strain (~30%) at very low SWNT loading levels (0.1–0.2 wt%, Figure 2.10). The effects of various parameters have been investigated, including the degree of pre-alignment, the

Figure 2.10 IR actuation of a 0.2 wt% SWNT–LCE nanocomposite film. (a) Before the IR was turned on; (b) after the IR was turned on for 4 s; (c) after the IR was turned on for 10 s; (d) the film returned to its original length when the IR beam was turned off. Reprinted by permission from Wiley-VCH Verlag GmbH & Co. KGaA.[95]

loading level of the SWNTs, and the curing time. In SWNT–LCE composites, semiconducting SWNTs can efficiently absorb and transform IR light into thermal energy, thereby serving as numerous nanoscale heaters being uniformly embedded in the LCE matrix. The absorbed thermal energy, if sufficient, then induces the LCE nematic–isotropic phase transition, which leads to the shape change of the nanocomposite film (Figures 2.10 and 2.11). The IR strain response of SWNT–LCE nanocomposites increases with an increasing SWNT loading level and a higher degree of hot-drawing, and decreases with longer photocuring.[95]

Terentjev and co-workers[96,97] have developed an alternative approach to non-covalent functionalization of CNTs for the preparation of CNT–LCE composites. A pyrene-containing main-chain (MC) thermotropic liquid crystal polymer [pyrene-MC (PyMC), Figure 2.12] has been synthesized and utilized for dispersing CNTs and fabrication of nematic CNT–LCE composites. The nematic LCE material with a polysiloxane backbone was prepared using the procedure described by Küpfer and Finkelmann (Figure 2.12).[98] They demonstrated that a small amount of added MWNTs does not change the thermal response of the polymer, and only a very small concentration (0.1%) of MWNTs is required in order for the LCE to become responsive to IR or visible light.[97] Their findings support the hypothesis that added CNTs make the LCE–CNT composites actuatable simply by providing local heat efficiently to the LCE thermal actuator.

2.3.2 Electroactive Shape-changing CNT–Polymer Composites

CNT–polymer composites at very low nanotube loadings exhibit substantial electrostrictive strains when exposed to an electric field that is dramatically lower than that required by neat polymers.[99,100] Zhang et al.[99] have shown that the crystallinity, Young's modulus, dielectric constant, electrostrictive strain, and elastic energy density of electrostrictive poly(vinylidene fluoride-trifluoroethylene-chlorofluoroethylene) [P(VDF-TrFE-CFE)] can be simultaneously improved by inclusion of only 0.5 wt% of MWNTs. At an applied electric field of 54 MV m^{-1}, the 0.5 wt% nanocomposite generates a strain of 2%, which nearly doubles that of pure P(VDF-TrFE-CFE) polymer.

Park et al.[100] have investigated the electromechanical properties of SWNT/LaRC-EAP (Langley Research Center-ElectroActive Polyimide) composites, which forms an intrinsic unimorph during the fabrication process that can actuate without the need for additional inactive layers (Figure 2.13). The 0.05% SWNT/LaRC-EAP exhibits a strain of 2.6% at a relatively low driving voltage (0.8 MV m^{-1}) while possessing excellent mechanical and thermal properties. The out-of-plane strain is proportional to the square of the electric field E, which suggests that the actuation mechanism of the SWNT/LaRC-EAP composite is primarily due to electrostriction.[101]

Electrical field-driven pure nematic LCE actuators have not been developed because the characteristic rotation energy density is too low compared with the

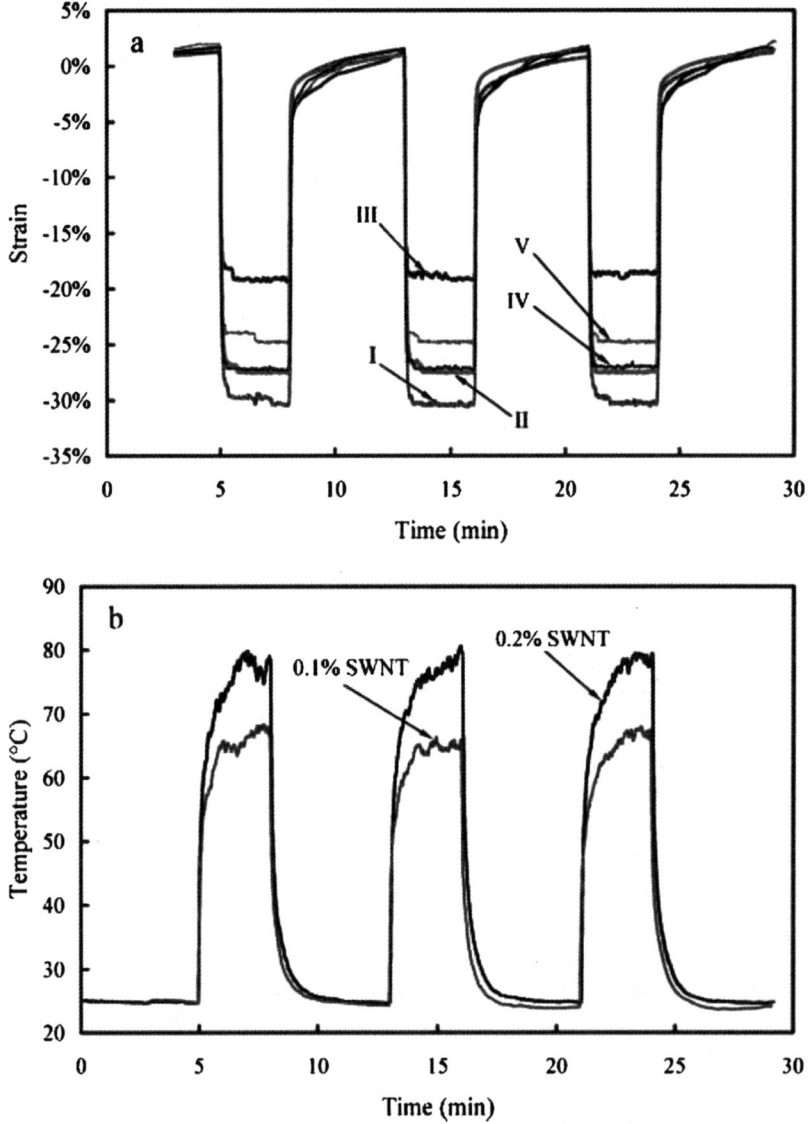

Figure 2.11 Change in strain (a) and temperature (b) of SWNT–LCE nanocomposite films under IR stimulus. (a) The strain responses to an IR stimulus in five different SWNT–LCE nanocomposite films: 0.2 wt% SWNT–LCE nanocomposite films subjected to 16 min of curing and different percentages of hot-drawing (I. 160%, II. 130%, and III. 70%) and 0.1 wt% SWNT–LCE nanocomposite films subjected to 200% hot-drawing and different curing times (IV. 16 min, and V. 20 min). (b) The temperature responses to an IR stimulus in SWNT–LCE nanocomposite films with 0.1 and 0.2 wt% PPE–SWNT loading levels. Reprinted by permission from Wiley-VCH Verlag GmbH & Co. KGaA.[95]

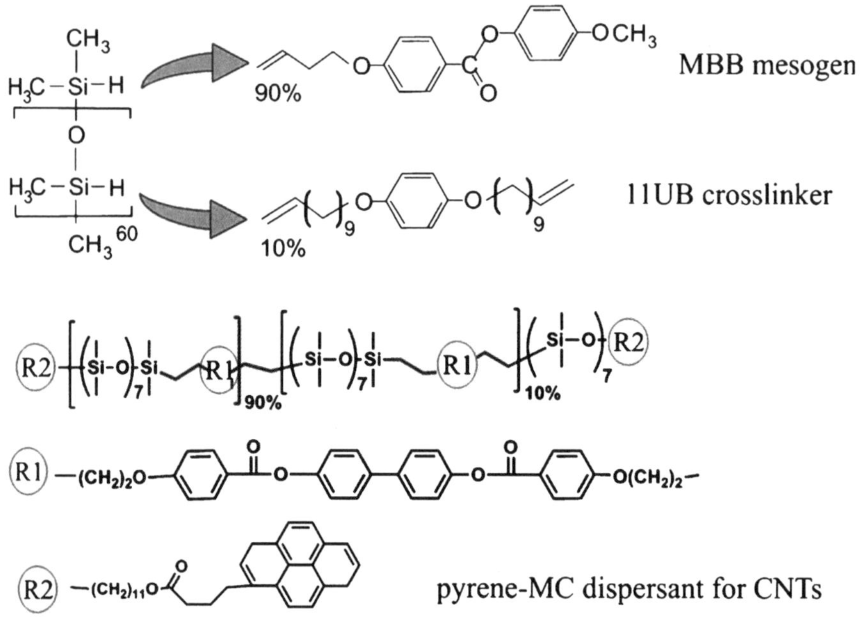

Figure 2.12 Scheme showing the components used to create LCE–CNT materials. The mesogens (MBB) and cross-linking component (11UB) are attached to the polysiloxane chain using a hydrosylation reaction with a platinum catalyst. The pyrene-MC (PyMC) is a dispersant for CNTs. Reprinted by permission from the Royal Society of Chemistry, Thomas Graham House.[97]

characteristic resistance energy density from the rubbery elastic network due to the low anisotropic dielectric permittivity of the LCEs, even under a very high electric field.[87] However, introducing highly polarizable anisotropic CNTs into the LCE matrix could lead to a significantly enhanced electromechanical effect and make electrical field-driven LCE actuators possible. Courty et al.[87] have demonstrated a relatively large electromechanical response with uniaxial stress of ~1 kPa in response to a constant field of ~1 MV m^{-1} in a nematic LCE filled with a very low (~0.0085 wt%) concentration of MWNTs. Surprisingly, the response stress in the isostrain conditions shows a linear dependence on the applied field E instead of E^4 that would be expected from a CNT rotation-induced electromechanical response.

Coupling of highly conductive CNT network with a soft polymer matrix with large coefficient of thermal expansion (CTE) leads to a new class of electrothermal CNT–polymer composite actuators,[102–106] which generates significant strains reversibly at applied electric fields at least two orders of magnitude lower than those reported for electrostrictive polymer nanocomposites.

Chen et al.[102] have reported electrothermal actuation of MWNT–PDMS composites where randomly oriented nanotubes form a conductive network in

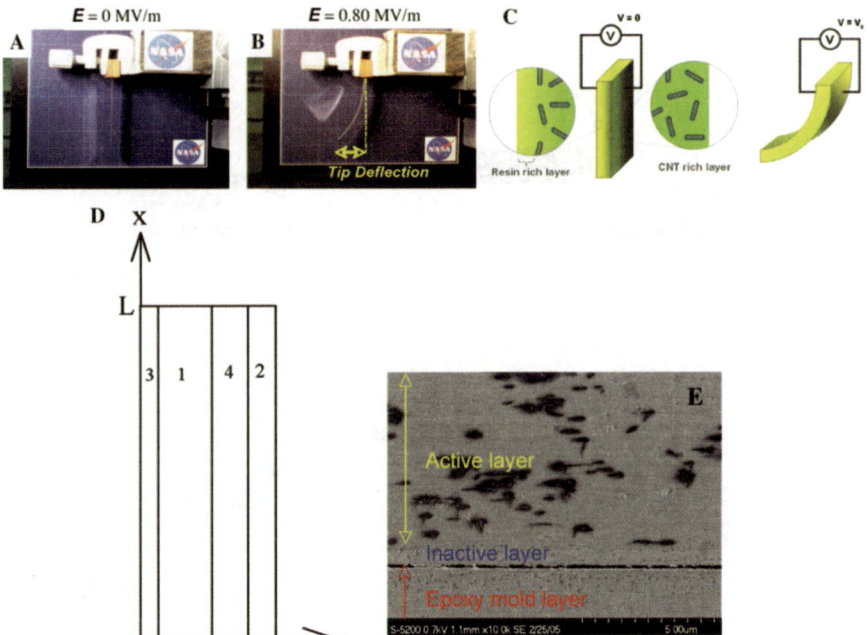

Figure 2.13 Electromechanical actuation of a SWNT/LaRC-EAP composite. Top: bending actuation of a cantilevered intrinsic unimorph (0.05% SWNT/LaRC-EAP) (side view): A) without and B) with an electric field (0.8MV m^{-1}), and C) Schematic structure of a cross-section of the intrinsic unimorph of SWNT/LaRC-EAP composite film without and with an electric field. Bottom: D) Schematic diagram of a four-layer electrostrictive bending prototype in passive state (side view). Layer widths and thicknesses are not to scale. Layer 1 is the active electrostrictive layer and layer 4 is the inactive layer. When activated, a voltage V is applied between metallic layers 2 and 3. E) Scanning electron microscopy (SEM) image of a side view of 0.05% SWNT/LaRC-EAP composite film. The top layer is a part of the SWNT-rich active composite layer under which can be found the polymer-rich inactive layer, which lies just above the epoxy mold. Reprinted by permission from Wiley-VCH Verlag GmbH & Co. KGaA.[100]

the elastomer matrix. The CTE (3×10^{-4} K^{-1}) of a 5 wt% MWNT–PDMS composite is nearly the same as pure insulating PDMS,[107] whereas the electrical conductivity of the nanocomposite is dramatically improved to approximately 0.1 S cm^{-1}. When a low electric field (1.5 KV m^{-1}) is applied, the 5 wt% MWNT–PDMS exhibits a significant maximum strain of 4.4% (Figure 2.14). In a later study, Chen *et al.*[103] have demonstrated a new type of bending actuator consisting of a thin layer (20 μm) of ~5 wt% super-aligned MWNT–PDMS composite and a thick layer (0.75 mm) of pristine PDMS. Compared with a 5 wt% random MWNT–PDMS composite,[102] the 5 wt% super-aligned MWNT-PDMS composite layer shows approximately two orders of magnitude decrease in CTE (from 3×10^{-4} K^{-1} to 6×10^{-6}

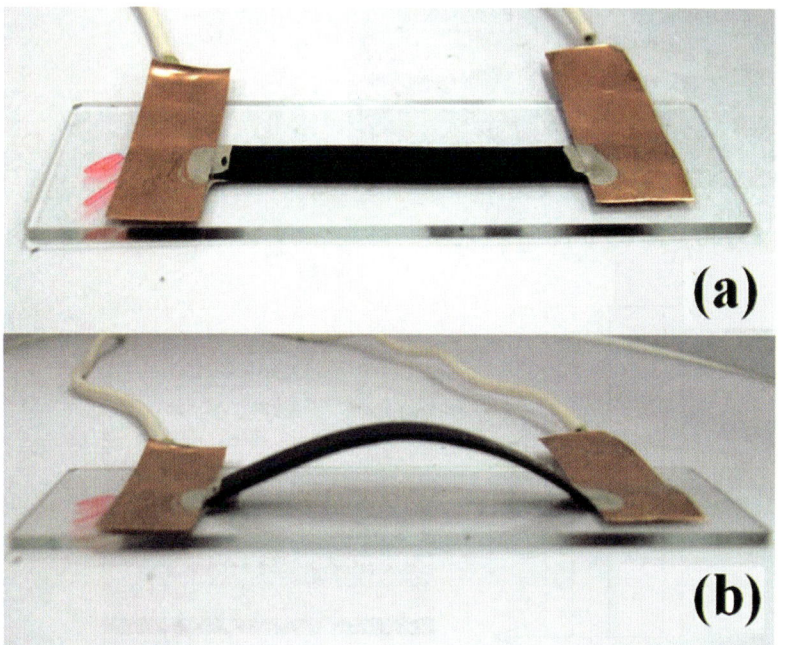

Figure 2.14 Photographs of a 5 wt% MWNT-PDMS composite before (a) and after (b) an electric field of 1.5 KV m^{-1} is applied. Reprinted by permission from the American Institute of Physics.[102]

K^{-1}) and two orders of magnitude increase in electrical conductivity (from 0.1 to 23 S cm^{-1}).[103] In the electrothermal actuator (Figure 2.15), resistive heating of the CNT-rich layer heats up the entire actuator. As the CTE mismatch between the CNT-rich layer in the direction parallel to nanotube alignment and the pure PDMS layer is enormous, the same temperature rise due to the resistive heating renders a much larger expansion of pure PDMS layer than that of the CNT-rich layer, which results in the bending actuation of the structure at very low electric fields (<700 V m^{-1}).

In electrothermal actuation, thermomechanical responses are induced through the inclusion of conductive nanoparticles such as CNTs that co-operatively behave as an internal resistive heating element. In addition to thermal expansion, which can be effective above or below the glass transition (T_g), Sellinger et al.[104] have shown that mechanical softening associated with the reversible phase transition at T_g can also be utilized, yielding substantial strain increases over a small temperature increase. For example, the reversible softening associated with temperature transitions through T_g in a 5 vol% SWNT–polyimide (CP2) nanocomposite provides ∼4% of strain increases over relatively small changes in an applied electric field ($\Delta E \approx 0.01$ MV m^{-1}) (Figure 2.16).[104,105]

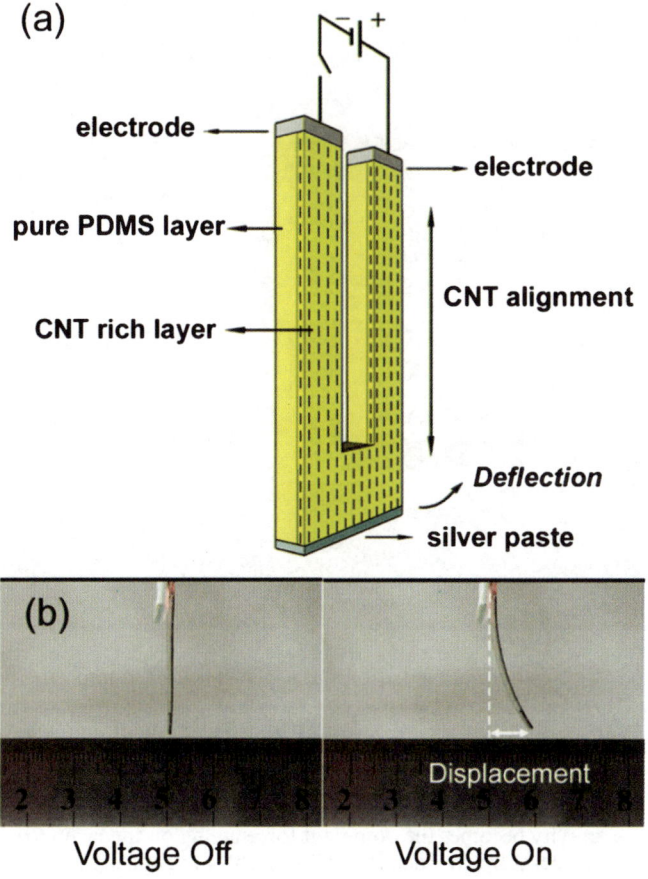

Figure 2.15 Electrothermal bending of a MWNT–PDMS composite film. (a) Schematic structure of a U-shaped actuator consisting of a thin layer of super-aligned MWNT–PDMS composite and a thick layer of pristine PDMS. The dashed lines represent the direction of CNT alignment. (b) Photographs of the actuator without (left) and with (right) an applied DC voltage of 40 V. Reprinted by permission from the American Chemical Society.[103]

Hu et al.[106] have reported the actuation behavior of an electroactive 25 wt% SWNT–chitosan composite. Under low alternating wave voltage, the SWNT–chitosan composite strip simultaneously vibrates in atmospheric conditions. The frequency and waveform of the motion are consistent with those of the applied voltage. It is believed that the vibrations are mostly controlled by thermal expansion and contraction of the polymer matrix, which is caused by periodical heating when an alternating current is passed through the nanotube conductive network.

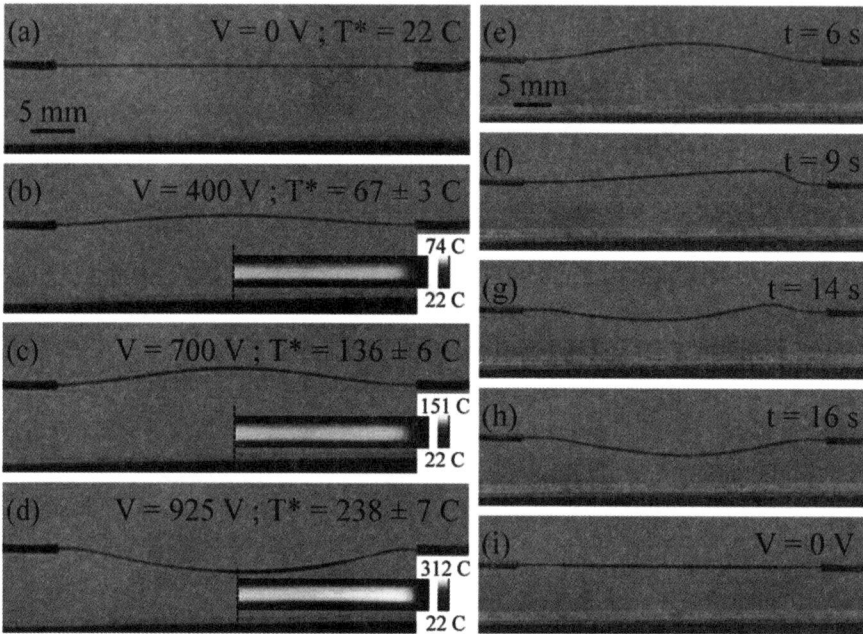

Figure 2.16 Electrothermal actuation of a SWNT–CP2 composite beam. (a–d) Deflection of a 0.5 vol% SWNT–CP2 beam ($L = 4$ cm; $d = 140$ μm; $w = 1.9$ mm) as a function of applied voltage and temperature. For $T^* < T_g$ (~220 °C), thermal expansion induces a positive deflection in the beam due to a buckling instability (b, c). At $T^* \approx T_g$, mechanical softening promotes a deflection inversion as gravity becomes the dominant force acting on the beam (d). The sample temperature was determined at each voltage using the thermal images shown (insets). (e–i) Deflection of a 0.5 vol% SWNT–CP2 beam as a function of time at V_{th} (voltage where T^* is first observed to exceed T_g). The sample initially buckles as a result of thermal expansion (e); however, as the sample temperature continues to increase, the sample kinks (f, g), and then eventually sags under its own weight at T_g (h). Once the voltage is removed, the beam reversibly actuates back to its original form (i). Reprinted by permission from Wiley-VCH Verlag GmbH & Co. KGaA.[104]

2.4 CNT–Polymer Composite IR Sensors

Organic electronic materials offer ease of material processing and integration, low cost, physical flexibility, and large device area as compared with traditional inorganic semiconductors. Optoelectronic materials that are responsive at wavelengths in the near-IR region (*e.g.*, 800–2000 nm) are highly desirable for various demanding applications such as telecommunication, thermal imaging, remote sensing, thermal photovoltaics, and solar cells.[108] SWNTs have strong and specific absorptions in the near-IR region owing to the first optical transition (S_{11}) of semiconducting nanotubes

Actuators and Infrared Sensors Based on Carbon Nanotube–Polymer Composites 43

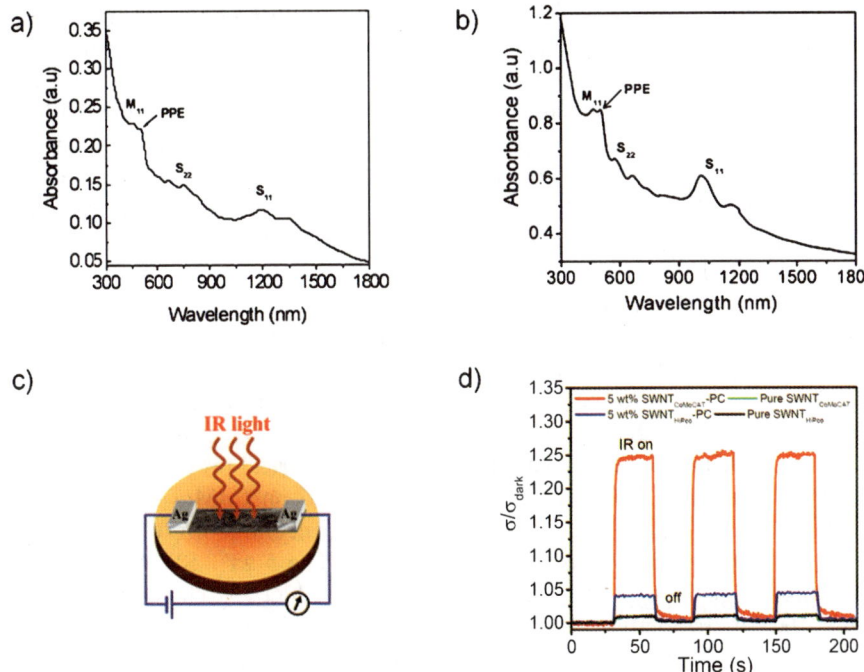

Figure 2.17 CNT–polymer composite IR sensors. UV-Vis-near-IR spectra of (a) 5 wt% $SWNT_{HiPco}$–PC and (b) 5 wt% $SWNT_{CoMoCAT}$–PC composite thin films. M_{11}, S_{11}, and S_{22} represent optical transitions in metallic and semiconducting SWNTs, respectively.(c) Schematic experimental setup. Near-IR light covers all of the sample area. (d) Relative conductivity (σ/σ_{dark}) responses of SWNT–PC nanocomposites to the on/off IR illumination (power intensity: 7 mW mm^{-2}).

(Figures 2.17a and 2.17b).[33–35] The inverse diameter dependence of the S_{11} optical transition energy enables the wavelength-tuning of the near-IR absorptions of SWNTs (Figure 2.1), which, coupled with rapid progress in nanotube separation,[38–40] will ultimately allow the development of SWNT IR sensors tailored to specific regions of the near-IR spectrum. There are a number of reports on the IR photoelectrical property of SWNT films. Levitsky and Euler[109] have demonstrated that the arc-produced SWNT ($SWNT_{arc}$) film is capable of generating a very weak photocurrent upon continuous-wave IR illumination (12 mW mm^{-2}) in the air at room temperature, and the current increase upon IR illumination is only approximately 0.2%. In addition, they have observed a dark current drift due to oxygen adsorption and a relatively slow rise/decay ($\sim$4–5 s relaxation time) of the photocurrent in response to the on/off irradiation. Itkis et al.[37] have reported that the IR photoresponse in the electrical conductivity of a $SWNT_{arc}$ film can be dramatically improved when the nanotube film is suspended in vacuum at low temperature (e.g., 50 K).

Chen and co-workers[91,92] have discovered that the IR photoresponse in the electrical conductivity of SWNTs is dramatically enhanced by embedding SWNTs in an insulating polymer matrix such as polycarbonate (PC) in the air at room temperature. Schematic experimental setup is shown in Figure 2.17(c). Near-IR light covers all sample area to minimize the localized position-sensitive effect.[110–115] In contrast to the gradual photoresponse and weak conductivity change (1.10%) observed in a HiPco-produced SWNT ($SWNT_{HiPco}$) film in the air at room temperature, the 5 wt% $SWNT_{HiPco}$–PC nanocomposite under the same IR illumination (power intensity: 7 mW mm^{-2}) shows a sharp photoresponse and strong conductivity change (4.26%, Figure 2.17d).[91] The dark current drift owing to oxygen adsorption is also minimized by embedding nanotubes in the polymer matrix. Semiconducting SWNTs are found to be critical to the IR photoresponse of nanotube materials.[91,92] Both SWNT types and nanotube–matrix polymer–nanotube junctions have profound impact on the IR photoelectrical property of SWNT–polymer composites.[92] Composite IR sensors based on CoMoCAT-produced SWNTs ($SWNTs_{CoMoCAT}$), which have more semiconducting tubes, significantly outperform those based on $SWNTs_{HiPco}$. The 5 wt% $SWNT_{CoMoCAT}$–PC nanocomposite demonstrates a very strong conductivity change of 23.45% upon the IR irradiation (7 mW mm^{-2}) in the air at room temperature (Figure 2.17d), which is 5.5 times of that (4.26%) observed in the 5 wt% $SWNT_{HiPco}$–PC composite film, nearly 27 times of that (0.88%) observed in the pure $SWNT_{CoMoCAT}$ film, 21 times of that (1.10%) observed in the pure $SWNT_{HiPco}$ film, and 117 times of that (0.2%) observed in the pure $SWNT_{arc}$ film in the air at room temperature (12 mW mm^{-2}). In addition, the 5 wt% $SWNT_{CoMoCAT}$–PC composite film shows a detectable IR photoresponse at a light intensity as low as 23.4 μW mm^{-2}.[92]

Levitsky and co-workers[116] have investigated isotropic and anisotropic SWNT–polystyrene composites, respectively, for mid-IR (2.5–20 μm) sensing applications. They have demonstrated that the composite alignment in conjunction with non-uniform distribution of nanotubes from the surface to the bulk in the polymer matrix enables a significant enhancement of the temperature coefficient of resistance, which leads to high mid-IR bolometric responsivity.

2.5 Conclusions

CNTs represent a rare class of materials, which demonstrate a number of outstanding properties in a single material system, such as high aspect ratio, small diameter, light weight, high mechanical strength, high electrical and thermal conductivities, and tunable near-IR optical and optoelectronic properties. Unlike traditional carbon fiber–polymer composites, which mainly exhibit passive mechanical, electrical, and thermal properties, CNT–polymer composites, if engineered appropriately, display both passive and active material functions. The synergy between multifunctional CNTs and the polymer matrix has been judiciously exploited to create highly desirable smart polymer composite devices such as shape-memory and shape-changing actuators, and IR sensors, which have demonstrated either new or dramatically enhanced actuation and sensing properties. Although not highlighted in this Chapter, other promising smart material functions have also been harnessed in CNT–polymer composites, including thermoelectric power devices that convert heat into electricity,[117–121] and dielectric polymer composites using core-shell MWNTs as fillers for electrical energy storage.[122,123] Rapid advances in the manipulation of CNTs in a polymer matrix[17,64,73,87,93,95,103,124–134] and nanotube separation[38–40,135] could ultimately enable precise control in both the architecture and properties of the CNT network in the polymer matrix. Although active CNT–polymer composites have already shown tremendous potentials, great challenges remain, including a better understanding of CNT–polymer interfaces and CNT–polymer–CNT junctions and their time- and space-resolved responses towards stimuli such as light, electric, or magnetic fields.

Acknowledgements

I am greatly indebted to my students and collaborators, and my colleagues in this field, whose names are cited in the references. Financial support from the National Science Foundation (DMI-06200338, CMMI-0625245, and CMMI-0856162), UWM start-up fund, UWM Research Growth Initiative award, and the Lynde and Harry Bradley Foundation is gratefully acknowledged.

References

1. A. Thess, R. Lee, P. Nikolaev, H. J. Dai, P. Petit, J. Robert, C. H. Xu, Y. H. Lee, S. G. Kim, A. G. Rinzler, D. T. Colbert, G. E. Scuseria, D. Tomanek, J. E. Fischer and R. E. Smalley, *Science*, 1996, **273**, 483.
2. I. Yakobson and R. E. Smalley, *Am. Sci.*, 1997, **85**, 324.
3. R. Saito, G. Dresselhaus and M. S. Dresselhaus, *Physical Properties of Carbon Nanotubes*, Imperial College Press, London, 1998.
4. P. M. Ajayan, *Chem. Rev.*, 1999, **99**, 1787.
5. R. H. Baughman, A. A. Zakhidov and W. A. de Heer, *Science*, 2002, **297**, 787.
6. P. M. Ajayan, L. S. Schadler, C. Giannaris and A. Rubio, *Adv. Mater.*, 2000, **12**, 750.
7. M. J. Biercuk, M. C. Llaguno, M. Radosavljevic, J. K. Hyun, A. T. Johnson and J. E. Fischer, *Appl. Phys. Lett.*, 2002, **80**, 2767.
8. C. Park, Z. Ounaies, K. A. Watson, R. E. Crooks, J. Smith, S. E. Lowther, J. W. Connell, E. J. Siochi, J. S. Harrison and T. L. St. Clair, *Chem. Phys. Lett.*, 2002, **364**, 303.
9. R. Andrews, D. Jacques, M. Minot and T. Rantell, *Macromol. Mater. Eng.*, 2002, **287**, 395.

10. A. Mamedov, N. A. Kotov, M. Prato, D. M. Guldi, J. P. Wicksted and A. Hirsch, *Nat. Mater.*, 2002, **1**, 190.
11. A. Mitchell, J. L. Bahr, S. Arepalli, J. M. Tour and R. Krishnamoorti, *Macromolecules*, 2002, **35**, 8825.
12. J. Zhu, J. Kim, H. Peng, J. L. Margrave, V. N. Khabashesku and E. V. Barrera, *Nano Lett.*, 2003, **3**, 1107.
13. Y. Lin, B. Zhou, K. A. Shiral Fernando, P. Liu, L. F. Allard and Y.-P. Sun, *Macromolecules*, 2003, **36**, 7199.
14. G. Viswanathan, N. Chakrapani, H. Yang, B. Wei, H. Chung, K. Cho, C. Y. Ryu and P. M. Ajayan, *J. Am. Chem. Soc.*, 2003, **125**, 9258.
15. Z. Yao, N. Braidy, G. A. Botton and A. Adronov, *J. Am. Chem. Soc.*, 2003, **125**, 16015.
16. R. Ramasubramaniam, J. Chen and H. Liu, *Appl. Phys. Lett.*, 2003, **83**, 2928.
17. J. C. Grunlan, A. R. Mehrabi, M. V. Bannon and J. L. Bahr, *Adv. Mater.*, 2004, **16**, 150.
18. J. Chen, R. Ramasubramaniam, C. Xue and H. Liu, *Adv. Funct. Mater.*, 2006, **16**, 114.
19. B. R. Sankapal, K. Setyowati, J. Chen and H. Liu, *Appl. Phys. Lett.*, 2007, **91**, 173103.
20. T. Thostenson, Z. Ren and T.-W. Chou, *Compos. Sci. Technol.*, 2001, **61**, 1899.
21. R. Andrews and M. C. Weisenberger, *Curr. Opin. Solid State Mater. Sci.*, 2004, **8**, 31.
22. M. Moniruzzaman and K. I. Winey, *Macromolecules*, 2006, **39**, 5194.
23. K. I. Winey, T. Kashiwagi and M. Mu, *MRS Bull.*, 2007, **32**, 348.
24. J. Chen, M. A. Hamon, H. Hu, Y. Chen, A. M. Rao, P. C. Eklund and R. C. Haddon, *Science*, 1998, **282**, 95.
25. Y.-P. Sun, K. Fu, Y. Lin and W. Huang, *Acc. Chem. Res.*, 2002, **35**, 1096.
26. J. L. Bahr, J. Yang, D. V. Kosynkin, M. J. Bronikowski, R. E. Smalley and J. M. Tour, *J. Am. Chem. Soc.*, 2001, **123**, 6536.
27. M. J. O'Connell, P. Boul, L. M. Ericson, C. Huffman, Y. H. Wang, E. Haroz, C. Kuper, J. Tour, K. D. Ausman and R. E. Smalley, *Chem. Phys. Lett.*, 2001, **342**, 265.
28. B. Dalton, C. Stephan, J. N. Coleman, B. McCarthy, P. M. Ajayan, S. Lefrant, P. Bernier, W. J. Blau and H. J. Byrne, *J. Phys. Chem. B*, 2000, **104**, 10012.
29. A. Star, J. F. Stoddart, D. Steuerman, M. Diehl, A. Boukai, E. W. Wong, X. Yang, S.-W. Chung, H. Choi and J. R. Heath, *Angew. Chem. Int. Ed.*, 2001, **40**, 1721.
30. R. J. Chen, Y. Zhang, D. Wang and H. Dai, *J. Am. Chem. Soc.*, 2001, **123**, 3838.
31. J. Chen, H. Liu, W. A. Weimer, M. D. Halls, D. H. Waldeck and G. C. Walker, *J. Am. Chem. Soc.*, 2002, **124**, 9034.
32. D. Tasis, N. Tagmatarchis, A. Bianco and M. Prato, *Chem. Rev.*, 2006, **106**, 1105.

33. J. W. G. Wildoer, L. C. Venema, A. G. Rinzler, R. E. Smalley and C. Dekker, *Nature*, 1998, **391**, 59.
34. T. W. Odom, J.-L. Huang, P. Kim and C. M. Lieber, *Nature*, 1998, **391**, 62.
35. M. A. Hamon, M. E. Itkis, S. Niyogi, T. Alvaraez, C. Kuper, M. Menon and R. C. Haddon, *J. Am. Chem. Soc.*, 2001, **123**, 11292.
36. F. Wang, G. Dukovic, L. E. Brus and T. F. Heinz, *Science*, 2005, **308**, 838.
37. M. E. Itkis, F. Borondics, A. Yu and R. C. Haddon, *Science*, 2006, **312**, 413.
38. A. Nish, J.-Y. Hwang, J. Doig and R. J. Nicholas, *Nat. Nanotech.*, 2007, **2**, 640.
39. M. C. Hersam, *Nat. Nanotech.*, 2008, **3**, 387.
40. X. Tu, S. Manohar, A. Jagota and M. Zheng, *Nature*, 2009, **460**, 250.
41. R. Vaia and J. Baur, *Science*, 2008, **319**, 420.
42. A. Lendlein and S. Kelch, *Angew. Chem. Int. Ed.*, 2002, **41**, 2034.
43. A. Lendlein, H. Jiang, O. Jünger and R. Langer, *Nature*, 2005, **434**, 879.
44. C. Liu, H. Qin and P. T. Mather, *J. Mater. Chem.*, 2007, **17**, 1543.
45. A. Rousseau, *Polym. Eng. Sci.*, 2008, **48**, 2075.
46. D. Ratna and J. Karger-Kocsis, *J. Mater. Sci.*, 2008, **43**, 254.
47. P. T. Mather, X. Luo and I. A. Rousseau, *Annu. Rev. Mater. Res.*, 2009, **39**, 445.
48. J. Leng, H. Lu, Y. Liu, W. M. Huang and S. Du, *MRS Bull.*, 2009, **34**, 848.
49. M. Behl, M. Y. Razzaq and A. Lendlein, *Adv. Mater.*, 2010, **22**, 3388.
50. M. Behl, J. Zotzmann and A. Lendlein, *Adv. Polym. Sci.*, 2010, **226**, 1.
51. S. A. Madbouly and A. Lendlein, *Adv. Polym. Sci.*, 2010, **226**, 41.
52. M. Yakacki and K. Gall, *Adv. Polym. Sci.*, 2010, **226**, 147.
53. C. Wischke, A. T. Neffe and A. Lendlein, *Adv. Polym. Sci.*, 2010, **226**, 177.
54. J. Hu and S. Chen, *J. Mater. Chem.*, 2010, **20**, 3346.
55. W. Small IV, P. Singhal, T. S. Wilson and D. J. Maitland, *J. Mater. Chem.*, 2010, **20**, 3356.
56. T. Xie, *Polymer*, 2011, **52**, 4985.
57. J. Leng, X. Lan, Y. Liu and S. Du, *Prog. Mater. Sci.*, 2011, **56**, 1077.
58. H. Y. Jiang, S. Kelch and A. Lendlein, *Adv. Mater.*, 2006, **18**, 1471.
59. H. Koerner, T. J. White, N. V. Tabiryan, T. J. Bunning and R. A. Vaia, *Mater. Today*, 2008, **11**, 34.
60. H. Koerner, G. Price, N. A. Pearce, M. Alexander and R. A. Vaia, *Nat. Mater.*, 2004, **3**, 115.
61. Q. Q. Ni, C. S. Zhang, Y. Fu, G. Dai and T. Kimura, *Compos. Struct.*, 2007, **81**, 176.
62. U. Khan, P. May, A. O'Neill, J. J. Vilatela, A. H. Windle and J. N. Coleman, *Small*, 2011, **7**, 1579.
63. H. Xia and M. Song, *J. Mater. Chem.*, 2006, **16**, 1843.

64. U. Khan, F. M. Blighe and J. N. Coleman, *J. Phys. Chem. C*, 2010, **114**, 11401.
65. R. Mohr, K. Kratz, T. Weigel, M. Lucka-Gabor, M. Moneke and A. Lendlein, *Proc. Natl. Acad. Sci. U.S.A.*, 2006, **103**, 18043.
66. Z. He, N. Satarkar, T. Xie, Y.-T. Cheng and J. Z. Hilt, *Adv. Mater.*, 2011, **23**, 3192.
67. H. Koerner, W. Liu, M. Alexander, P. Mirau, H. Dowty and R. A. Vaia, *Polymer*, 2005, **46**, 4405.
68. W. Cho, J. W. Kim, Y. C. Jung and N. S. Goo, *Macromol. Rapid Commun.*, 2005, **26**, 412.
69. P. Miaudet, A. Derré, M. Maugey, C. Zakri, P. M. Piccione, R. Inoubli and P. Poulin, *Science*, 2007, **318**, 1294.
70. Y. Liu, H. Lv, X. Lan, J. Leng and S. Du, *Compos. Sci. Technol.*, 2009, **69**, 2064.
71. L. Viry, C. Mercader, P. Miaudet, C. Zakri, A. Derré, A. Kuhn, M. Maugeya and P. Poulin, *J. Mater. Chem.*, 2010, **20**, 3487.
72. X. Luo and P. T. Mather, *Soft Matter*, 2010, **6**, 2146.
73. K. Yu, Z. Zhang, Y. Liu and J. Leng, *Appl. Phys. Lett.*, 2011, **98**, 074102.
74. Y. Bar-Cohen, *Electroactive Polymer (EAP) Actuators as Artificial Muscles-Reality, Potential and Challenges*, SPIE Press, Bellingham, WA, 2001.
75. J. D. Madden, N. A. Vandesteeg, P. A. Anquetil, P. G. A. Madden, A. Takshi, R. Z. Pytel, S. R. Lafontaine, P. A. Wieringa and I. W. Hunter, *IEEE J. Ocean. Eng.*, 2004, **29**, 706.
76. S. Ahir and E. M. Terentjev, *Nat. Mater.*, 2005, **4**, 491.
77. R. Vaia, *Nat. Mater.*, 2005, **4**, 429.
78. S. V. Ahir, Y. Y. Huang and E. M. Terentjev, *Polymer*, 2008, **49**, 3841.
79. H. Finkelmann, H.-J. Kock and G. Rehage, *Makromol. Chem. Rapid Commun.*, 1981, **2**, 317.
80. M. Warner and E. M. Terentjev, *Liquid Crystal Elastomers*, Oxford University Press, Oxford, 2007.
81. X. Ping and R. Zhang, *J. Mater. Chem.*, 2005, **15**, 2529.
82. T. Ikeda, J. Mamiya and Y. Yu, *Angew. Chem. Int. Ed.*, 2007, **46**, 506.
83. *Cross-linked Liquid Crystalline Systems: From Rigid Polymer Networks to Elastomers*, ed. D. Broer, G. Crawford and S. Žumer, CRC Press, Boca Raton, FL, 2011.
84. L. Thomsen, III, P. Keller, J. Naciri, R. Pink, H. Jeon, D. Shenoy and B. Ratna, *Macromolecules*, 2001, **34**, 5868.
85. J. Naciri, A. Srinivasan, H. Jeon, N. Nikolov, P. Keller and B. R. Ratna, *Macromolecules*, 2003, **36**, 8499.
86. S. V. Ahir, A. M. Squires, A. R. Tajbakhsh and E. M. Terentjev, *Phys. Rev. B*, 2006, **73**, 085420.
87. S. Courty, J. Mine, A. R. Tajbakhsh and E. M. Terentjev, *Europhys. Lett.*, 2003, **64**, 654.
88. C. Li, Y. Liu, C.-W. Lo and H. Jiang, *Soft Matter*, 2011, **7**, 7511.

89. J. Chen, C. Xue, R. Ramasubramaniam and H. Liu, *Carbon*, 2006, **44**, 2142.
90. K. Setyowati, M. J. Piao, J. Chen and H. Liu, *Appl. Phys. Lett.*, 2008, **92**, 043105.
91. B. Pradhan, K. Setyowati, H. Liu, D. H. Waldeck and J. Chen, *Nano Lett.*, 2008, **8**, 1142.
92. B. Pradhan, R. R. Kohlmeyer, K. Setyowati, H. A. Owen and J. Chen, *Carbon*, 2009, **47**, 1686.
93. B. Pradhan, R. R. Kohlmeyer and J. Chen, *Carbon*, 2010, **48**, 217.
94. R. R. Kohlmeyer, M. Lor, J. Deng, H. Liu and J. Chen, *Carbon*, 2011, **49**, 2352.
95. L. Yang, K. Setyowati, A. Li, S. Gong and J. Chen, *Adv. Mater.*, 2008, **20**, 2271.
96. Y. Ji, Y. Y. Huang, R. Rungsawang and E. M. Terentjev, *Adv. Mater.*, 2010, **22**, 3436.
97. J. E. Marshall, Y. Ji, N. Torras, K. Zinoviev and E. M. Terentjev, *Soft Matter*, 2012, **8**, 1570.
98. J. Küpfer and H. Finkelmann, *Makromol. Chem. Rapid Commun.*, 1991, **12**, 717.
99. S. Zhang, N. Zhang, C. Huang, K. Ren and Q. Zhang, *Adv. Mater.*, 2005, **17**, 1897.
100. C. Park, J. H. Kang, J. S. Harrison, R. C. Costen and S. E. Lowther, *Adv. Mater.*, 2008, **20**, 2074.
101. J. F. Nye, *Physical Properties of Crystals*, Oxford University Press, Oxford, UK, 1976.
102. L. Z. Chen, C. H. Liu, C. H. Hu and S. S. Fan, *Appl. Phys. Lett.*, 2008, **92**, 263104.
103. L. Chen, C. K. Liu, C. Meng, C. Hu, J. Wang and S. Fan, *ACS Nano*, 2011, **5**, 1588.
104. T. Sellinger, D. H. Wang, L.-S. Tan and R. A. Vaia, *Adv. Mater.*, 2010, **22**, 3430.
105. J. Arlen, D. Wang, J. D. Jacobs, R. Justice, A. Trionfi, J. W. P. Hsu, D. Schaefer, L. S. Tan and R. A. Vaia, *Macromolecules*, 2008, **41**, 8053.
106. Y. Hu, W. Chen, L. Lu, J. Liu and C. Chang, *ACS Nano*, 2010, **4**, 3498.
107. A. Govindaraju, A. Chakraborty and C. Luo, *J. Micromech. Microeng.*, 2005, **15**, 1303.
108. S. A. McDonald, G. Konstantatos, S. Zhang, P. W. Cyr, E. J. D. Klem, L. Levina and E. H. Sargent, *Nat. Mater.*, 2005, **4**, 138.
109. A. Levitsky and W. B. Euler, *Appl. Phys. Lett.*, 2003, **83**, 1857.
110. H. Lien, W. K. Hsu, H. W. Zan, N. H. Tai and C. H. Tsai, *Adv. Mater.*, 2006, **18**, 98.
111. S. X. Lu and B. Panchapakesan, *Nanotechnology*, 2006, **17**, 1843.
112. C. A. Merchant and N. Markovic, *Appl. Phys. Lett.*, 2008, **92**, 3.
113. P. Stokes, L. W. Liu, J. H. Zou, L. Zhai, Q. Huo and S. I. Khondaker, *Appl. Phys. Lett.*, 2009, **94**, 3.

114. B. C. St-Antoine, D. Ménard and R. Martel, *Nano Lett.*, 2009, **9**, 3503.
115. B. K. Sarker, M. Arif and S. I. Khondaker, *Carbon*, 2010, **48**, 1539.
116. Y. Glamazda, V. A. Karachevtsev, W. B. Euler and I. A. Levitsky, *Adv. Funct. Mater.*, 2012, **22**, 2177.
117. C. Yu, Y. S. Kim, D. Kim and J. C. Grunlan, *Nano Lett.*, 2008, **8**, 4428.
118. D. Kim, Y. Kim, K. Choi, J. C. Grunlan and C. Yu, *ACS Nano*, 2010, **4**, 513.
119. C. Hu, C. Liu, L. Chen, C. Meng and S. Fan, *ACS Nano*, 2010, **4**, 4701.
120. C. Yu, K. Choi, L. Yin and J. C. Grunlan, *ACS Nano*, 2011, **5**, 7885.
121. C. A. Hewitt, A. B. Kaiser, S. Roth, M. Craps, R. Czerw and D. L. Carroll, *Nano Lett.*, 2012, **12**, 1307.
122. R. R. Kohlmeyer, A. Javadi, B. Pradhan, S. Pilla, K. Setyowati, J. Chen and S. Gong, *J. Phys. Chem. C*, 2009, **113**, 17626.
123. T. Zhou, J.-W. Zha, Y. Hou, D. Wang, J. Zhao and Z.-M. Dang, *ACS Appl. Mater. Interfaces*, 2011, **3**, 4557.
124. L. Jin, C. Bower and O. Zhou, *Appl. Phys. Lett.*, 1998, **73**, 1197.
125. T. Kimura, H. Ago, M. Tobita, S. Ohshima, M. Kyotani and M. Yumura, *Adv. Mater.*, 2002, **14**, 1380.
126. F. Du, J. E. Fischer and K. I. Winey, *Phys. Rev. B*, 2005, **72**, 121404.
127. B. S. Shim and N. A. Kotov, *Langmuir*, 2005, **21**, 9381.
128. C. Park, J. Wilkinson, S. Banda, Z. Ounaies, K. E. Wise, G. Sauti, P. T. Lillehei and J. S. Harrison, *J. Polym. Sci., Part B: Polym. Phys.*, 2006, **44**, 1751.
129. N. Akima, Y. Iwasa, S. Brown, A. M. Barbour, J. Cao, J. L. Musfeldt, H. Matsui, N. Toyota, M. Shiraishi, H. Shimoda and O. Zhou, *Adv. Mater.*, 2006, **18**, 1166.
130. G. Yu, A. Cao and C. M. Lieber, *Nat. Nanotech.*, 2007, **2**, 372.
131. N. Adachi, T. Fukawa, Y. Tatewaki, H. Shirai and M. Kimura, *Macromol. Rapid Commun.*, 2008, **29**, 1877.
132. S. Shoji, H. Suzuki, R. P. Zaccaria, Z. Sekkat and S. Kawata, *Phys. Rev. B*, 2008, **77**, 153407.
133. Q. Wang, J. Dai, W. Li, Z. Wei and J. Jiang, *Compos. Sci. Technol.*, 2008, **68**, 1644.
134. W. Wang, X. Sun, W. Wu, H. Peng and Y. Yu, *Angew. Chem. Int. Ed.*, 2012, **51**, 4644.
135. H. Liu, D. Nishide, T. Tanaka and H. Kataura, *Nat. Commun.*, 2011, **2**, 309.

CHAPTER 3

Photoelectrical Responses of Carbon Nanotube–Polymer Composites

YUMENG SHI*[a] AND LAIN-JONG LI*[b]

[a] SUTD-MIT International Design Centre, Singapore University of Technology and Design, 20 Dover Drive, 138682, Singapore; [b] Institute of Atomic and Molecular Sciences, Academia Sincia, Taipei, 10617, Taiwan
* E-mail: yumeng@sutd.edu.sg or lanceli@gate.sinica.edu.tw

3.1 Introduction

Because of the unique electrical properties and high charge sensitivities of single-walled carbon nanotubes (SWNTs)[1] it has been considered and demonstrated as a potential material for a variety of applications such as field-effect transistors (FETs),[2] memory devices,[3] photovoltaic devices,[4–6] and chemical/biological sensors.[7,8] SWNTs are also attractive in optoelectronic applications due to their unique electronic structures. The pseudo one-dimensional geometry of SWNTs, leading to their unique band structures and large Coulomb interaction between electrons and holes,[9] strongly dominates their optical and electro-optical characteristics. It is therefore important to study the fundamentals of electrical-to-optical or optical-to-electrical conversion for SWNT-based devices. The photoconductivity of SWNTs has been attributed to the direct excitation through series of van Hove singularities of SWNTs.[10–12] The photocurrent is due to the photon-induced exciton generation and subsequent charge separation by an electric field. The effect

---

RSC Nanoscience & Nanotechnology No. 27
Carbon Nanotube-Polymer Composites
Edited by Dimitrios Tasis
© The Royal Society of Chemistry 2013
Published by the Royal Society of Chemistry, www.rsc.org

of metal electrodes on the photoconductivity of nanotubes through the modulation of Schottky barrier height has also been revealed.[13] On the other hand, photosensitive polymers are widely used in organic optoelectronic devices such as organic light-emitting diodes (OLEDs) and organic photovoltaic cells (OPVs). The interaction between photosensitive polymers and carbon nanotube (CNT) transistors has been studied. The observed optoelectronic switching behaviors of the SWNT–polymer composite devices have been proposed to serve as photodetector, photovoltaic, and memory devices.[14,15] In this Chapter, we mainly discuss the photoelectrical responses from the photosensitive polymer-coated SWNTs. The SWNTs can be functionalized with photosensitive polymers to form CNT–polymer composite films, and these composites become sensitive to visible or ultraviolet (UV) wavelength, which largely extends the photosensitivity and detetion ranges.

A SWNT can be treated as a rolled-up cylinder from a graphene sheet along a certain direction of graphene crystal. Depending on the roll-up chirality, determined by the roll-up angle and diameter, SWNTs show either metallic or semiconducting electronic structures. Benefiting from the unique one-dimensional and defect-free structure, the electron transport in SWNTs exhibits ultra-high intrinsic mobility up to 100 000 cm^2 $(V\ s)^{-1}$.[16–18] The ballistic electron transport property gives SWNTs a high potential for nano-electronic applications. The simple tight-binding calculation results for a two-dimensional graphene sheet gives a good understanding on how to derive the electronic structure of SWNTs and how it deviates from that of graphene.[19–21] Here, we briefly describe the electronic structure of SWNTs and show that SWNTs can either be n- or p-type depending on their doping level, which means that both holes and electrons can transport within the conductive channel of the SWNTs. Most importantly, the direct band gap property of semiconducting SWNTs makes it possible to achieve light emission from the pristine SWNTs in which the electron and hole can recombine through its electronic gap radiatively.[22]

3.2 Band Structure and Chirality Dependence

A van Hove singularity is a discontinuity in the density of states (DOS) of a solid. SWNTs show a typical DOS of one-dimensional crystals, as illustrated in Figure 3.1. The electronic and optical properties of SWNTs can be predicted from the DOS of SWNTs. Electron energy zero presents the Fermi energy level for pristine SWNTs. Depending on the chirality, there is either an absence of DOS at Fermi energy level for semiconducting tubes (as shown in Figure 3.1, left) or a presence of DOS at Fermi energy level for metallic tubes (as shown in Figure 3.1, right). These determine the electrical properties of the respective SWNTs.

The band structure of SWNTs as a dependence of chirality was first calculated by Hiromichi Kataura and co-workers in 1999.[24] The theoretical graph based on Kataura's calculation shows the relationship between the

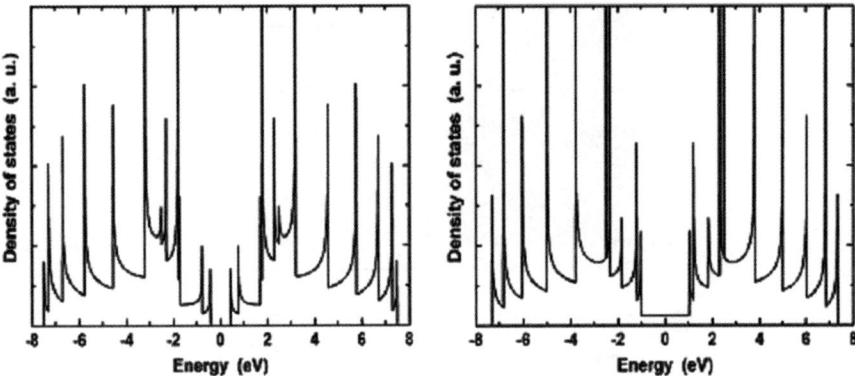

Figure 3.1 DOS for (11,0) semiconducting (left) and (12,0) metallic (right) CNTs showing van Hove singularities based on a tight-binding model. Reprinted (adapted) with permission from Anantram, M. P., and Léonard, F.,[23] *Physics of carbon nanotube electronic devices.* Rep. Prog. Phys., 2006. **69**: pp. 507–561. Copyright (2006) Institute of Physics and IOP Publishing Limited.

nanotube diameter and its band gap energy for all nanotubes in a certain diameter range. This theoretical graph was given the name "Kataura Plot" from which it is convenient to find the band gap energy of a SWNT with a defined tube chirality.

3.3 Band-to-band Transition of SWNTs

The excitation-generated electrons and holes in metallic SWNTs can immediately and non-radiatively recombine due to the absence of a band gap. Hence, the majority of optical studies have been focused on semiconducting SWNTs because the electron and hole can recombine through its electronic gap radiatively. It is noted that the transfer curve of a semiconducting SWNT exhibits an ambipolar behavior as the Fermi energy of the SWNT can be simply tuned by a gate voltage, which means that, by changing the applied external electrical field, either electron or hole charge carriers can transport within SWNTs. Therefore, the electrical luminescence can be achieved by intentionally balancing the hole and electron carrier density to generate enough hole–electron pairs *via* the application of a bias voltage across a semiconducting SWNT.[22,23] Figure 3.2 shows the example of the infrared (IR) emission caused by the recombination of electrons and holes through the 1st van Hove transition (E_{11}) in a semiconducting SWNT, in which the electrons and holes are accelerated and moving towards each other under an external bias across the tube. It can be clearly seen from Figure 3.2(A) that the emission is localized at the position of the CNT. Figure 3.2(B) further demonstrates that the emission intensity exhibits a good relation to the transfer curve (drain current *versus* gate voltage) of the SWNT transistor. Meanwhile, the inset

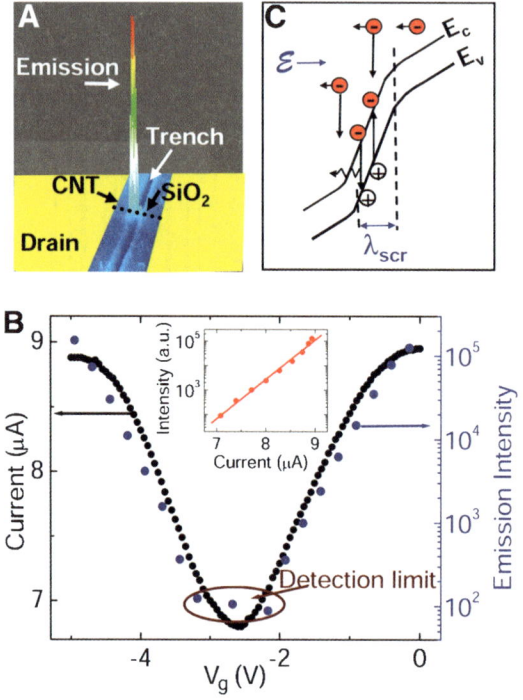

Figure 3.2 Infrared (IR) emission from a single wall carbon nanotube. (A) E_{11} IR emission overlaid with an optical image of a partially suspended carbon nanotube field effect transistor (CNTFET) with 200 nm SiO_2, Pd source–drain contacts and a Si back gate. The length of the channel is 26 μm, the width of the trench is 5 μm, and the CNT diameter is ∼2 nm. (B) Drain current (black solid circles) and emission intensity (Blue solid circles) *versus* gate bias at 1.6 μm of a partially suspended CNTFET at $V_d = -5$ V. One unit of emission intensity corresponds to approximately 240 photons s^{-1} (4π solid angle)$^{-1}$. The inset shows that the emission intensity increases exponentially with drive current. The solid line is meant as a guide to the eyes. (C) Schematics of the impact excitation process. An incoming hot electron is accelerated by the band-bending at the suspended/supported interface to energies larger than the band gap and generates an exciton that decays radiatively. Subsequently, the "cooled" electron picks up more energy from the electrical field and continues this process. Reprinted (adapted) with permission from Chen, J. et al.,[22] *Bright infrared emission from electrically induced excitons in carbon nanotubes.* Science, 2005. **310**(5751): pp. 1171–1174. Copyright (2005) The American Association for the Advancement of Science.

shows that the emission intensity increases exponentially with the driving current. Due to the band structure dependence of SWNTs on their chirality, the electronic and optical properties of SWNTs can be predicted from their diameter and rolling angle. The research work of Chen *et al.*[22] on the photoemission from SWNTs suggests that light emission with a desired wavelength could be achieved through the structure design of SWNTs.

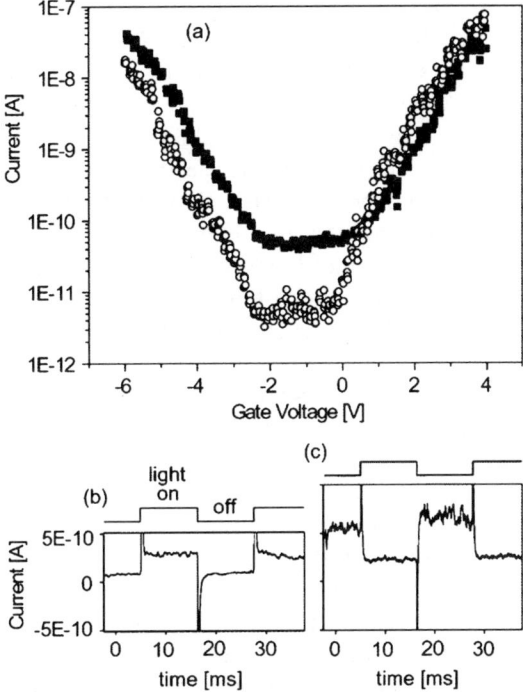

Figure 3.3 Photoconductivity of carbon nanotubes induced by IR laser-excitation. (a) Gating characteristics of an ambipolar CNT-FET with (■) and without (○) IR illumination. (The drain voltage is −0.5 V.) Time traces of the current with the light chopped, taken at gate voltages of (b) −3 V (hole conduction) and (c) +1.5 V (electron conduction). The sign of the current spikes is independent of gate voltage. The plateaus correspond to the data in (a). Reprinted (adapted) with permission from Freitag, M. et al.,[25] Photoconductivity of single carbon nanotubes. Nano Letters, 2003. **3**(8): pp. 1067–1071. Copyright (2003) American Chemical Society.

The photon emission is from the recombination of electron and hole pairs. For a reverse process, the photons can be also dissociated into electrons and holes in SWNTs. In addition to the direct observation of electrically induced photoemission in SWNTs, the IR laser-excited photoconductivity has also been observed in a SWNT-FET.[24] Figure 3.3(a) shows that the transfer curve of a semiconducting SWNT is shifted to the right-hand side with the IR illumination, meaning that the photon-generated carriers impose effective p-doping to the SWNT-FET device. Hence, the photoconductivity is opposite when the gate voltage is tuned to the p- and n-channels, as shown in Figures 3.3(b) and 3.3(c). From the dependence of wavelength and polarization of the photocurrent on the SWNTs tube, it is concluded that the photocurrent generation originates from the resonant excitation of the second exciton state of the semiconducting tube. The excited state decays to electron–hole pair at E_{11}, which further contributes to the photocurrent of SWNTs.[25]

Recently, multiple electron–hole pair generation in carbon nanotubes has also been reported.[26,27] In the p–n junction of SWNTs, a single photon with energy larger than the second electronic sub-band (E_{22}) generates multiple electron–hole pairs, which is quite different from the standard limit of photovoltaic efficiency established by Shockley and Queisser,[28] in which a single photon can only convert into one electron–hole pair and the excess energy from photons with energies greater than the band gap is lost through phonon emission. Therefore, the multiple exciton generation process in SWNTs may lead to an enhanced photocurrent generation and improved efficiency in the carbon-based photovoltaic devices.[4]

As mentioned above, the band structure of SWNTs strongly depends on the chirality of SWNTs; therefore, it is possible to utilize a desired band-to-band transition energy by selecting a SWNT with appropriate chirality. However, as SWNTs typically obtained are an ensemble of tubes with various diameters and chiralities, it is still very challenging to extract or select a specific type of SWNTs from the mixture. For instance, enrichment of semiconducting SWNTs from the as-obtained (or as-synthesized) nanotubes by all means is the major challenge for realizing the electronic applications based on SWNT-FET owing to the coexistence of metallic and semiconducting SWNTs. Despite tremendous efforts on development of the growth method of SWNTs, the selective SWNT growth technique for an effective production of an identical population of SWNTs with desired properties still requires more investigation. Meanwhile, the optical electronic application of SWNTs based on the band-to-band transition is limited to the IR or near-infrared (NIR) wavelengths. It is necessary to extend the working wavelength of the SWNT to meet the commonly used visible range for optoelectronic devices. Hence, the method of surface modifications with various functional groups to SWNTs has been widely adapted. Photosensitive polymers are extensively used in the organic optoelectronics such as OPVs and OLEDs. Carbon nanotube and photosensitive polymer composites have also been used as the photoactive layer in solar cells.[29–33] By the functionalization of SWNTs with a photosensitive polymer, the photoresponsive range of SWNTs can also be further extended to a visible wavelength of light, which opens a great avenue for the application of SWNTs in photodetector, photovoltaic, and memory devices.[14,15]

3.4 Wrapping SWNTs with Polymers

Before discussing the properties of SWNTs/polymer composites and their applications in optoelectronics, we will first introduce the interactions between SWNT and some unique polymers. SWNTs in solution phase are much easier to be utilized for device fabrication *via* wet-coating processes, especially for low cost production, compared with its solid state counterpart.[4] During the past few years, many efforts have been devoted to the searching of methods for solubilizing SWNTs in solutions. However, due to the strong tube–tube interaction, SWNTs tend to aggregate to form large bundles, resulting in poor solubility in solutions.[34,35] To achieve a stable SWNT dispersion, chemical functionalization of SWNT surfaces has been reported.[36,37] For example, the covalent functionalization of SWNTs can be done by strong acid treatment, and this oxidation process introduces hydrophilic groups along the tube sidewall. However, an obvious drawback is that the chemical functionalization unavoidably destroys the electronic properties of pristine SWNTs. Alternatively, the non-covalent surface modification such as wrapping SWNTs with surfactants or polymers can also enhance the solubility of SWNTs.[38] It has been found that many types of polymers,[39–41] including poly(vinyl pyrrolidone), poly(phenylene vinylene)s, polysaccharides, poly(9,9-dioctylfluoreny-2,7-diyl) (PFO), and DNA,[42,43] assist the dispersion of SWNTs in organic solvents. The polyfluorene-based conjugating polymer could be one of the most interesting wrapping agents, not only due to their good solubility to carbon materials without degrading the properties of SWNTs, but also for the specific wrapping selectivity for SWNTs. The stabilization energy of polyfluorene-wrapped SWNTs depends strongly on the corrugations between the surface of the SWNTs and the polymer. Experimentally, Nicholas and co-workers[44] have demonstrated that a polymer PFO exhibits high wrapping selectivity for specific diameters or chiral angles. Besides a high selectivity for large chiral angle SWNTs, enrichment of (7,5) species from the (6,5)-dominated SWNT ensemble (CoMoCAT) *via* PFO wrapping also corroborates the high selectivity on specific chirality of tubes. Similar results have also been achieved by Li and co-workers[45] when PFO is used to disperse the SWNT produced by chemical vapor deposition (CVD) based on CO disproportionation on Co-MCM-41 catalysts (Co-MCM SWNTs), where the PFO extracts a high purity ($\sim$79%) of the (7,5) species from the semiconducting SWNT ensemble. Figure 3.4 displays the photo-luminescence excitation (PLE) mappings for the SWNT ensemble dispersed in a typical sodium dodecylbenzene sulfonate (SDBS) surfactant (without tube selectivity) and that wrapped by PFO polymers.[32] The high selectivity of PFO

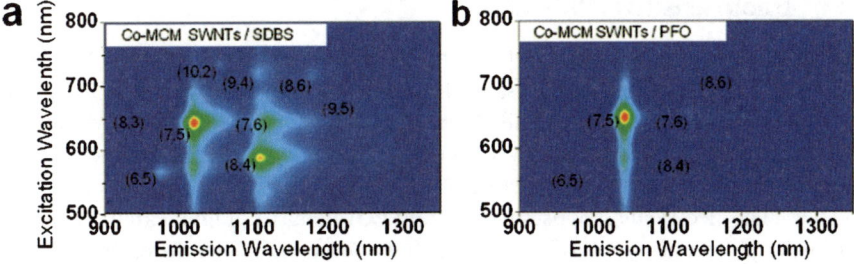

Figure 3.4 PLE mappings for the CoMCM SWNTs dispersed in (a) SDBS/D$_2$O and in (b) PFO/toluene, respectively. Reprinted (adapted) with permission from Chen, F. *et al.*,[45] *Toward the Extraction of Single Species of Single-Walled Carbon Nanotubes Using Fluorene-Based Polymers.* Nano Letters, 2007. **7**(10): pp. 3013–3017. Copyright (2007) American Chemical Society.

with a specific diameter is attributed to the formation of self-assembled intermolecular structures with an n-fold symmetry on SWNT surfaces.

Not only limited to the PFO polymer, another fluorene-based polymer poly(9,9-dioctylfluorene-*alt*-benzothiadiazole) (F8BT) is also able to discriminately wrap SWNTs with larger diameters such as a (15,4) tube.[46] It has been suggested that a stable exciplex between poly[(9,9-dioctyl-fluorenyl-2,7-diyl)-*co*-(bithiophene)] (F8T2) and SWNTs exists, in which the energy level matching between SWNTs and fluorene-based polymers is responsible for the chirality selectivity.

3.5 Energy Transfer from Photosensitive Polymers to SWNTs

Photon absorption in photosensitive polymers forms bound-state electron–hole pairs which are called excitons.[35] These excitons can radiatively recombine and emit photons through band-to-band transition of the polymer, which is known as photoluminescence. The electron–hole pairs can also transfer from polymers to SWNTs, which is called energy transfer. Direct evidence of the PLE mapping proves that the photon-generated excitons in aromatic polymers are transferred from conjugated polymers to SWNTs, resulting in NIR photoemission from the semiconducting SWNTs.[47,48] The enhanced band-gap emission of SWNTs in the NIR region is originated from the polymer wrapping layer which acts as a light-harvesting coating. Due to the strong interaction between the polymer and SWNTs, the absorbed photons can be utilized as the excitation light source for SWNTs band to band transition *via* energy transfer between SWNTs and polymer. The effect of polymer concentration on the energy transfer process from fluorene-based polymers to SWNTs has also been studied,[49] where the polymer concentration governs the polymer aggregation on SWNTs which directly affects the energy transfer process.

3.6 Photoelectric Responses from the SWNTs Coated with Photosensitive Polymers

In addition to energy transfer from photosensitive polymers to SWNTs, the electron–hole pairs can also dissociate at the interface between SWNTs and polymers, which process is called charge transfer and is governed by the electronic energy alignment between SWNTs and polymers. The charge transfer process can be detected by the conductance change in SWNT-based electronic devices. The electrical responses to different wavelengths of light can be tuned by choosing different types of photosensitive polymers. Therefore, the photosensitivity of the photosensitive polymer greatly extends the photoelectrical responses of SWNTs. The charge transfer from photosensitive polymers to SWNTs is affected by many factors, such as Fermi energy alignment between polymer and SWNTs, the charge trapping, and the electrostatic gating. Furthermore, the one-dimensional geometry of individual SWNTs gives a low photon-capture cross-section which could limit the light harvesting. The large electron–hole recombination probability restricts the applications of SWNTs in photodetection. To solve this problem, SWNT networks have been used as the channel materials to increase the photon capture probability in the device. Self-assembled SWNT networks using an organic/inorganic phase separation method is one of the most practical way for SWNT network formation.

In this Section, we introduce the charge transfer process from photosensitive polymer to SWNTs in the following specific aspects: (1) intrinsic charge transport in polymer-coated SWNTs; (2) extrinsic effects such as electrostatic gating from photo-generated charges, and photo-gating effect from substrates; and (3) device fabrication method and its influence on the optoelectronic properties.

3.6.1 SWNT Optoelectronic Devices Based on Photosensitive Polymers

Modifying the surface of SWNTs with functional materials is one way to engineer the photoelectrical properties of SWNTs. Hence, chemical functionalization of SWNTs has been a subject attracting much attention.[50–53] Compared with the covalent functionalization of SWNTs, the non-covalent functionalization method is particularly attractive as this method presents the possibility of attaching chemical handles to SWNTs without severely disrupting the bonding network in SWNTs. The functionalization of SWNTs with chromophores has been reported,[54] while other research groups have used photosensitive polymers to achieve the modulation of the conductance in SWNTs.[14,15,55]

Photosensitive polymer-wrapped SWNTs can be used as the conductance channel for optoelectronic devices. The optoelectronic devices based on polymer-wrapped SWNTs reveal a photo-gating effect on charge transport which can rectify or amplify the current flow through the tubes. For example, Steuerman and co-workers[55] reported the interaction between conjugated polymers and SWNT ropes (or bundles), where the SWNTs assist in the photo-generated charge stabilization. Two structurally similar polymers, poly{(m-phenylenevinylene)-co-[(2,5-dioctyloxy-p-phenylene)vinylene]} (PmPV), and poly{(2,6-pyridinylenevinylene)-co-[(2,5-dioctyloxy-p-phenylene)vinylene]} (PPyPV), were investigated in their study. Both of these two polymer-wrapped SWNT devices show that the wavelength dependence of the photo-gating effect correlates to the absorption spectrum of the corresponding polymer. Star and co-workers[14] have demonstrated photoelectrical device structures based on the film of the photosensitive polymer–SWNT mixture (as shown in Figure 3.5). In these devices, the photo-generated holes are transferred to the SWNTs because of the alignment of the nanotube and polymer valence bands. The positive shift in the threshold voltage upon light exposure is the direct evidence for this effect.

The second type of photosensitive SWNT-FET devices adopts SWNT networks as the transistor channel. A polymer layer is then deposited over the contacts and nanotubes by a drop-casting or spin-coating method. As shown in Figure 3.6, the coating of photosensitive polymer doesn't significantly

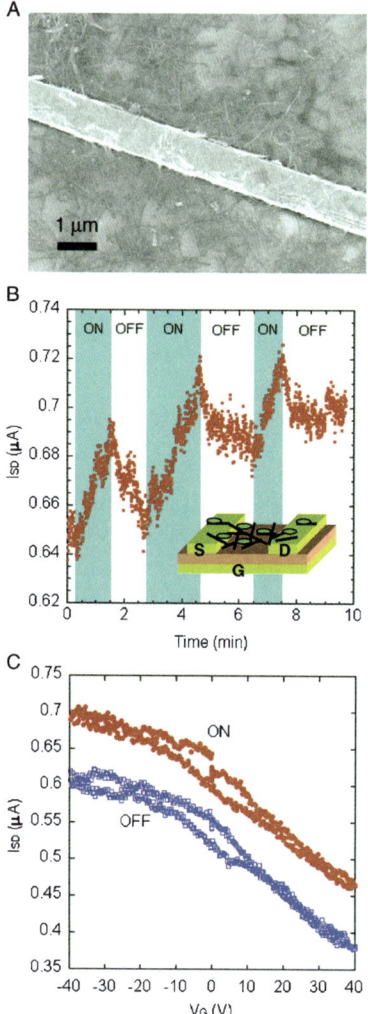

Figure 3.5 Response of PmPV polymer–SWNT composite to UV light illumination. (A) Scanning electron microscopy (SEM) image of PmPV–SWNT composite deposited on a Si wafer with optically patterned Au electrodes. One metal electrode (1 μm wide) is shown. The polymer–SWNT composite form a dense network on the wafer. (B) Current measured during UV illumination cycles in air. The shaded and unshaded regions mark the UV-on and -off periods, respectively. The inset shows the solution-deposited SWNT-FET device geometry where source (S), drain (D), and gate (G) gold electrodes were patterned on a Si wafer prior polymer–SWNT composite deposition. (C) I_{SD}–V_G curves of the resulting polymer–nanotube device recorded under bias voltage (V_{SD}) of 2 V in air at UV-off (blue curve) and UV-on (red curve). Reprinted (adapted) with permission from Star, A. et al.,[14] *Nanotube Optoelectronic Memory Devices.* Nano Letters, 2004. **4**(9):pp. 1587–1591. Copyright (2004) American Chemical Society.

change the transport properties of SWNT-FETs. Also, the photon induced current can remain as long as several hours after turning off the light, suggesting that the light induced photoresponse in SWNTs-FETs may be used as a charge storage reservoir or an electro-optical memory.

Instead of using SWNT networks, Borghetti and co-workers[15] developed an optoelectronic switch and memory device, where they used the preselected semiconducting SWNTs as the conductive channel and polymers are coated on the device area. Due to the exclusion of metallic SWNTs in the charge

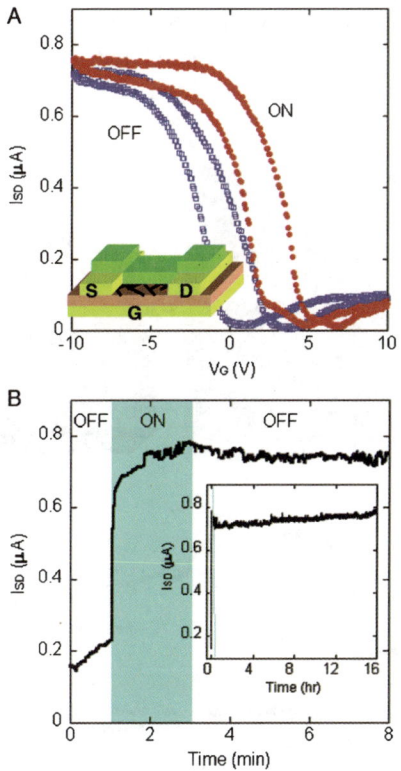

Figure 3.6 Response of PmPV polymer-coated CVD-grown SWNT-FET device to UV light ($\lambda = 365$ nm). (A) The source–drain current (I_{SD}) *versus* the gate voltage (V_G) of the device in air ($V_{SD} = 1$ V) at UV-off (blue curves) and UV-on (red curves) conditions. The reversible hysteresis (forward I_{SD} – reverse I_{SD}) in the device measured in the range of 20 V (–10 V to +10 V) at the sweep rate of 4 Hz. The inset shows the polymer-coated CVD-grown SWNT-FET device geometry. (B) Current (I_{SD}) *versus* time response to UV illumination of PmPV-coated SWNT-FET device in air at room temperature ($V_G = 4$ V, $V_{SD} = 1$ V). The inset shows no apparent recovery in the device conductance after 16 h at fixed V_G conditions. Shaded and unshaded regions mark the UV-on and -off periods, respectively. Reprinted (adapted) with permission from Star, A. *et al.*,[14] *Nanotube Optoelectronic Memory Devices.* Nano Letters, 2004. **4**(9): pp. 1587–1591. Copyright (2004) American Chemical Society.

transport from polymer to SWNTs, their results clearly demonstrated the electrical responses are different at different applied gate voltage. Figure 3.7 shows the different photoresponses of the SWNT–polymer system working on either a positive or negative gate bias. These two different dynamics enable two kinds of applications: at $V_{GS} = -3$ V the devices can be used as an optical switch, where the current can be modulated by light illumination, whereas at $V_{GS} = +3$ V, due to the slow current relaxation, it can be seen as a non-volatile optoelectronic memory. The photoresponse dynamics are very similar to that for a typical polymer transistor without the presence of SWNTs. The difference is that for the SWNT-FETs, the current level is determined by the number of holes in the channel which is related to the electrostatic potential defined by the trapped electrons at the interface of dielectric layer. Therefore,

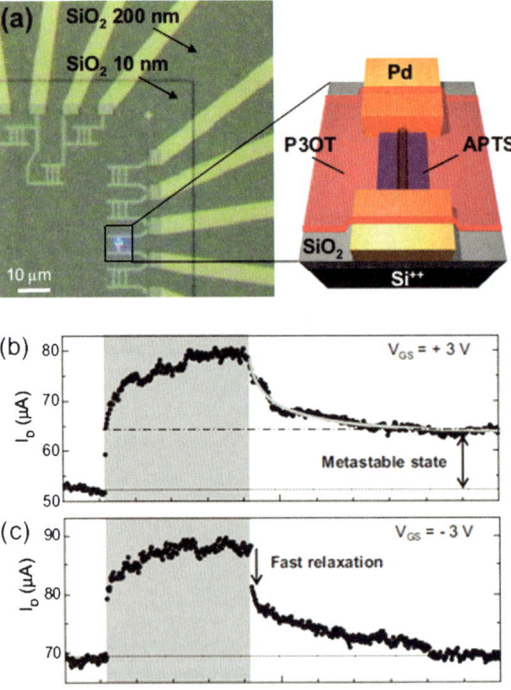

Figure 3.7 Photoresponse of polymer-functionalized carbon nanotube transistor. (a) Optical microscope image of the SWNT-FETs device coated with a photosensitive polymer [poly(3-octylthiophene) (P3OT)] and the schematic device structure; (b, c) photoresponse of the device at different fate voltages. For the device working at the positive gate voltage, turning off the light results a persistent photoconductivity with a slow current decay. While for the device applied with the negative gate voltage, the current relaxation time is much faster. Reprinted (adapted) with permission from Borghetti, J. *et al.*,[15] *Optoelectronic Switch and Memory Devices Based on Polymer-Functionalized Carbon Nanotube Transistors.* Advanced Materials, 2006. **18**(19): pp. 2535-2540. Copyright (2006) John Wiley and Sons.

the possible mechanisms responsible for the different photoresponse dynamics are described as follows. At a positive gate voltage, the SWNT devices work at a depletion mode and the photo-generated electrons can be trapped at the interface of the dielectric layer. Once the light illumination is turned off, a final metastable state can be reached because these electrons remain trapped by the positive back-gate potential hindering the complete de-trapping. On the other hand, for a negative back-gate voltage, the SWNT-FET turned to work at accumulation mode with sufficient holes to recombine with the trapped electrons. Moreover, the negative gate potential can assist in the de-trapping process and result in a fast current relaxation upon turning off the light.

3.6.2 Electrostatic Force Microscopy (EFM) Measurement of SWNT–Polymer

The photo-switching behaviors of the polymer–SWNTs have been introduced and the mechanism was attributed to the photo-induced electrostatic gating effect, meaning that the photocurrent was caused by the gating effect of the trapped photoelectrons at the dielectric surfaces. The technique of EFM has been widely used to study the electrical properties at the nanometer scale.[56–58] The EFM has also been applied to study the charge trapping in the polymer–SWNT system. EFM is a dual-pass scan-probing technique. Two scans are conducted in a tapping mode, in which the tip is mechanically driven around its resonance frequency. During the first scan, topography information is acquired. The tip is then lifted with a certain distance from the sample surface (typically ~ 10 nm), and the line scan is repeated at this constant distance from the surface based on the recorded profile. During the second scan (interleave scan), a DC voltage is applied to the tip. The long-range electrostatic force between the tip and the sample surface alters the tip resonance frequency, inducing a change in both the phase and amplitude signals. Attractive and repulsive forces will give rise to an opposite phase shift. Recording the phase shift reveals information about charge/potential distribution on the sample surface.

Figure 3.8 shows the typical EFM results carried out for the F8T2-coated SWNTs and the corresponding atomic force microscopy (AFM) images show the topography of these samples. As seen from the image, AFM shows no changes upon light illumination, whereas the EFM images exhibit dramatic response to the light illumination. This phenomenon suggests that the photo-generated charges from the polymer–SWNT change the surface potential distribution significantly. Figure 3.8(a) demonstrates that the phase lag for F8T2-coated SiO_2/Si substrates is positively shifted when a tip voltage bias (V_{tip}) of –3 V is applied. By contrast, the phase lag is negatively shifted when V_{tip} is applied with 3 V. The result suggests that the surface of the F8T2 polymers is positively charged after light exposure, as the more attractive force between the tip and polymer surface results in a positively shift in the phase lag in this particular EFM system. Figure 3.8(b) shows the result of parallel EFM measurements for F8T2-coated SWNTs, where the location of SWNTs under

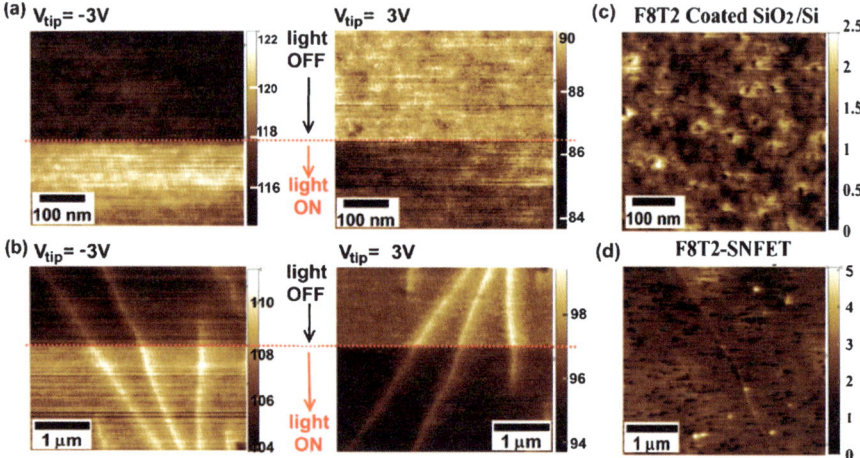

Figure 3.8 EFM phase images upon light switching for (a) photosensitive polymer-coated SiO$_2$/Si substrates and (b) polymer-coated SWNTs with a tip bias at −3 and +3 V, respectively. The conventional corresponding AFM height images are shown in (c) and (d). Reprinted (adapted) with permission from Shi, Y. et al.,[59] *Photoconductivity from Carbon Nanotube Transistors Activated by Photosensitive Polymers.* The Journal of Physical Chemistry C, 2008. **112**(46): pp. 18201–18206. Copyright (2008) American Chemical Society.

the F8T2 polymers can be clearly identified. The phase shift direction at either the blanket polymer area or the location with SWNTs underneath is consistent with the observation in Figure 3.8(a), indicating that the phase shift is dominantly from the photon-induced change of the polymer F8T2.

Figure 3.9(a) shows that the change in phase shift along the trace with SWNTs underneath the polymer is larger than the trace without SWNTs underneath. From the photoelectrical measurement of polymer–SWNTs transistor, the photo-generated excitons in polymer dissociated into electrons and holes, where the electrons can be trapped at the interface of polymer and the dielectric layer, leading to the photo-gating effect. The EFM results provide direct evidence for Borghetti's proposal that the trapping of electrons at the SiO$_2$ interface in the direct vicinity of the SWNTs governs the electrical current in polymer-coated SWNT-FETs.[15] The EFM results also suggest that the excitons preferentially dissociate at the polymer–SWNT interface. The EFM results agree well with the argument for the photocurrent in polymer–SWNT systems. The dielectric interface plays an important part in the photo-response for the polymer–SWNT system.

3.6.3 The Ability for Hole and Electron Discrimination in SWNTs

In the above-mentioned polymer–SWNT system the holes are preferentially transferred to the nanotube and the electrons that are trapped at the dielectric layer

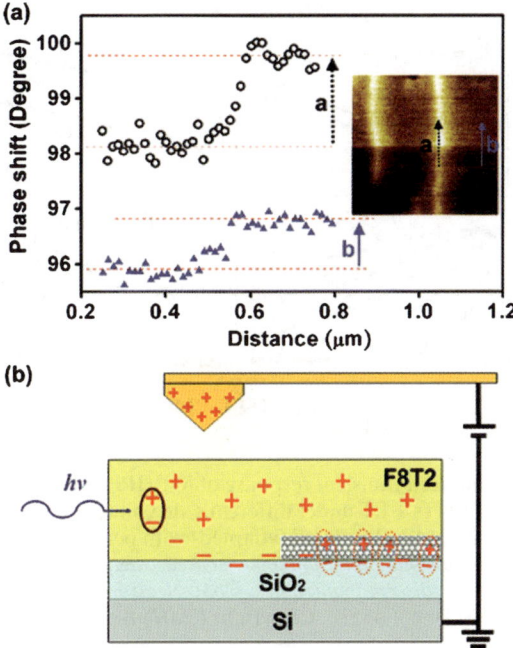

Figure 3.9 EFM study of the photogenerated charge distribution in polymer/carbon nanotube composite. (a) Light-induced phase shift for a F8T2-coated SWNTs in EFM measurements with tip voltage of 3 V, where the labels "a" and "b" represent the traces with and without SWNTs underneath the polymer. The inset shows the phase image change of EFM. (b) Schematic illustration for the distribution of photo-excited charges in the polymer layer and the charge trapping at the dielectric interface. Reprinted (adapted) with permission from Shi, Y. et al.,[59] *Photoconductivity from Carbon Nanotube Transistors Activated by Photosensitive Polymers*. The Journal of Physical Chemistry C, 2008. **112**(46): pp. 18201–18206. Copyright (2008) The American Chemical Society.

can effectively gate the SWNT transistors. However, due to the energy alignment of the nanotube and photosensitive polymer, it is also possible for the electrons to transfer to SWNTs and results in a negative photocurrent upon light illumination.

Chawla et al.[60] discussed the effect of two different polymers including the cyano derivative of poly(p-phenylene vinylene) (CNPPV) and poly[2-methoxy-5-(2'-ethylhexyloxy)-1,4-phenylenevinylene] (MEHPPV). The most interesting feature is that a current decrease was observed under a negative gate voltage for the CNPPV-coated SWNT-FETs as the CNPPV polymer chains have a higher electron affinity compared with MEHPPV, as shown in Figure 3.10. With light illumination, the electrons can efficiently transfer from CNPPV to p-type SWNTs. At negative gate voltage, the SWNT-FET works in the accumulation mode, therefore, the electron injection from polymer to SWNTs leads to the electron–hole recombination in the nanotube. By contrast, at a

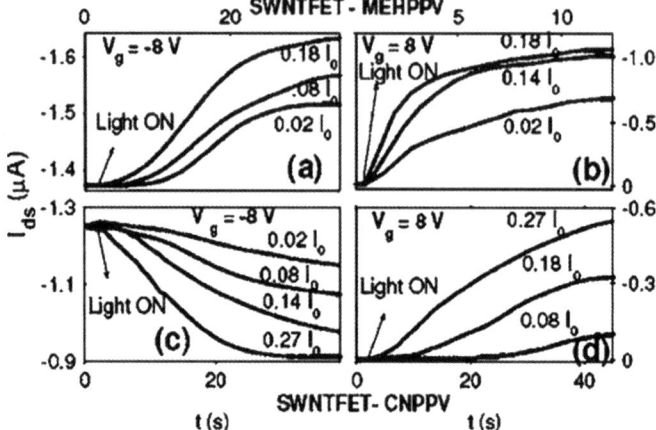

Figure 3.10 Light-induced transient response of MEHPPV- (a, b) and CNPPV-coated (c, d) SWNTs-FET under different gate voltages [–8 V for (a) and (c); +8 V for (b) and (d)]. Reprinted (adapted) with permission from Chawla, J.S. et al.,[60] Semiconducting polymer coated single wall nanotube field-effect transistors discriminate holes from electrons. Applied Physics Letters, 2007. **91**(4): p. 043510. Copyright (2007) American Institute of Physics.

positive gate voltage, the SWNT-FET works in a depletion mode. The rate of the current increase in CNPPV is slower than that of MEHPPV, which suggests the competing process of electron transfer across the interface accompanied by a reduced recombination in the depleted nanotube interface. However, this process is even more complicated, when taking the electron trapping effect at the dielectric layer into account.

To study the charge interaction between photosensitive polymers and SWNTs, it is easiest to select the device with very few surface charge traps from the dielectric layer. Shi et al.[61] compared the charge transfer from polymer to SWNTs using either transistor or resistor type of devices. F8T2 and F8BT were used as an electron acceptor and a donor respectively. Figures 3.11(a) and 3.11(b) show the charge transport characteristics of the FETs in dark before and after polymer coating. After F8T2 coating the source–drain current of the SWNT-FET increases at a V_g ranging from −20 to 20 V, whereas the F8BT coating slightly decreases the drain current when the applied V_g is below −15 V. These observations suggest that the electrostatic gating effect, due to electron trapping, plays an important role in current increases for the two polymers. While the polarity of the photocurrent at $V_g = -20$ V is likely to be determined by the polarity of the charge carriers received by the SWNTs, which is in consistance with the report of Chawla et al.[60] At such gate voltages, SWNT-FETs operate at the accumulation mode and the majority carriers are holes. The observed drain current increase in F8T2-coated devices is simply because additional (photo-generated) holes are transferred from F8T2 to SWNTs. In case of F8BT-coated devices, photo-generated electrons are donated to SWNTs,

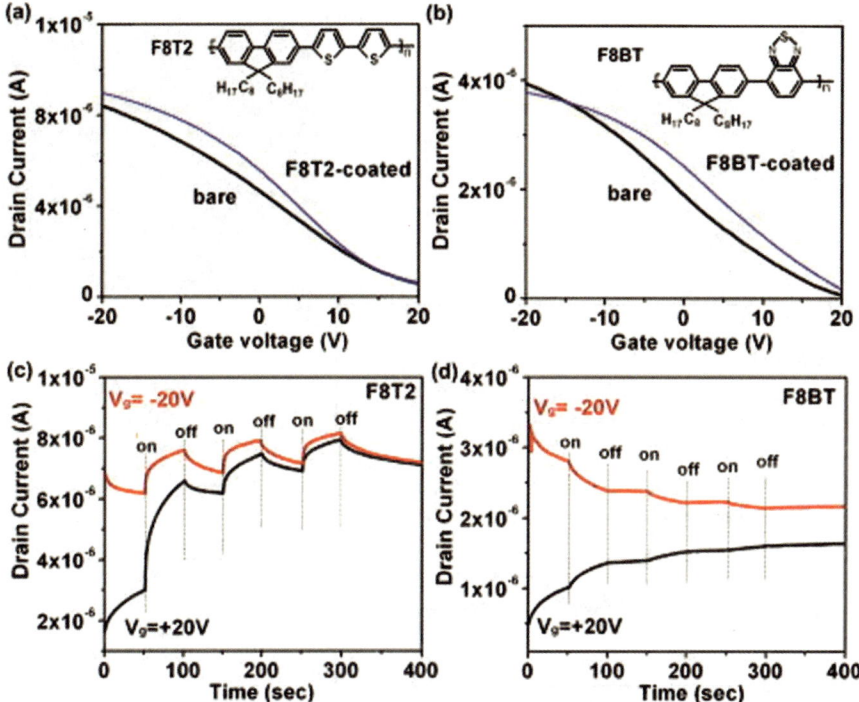

Figure 3.11 Photoresponse of F8T2 and F8BT coated SWNTs-FET. Transport curve of the SWNT-FET before and after coating with (a) F8T2 and (b) F8BT. The inset in each panel shows the chemical structure of each polymer. The photoresponses observed for (c) F8T2- and (d) F8BT-coated SWNTs-FET to the illumination cycles of the 450 nm wavelength of light. Reprinted (adapted) with permission from Shi, Y. et al.,[61] *Effects of substrates on photocurrents from photosensitive polymer coated carbon nanotube networks*. Applied Physics Letters, 2008. **92**(10): pp. 103310. Copyright (2008) American Institute of Physics.

leading to electron–hole recombination and hence reducing the source–drain current. When SWNT-FETs are biased at $V_g = 20$ V, hole carriers are largely depleted. Under depletion condition, the current increases in both devices with light exposure. These observations can be attributed to the same electrostatic gating effects from the trapped photo-induced charges on SiO_2 surfaces on which the nanotubes reside. Electrostatic gating from the trapped electrons increases the number of holes in SWNTs and hence the current. At sufficiently negative voltages, such electrostatic gating effects become less pronounced because of the abundance of current carriers. Therefore, the current modulates are basically determined by charge transfer from the donor polymers.

To minimize the surface charge trapping effect, a high quality dielectric layer is essential for the photocurrent response study in the polymer–SWNT system. The study on quartz substrates with very small amounts of surface defects

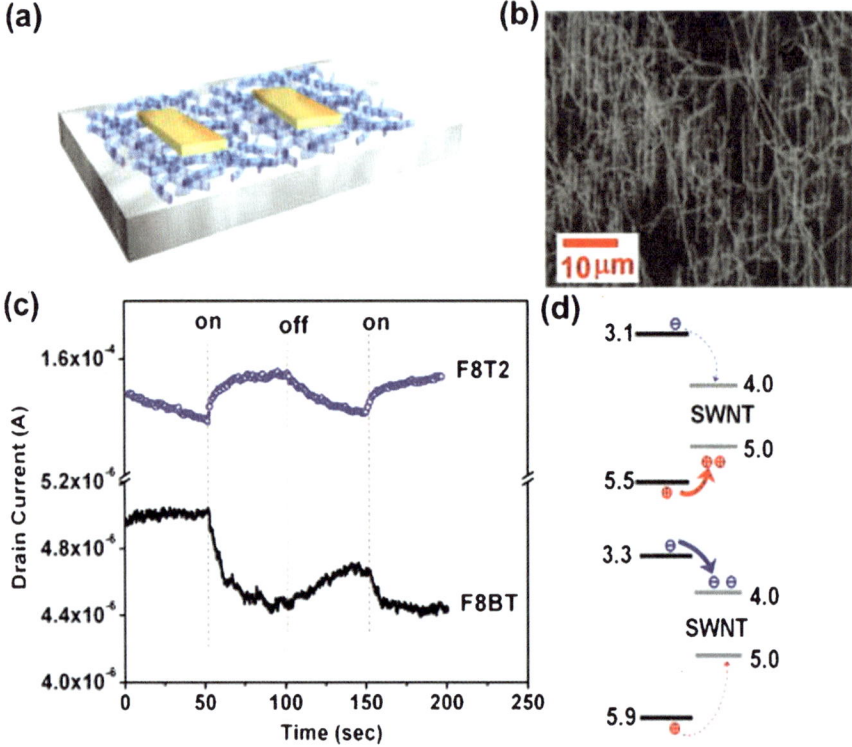

Figure 3.12 Photoresponse of polymer coated SWNT resistors on quartz substrates. (a) Schematic illustration of SWNT resistor devices on quartz. (b) SEM image of SWNT on quartz. (c) Photoresponses of both F8T2- and F8BT-coated SWNT networks on quartz (to the 450 nmlight). (d) Energy band alignment of these polymers with SWNTs. Reprinted (adapted) with permission from Shi, Y. et al.,[61] *Effects of substrates on photocurrents from photosensitive polymer coated carbon nanotube networks.* Applied Physics Letters, 2008. **92**(10): pp. 103310. Copyright (2008) American Institute of Physics.

gives more insight into the charge transfer process from polymer to SWNTs. As shown in Figure 3.12, a resistor type device using quartz as a substrate was applied for the charge transfer study. The F8T2-coated SWNTs resistor shows an increase in current, whereas the F8BT-coated one shows a decrease in current when the light is on. These distinct current responses are consistent with the charge transfer mechanism, indicating that the electrostatic gating effects due to the charge is not pronounced for quartz substrates. Figure 3.12(d) illustrates the energy band alignment of these polymers and SWNTs and the proposed charge transfer flow, assuming the average band gap for the chemical vapor deposition (CVD) grown nanotubes is approximately 1 eV. This model suggests that SWNTs may either be hole or electron acceptors, depending on their energy level alignment with photosensitive polymers.

3.7 Conclusions

Carbon nanotubes are promising materials for electronic applications. The optical properties of the SWNT-based devices are well-defined by the atomic structure of the carbon nanotube. However, the use of SWNTs in optoelectronic devices was restricted due to the physical constraints such as a low photon-capture area, large recombination probability, and other intrinsic reasons, which lead to low internal photon to current conversion efficiencies. The modification of SWNTs with fluorene-based polymers introduces a high selectivity on specific chirality of tubes, which makes it possible to separate SWNTs with certain energy gap. Furthermore, the polymer changes the optical and electro-optical characteristics of SWNTs, for example, photosensitive polymers greatly extend the capabilities of SWNT-FETs in optoelectronic devices including memory and photodetectors. The enhancement of photoresponses is due to charge transfer from polymers to SWNTs. The SWNTs in the polymer–SWNT composites can be either electron or hole acceptors depending on the energy levels of the polymers, and the knowledge is useful for the development of more delicate optoelectronic devices.

References

1. R. H. Baughman, A. A. Zakhidov and W. A. de Heer, *Science*, 2002, **297**, 787–792.
2. S. J. Wind, J. Appenzeller and P. Avouris, *Physical Review Letters*, 2003, **91**, 058301.
3. M. S. Fuhrer, B. M. Kim, T. Dürkop and T. Brintlinger, *Nano Letters*, 2002, **2**, 755–759.
4. V. C. Tung, J.-H. Huang, J. Kim, A. J. Smith, C.-W. Chu and J. Huang, *Energy & Environmental Science*, 2012, **5**(7), 7810–7818.
5. A. K. K. Kyaw, H. Tantang, T. Wu, L. Ke, C. Peh, Z. H. Huang, X. T. Zeng, H. V. Demir, Q. Zhang and X. W. Sun, *Appl. Phys. Lett.*, 2011, **99**, 021107–021103.
6. G. Gruner, *Journal of Materials Chemistry*, 2006, **16**, 3533–3539.
7. J. Kong, N. R. Franklin, C. Zhou, M. G. Chapline, S. Peng, K. Cho and H. Dai, *Science*, 2000, **287**, 622–625.
8. M. T. Martinez, Y.-C. Tseng, N. Ormategui, I. Loinaz, R. Eritja and J. Bokor, *Nano Letters*, 2009, **9**, 530–536.
9. T. Ando, *Phys. Soc. Jpn.*, 1997, **66**, 1066–1073.
10. M. Freitag, Y. Martin, J. A. Misewich, R. Martel and P. Avouris, *Nano Letters*, 2003, **3**, 1067–1071.
11. D. A. Stewart and F. Léonard, *Nano Letters*, 2005, **5**, 219–222.
12. K. Balasubramanian, Y. W. Fan, M. Burghard, K. Kern, M. Friedrich, U. Wannek and A. Mews, *Appl. Phys. Lett.*, 2004, **84**, 2400–2402.
13. D. H. Lien, W. K. Hsu, H. W. Zan, N. H. Tai and C. H. Tsai, *Adv. Mater.*, 2006, **18**, 98–103.

14. A. Star, Y. Lu, K. Bradley and G. Grüner, *Nano Letters*, 2004, **4**, 1587–1591.
15. J. Borghetti, V. Derycke, S. Lenfant, P. Chenevier, A. Filoramo, M. Goffman, D. Vuillaume and J. P. Bourgoin, *Adv. Mater.*, 2006, **18**, 2535–2540.
16. T. Durkop, S. A. Getty, E. Cobas, M. S. Fuhrer, *Nano Letters*, 2004, **4**, 35–39.
17. H. C. d'Honincthun, S. Galdin-Retailleau, J. See and P. Dollfus, *Appl. Phys. Lett.*, 2005, **87**(17), 172112.
18. Y. F. Chen and M. S. Fuhrer, *Phys. Rev. Lett.*, 2005, **95**(23), 236803.
19. R. Saito, M. Fujita, G. Dresselhaus and M. S. Dresselhaus, *Appl. Phys. Lett.*, 1992, **60**, 2204–2206.
20. S. Reich, J. Maultzsch, C. Thomsen and P. Ordejon, *Phys. Rev. B*, 2002, **66**(3), 035412.
21. M. S. Dresselhaus, G. Dresselhaus and R. Saito, *Carbon*, 1995, **33**, 883–891.
22. J. Chen, V. Perebeinos, M. Freitag, J. Tsang, Q. Fu, J. Liu and P. Avouris, *Science*, 2005, **310**, 1171–1174.
23. M. P. Anantram and F. Léonard, *Rep. Prog. Phys.*, 2006, **69**, 507–561.
24. H. Kataura, Y. Kumazawa, Y. Maniwa, I. Umezu, S. Suzuki, Y. Ohtsuka and Y. Achiba, *Synth. Met.*, 1999, **103**, 2555–2558.
25. M. Freitag, Y. Martin, J. A. Misewich, R. Martel and P. H. Avouris, *Nano Letters*, 2003, **3**, 1067–1071.
26. N. M. Gabor, Z. Zhong, K. Bosnick, J. Park and P. L. McEuen, *Science*, 2009, **325**, 1367–1371.
27. S. Wang, M. Khafizov, X. Tu, M. Zheng, T. D. Krauss, *Nano Letters*, 2010, **10**, 2381–2386.
28. W. Shockley and H. J. Queisser, *Journal of Applied Physics*, 1961, **32**, 510.
29. E. Kymakis, I. Alexandrou and G. A. J. Amaratunga, *Journal of Applied Physics*, 2003, **93**, 1764–1768.
30. S. Cataldo, P. Salice, E. Menna, B. Pignataro, *Energy & Environmental Science*, 2012, **5**, 5919–5940.
31. B. J. Landi, R. P. Raffaelle, S. L. Castro and S. G. Bailey, *Progress in Photovoltaics: Research and Applications*, 2005, **13**, 165–172.
32. E. Kymakis and G. A. J. Amaratunga, *Appl. Phys. Lett.*, 2002, **80**, 112–114.
33. C. Li, Y. Chen, Y. Wang, Z. Iqbal, M. Chhowalla and S. Mitra, *Journal of Materials Chemistry*, 2007, **17**, 2406–2411.
34. Y. Ryu, L. Yin and C. Yu, *Journal of Materials Chemistry*, 2012, **22**, 6959–6964.
35. C. W. Lee, C.-H. Weng, L. Wei, Y. Chen, M. B. Chan-Park, C.-H. Tsai, K.-C. Leou, C. H. P. Poa, J. Wang and L.-J. Li, *The Journal of Physical Chemistry C*, 2008, **112**, 12089–12091.
36. H.-Z. Geng, K. K. Kim, K. P. So, Y. S. Lee, Y. Chang and Y. H. Lee, *Journal of the American Chemical Society*, 2007, **129**, 7758–7759.
37. C. H. Lau, R. Cervini, S. R. Clarke, M. G. Markovic, J. G. Matisons, S. C. Hawkins, C. P. Huynh and G. P. Simon, *J. Nanoparticle Res.*, 2008, **10**, 77–88.
38. M. Shigeta, M. Komatsu, and N. Nakashima, *Chemical Physics Letters*, 2006, **418**, 115–118.
39. I. Musa, M. Baxendale, G. A. J. Amaratunga and W. Eccleston, *Synth. Met.*, 1999, **102**, 1250–1250.

40. J. N. Coleman, S. Curran, A. B. Dalton, A. P. Davey, B. McCarthy, W. Blau, R. C. Barklie, *Synth. Met.*, 1999, **102**, 1174–1175.
41. G. Z. Chen, M. S. P. Shaffer, D. Coleby, G. Dixon, W. Z. Zhou, D. J. Fray, A. H. Windle, *Adv. Mater.*, 2000, **12**, 522–526.
42. M. Zheng, A. Jagota, E. D. Semke, B. A. Diner, R. S. McLean, S. R. Lustig, R. E. Richardson and N. G. Tassi, *Nat Mater*, 2003, **2**, 338–342.
43. M. S. Strano, M. Zheng, A. Jagota, G. B. Onoa, D. A. Heller, P. W. Barone and M. L. Usrey, *Nano Letters*, 2004, **4**, 543–550.
44. A. Nish, J.-Y. Hwang, J. Doig, R. J. Nicholas, *Nature Nanotechnology*, 2007, **2**, 640–646.
45. F. Chen, B. Wang, Y. Chen and L.-J. Li, *Nano Letters*, 2007, **7**, 3013–3017.
46. M. Tange, T. Okazaki and S. Iijima, *Journal of the American Chemical Society*, 2011, **133**, 11908–11911.
47. T. Umeyama, N. Kadota, N. Tezuka, Y. Matano, H. Imahori, *Chemical Physics Letters*, 2007, **444**, 263–267.
48. A. Nish, J.-Y. Hwang, J. Doig, R. J. Nicholas, *Nanotechnology*, 2008, **19**, 095603.
49. F. Chen, W. Zhang, M. Jia, L. Wei, X.-F. Fan, J.-L. Kuo, Y. Chen, M. B. Chan-Park, A. Xia, L.-J. Li, *The Journal of Physical Chemistry C*, 2009, **113**, 14946–14952.
50. J. L. Bahr and J. M. Tour, *Journal of Materials Chemistry*, 2002, **12**, 1952–1958.
51. S. Banerjee, T. Hemraj-Benny and S. S. Wong, *Adv. Mater.*, 2005, **17**, 17–29.
52. D. Tasis, N. Tagmatarchis, A. Bianco and M. Prato, *Chemical Reviews*, 2006, **106**, 1105–1136.
53. P. Singh, S. Campidelli, S. Giordani, D. Bonifazi, A. Bianco and M. Prato, *Chemical Society Reviews*, 2009, **38**, 2214–2230.
54. X. Guo, L. Huang, S. O'Brien, P. Kim and C. Nuckolls, *Journal of the American Chemical Society*, 2005, **127**, 15045–15047.
55. D. W. Steuerman, A. Star, R. Narizzano, H. Choi, R. S. Ries, C. Nicolini, J. F. Stoddart and J. R. Heath, *The Journal of Physical Chemistry B*, 2002, **106**, 3124–3130.
56. D. C. Coffey and D. S. Ginger, *Nat Mater*, 2006, **5**, 735–740.
57. M. Zdrojek, T. Melin, H. Diesinger, D. Stievenard, W. Gebicki and L. Adamowicz, *Journal of Applied Physics*, 2006, **100**, 114326.
58. T. Sand Jespersen and J. Nygård, *Applied Physics A: Materials Science & Processing*, 2007, **88**, 309–313.
59. Y. Shi, X. Dong, H. Tantang, C.-H. Weng, F. Chen, C. Lee, K. Zhang, Y. Chen, J. Wang and L.-J. Li, *The Journal of Physical Chemistry C*, 2008, **112**, 18201–18206.
60. J. S. Chawla, D. Gupta, K. S. Narayan and R. Zhang, *Appl. Phys. Lett.*, 2007, **91**, 043510.
61. Y. Shi, H. Tantang, C. W. Lee, Y. Shi, C.-H. Weng, X. Dong, L.-J. Li and P. Chen, *Appl. Phys. Lett.*, 2008, **92**, 103310.

CHAPTER 4

Chemical Functionalisation of Carbon Nanotubes for Polymer Reinforcement

YURII K. GUN'KO*

School of Chemistry and CRANN, Trinity College Dublin, Dublin 2, Ireland.
*E-mail: igounko@tcd.ie

4.1 Introduction

The field of carbon nanotube (CNT)–polymer composites has been progressing extremely rapidly over the last two decades. CNTs theoretically have exceptional mechanical properties such as elastic modulus and strengths 10–100 times higher than the strongest steel at only a fraction of the weight. When individual nanotubes are tested, their mechanical properties reach those estimated for the theoretical values. In 2000, Yu *et al.*[1] measured the Young's modulus values for individual multi-walled nanotubes (MWNTs) of between 0.27–0.95 TPa, strengths in the 11–63 GPa range and a toughness of ~ 1240 J g^{-1}. For single-walled nanotubes (SWNTs), the Young's moduli were found to be in the range 0.32–1.47 TPa, and strengths between 10 and 52 GPa, with a toughness of ~ 770 J g^{-1}.[2] These really outstanding mechanical properties and performance make CNTs potentially ideal additives for the formation of polymer composites with improved mechanical properties.[3] This area is very topical and there are a huge number of publications and good reviews on the mechanical properties of CNT–polymer composites.[4–19]

With CNTs becoming easier to produce and much cheaper to buy, CNT-based manufacturing could potentially overtake that of the carbon fibre industry and result in a number of new commercial polymer composite products. However, there are still many challenges in this area including the development of functionalised CNTs as additives for polymer reinforcement.

The chemical functionalisation of CNTs has been a subject of several reviews.[20-24] Here, we will only focus on some recent developments of the functionalisation of CNTs for polymer composite formation. It is known that nanotube solubility, dispersion and stress transfer strongly depend on their surface functionalisation. Therefore, functionalisation of nanotubes is extremely important for their processing and potential applications in polymer composites. Unless the interface between nanotube and polymer is carefully engineered, poor load transfer between nanotubes, when in bundles, and between nanotubes and surrounding polymer chains may result in interfacial slippage.[25] In general, composites based on chemically modified nanotubes show the best mechanical results because functionalisation enables a significant improvement in both dispersion and stress–strain transfer. The treatment of CNTs by chemical functionalisation and/or ultrasonication is also widely used to increase the dispersion of nanotubes in solvents.[26-32] In this Chapter, we discuss the progress in the research on the functionalisation of CNTs for the production of reinforced polymer composites. Various functionalisation and fabrication approaches and their role in the preparation of CNT–polymer composites with improved mechanical properties will be discussed. We will also tabulate the typical values of the Young's modulus for various CNT–polymer composites and also compare the effectiveness of different processing techniques.

4.2 Non-covalent Functionalisation of CNTs

There are two major approaches to nanotube functionalisation: non-covalent supramolecular modifications and covalent functionalisation.[24,33,34]

The non-covalent functionalisation of nanotubes normally involves van der Waals, π–π, CH–π or electrostatic interactions between polymer molecules and the CNT surface.[34-36]

The advantage of non-covalent functionalisation is that it does not alter the structure of the nanotubes and therefore both the initial electrical and mechanical properties should also remain unchanged. However, the efficiency of the load transfer might decrease as the forces between the wrapping molecules and the nanotube surface may be relatively weak.

There are several non-covalent approaches for nanotube functionalisation such as surfactant-assisted dispersion, polymer wrapping and the polymerisation-filling technique (PFT).[33]

Surfactant-assisted dispersion is a very common technique that enables the transfer of nanotubes to the aqueous phase in the presence of surface-active molecules such as sodium dodecyl sulfate (SDS) or benzylalkonium chloride.

The presence of an aromatic group in the surfactant molecule allows for π–π stacking interactions with the graphitic sidewalls of the nanotubes, which results in their effective coating and dispersion.[36]

The polymer wrapping approach is particularly useful for the non-covalent functionalisation of nanotubes. This method involves the utilisation of conjugated and aromatic group containing polymers [*e.g.* poly(vinyl pyrrolidone), poly(phenylene vinylene), pyrene–poly(ethylene glycol) (PEG)], which can wrap around CNTs through π–π stacking and van der Waals interactions.[34,37,38] Coleman and Ferreira[39] showed that polymer wrapping can minimise energy for purely geometric reasons.

Molecular dynamic simulations (Figure 4.1) of the interfacial binding between SWNTs and selected conjugated polymers [polythiophene (PT), polypyrrole (PPy), poly(2,6-pyridinylenevinylene-*co*-2,5-dioctyloxy-*p*-phenylenevinylene) (PPyPV) and poly(*m*-phenylenevinylene-*co*-2,5-dioctyloxy-*p*-phenylenevinylene) (PmPV)] demonstrated that the intermolecular interaction in these systems is strongly influenced by the specific monomer structure of polymer and the nanotube radius, but the influence of temperature could be negligible. The simulations showed that the strongest interaction between the SWNTs and these conjugated polymers was observed, first for PT, then PPy and PmPV, and finally PPyPV. It was also suggested that an efficient load

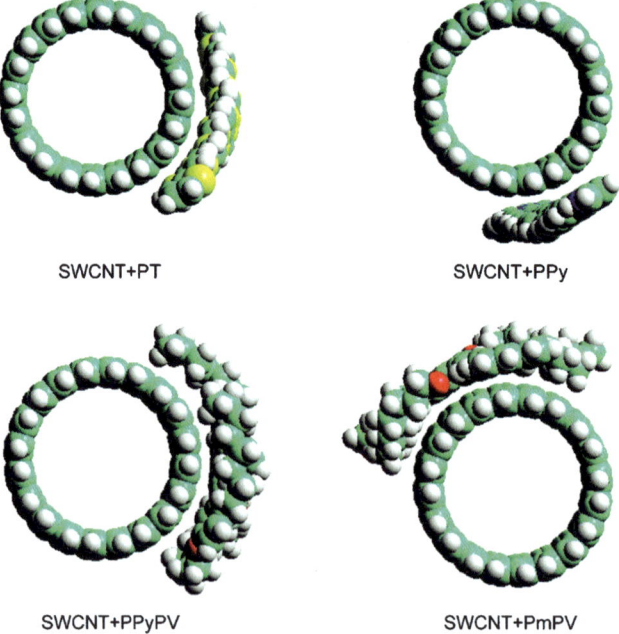

Figure 4.1 Molecular dynamic simulation snapshots of the wrapping of a SWNT (10,10) by different polymers (cross-sectional view). Reproduced with permission from Foroutan and Nasrabadi.[40]

transfer should exist in the interface between nanotubes and heterocyclic conjugated polymers, which plays a key role in the composite reinforcement practical applications.[40]

In order to achieve a good dispersion and coating of the nanotube surface with the polymer, Mandal and Nandi[41] have designed a new compatibilizer (P2) containing a thiophene moiety and a poly(dimethylamino ethyl methacrylate) group. This compatibilizer was prepared by atom transfer radical polymerisation (Figure 4.2). The dispersion of compatibilizer-coated MWNT in N,N-dimethylformamide was stable for more than for 3 months. The addition of the P2-coated MWNTs to the poly(vinylidene fluoride)

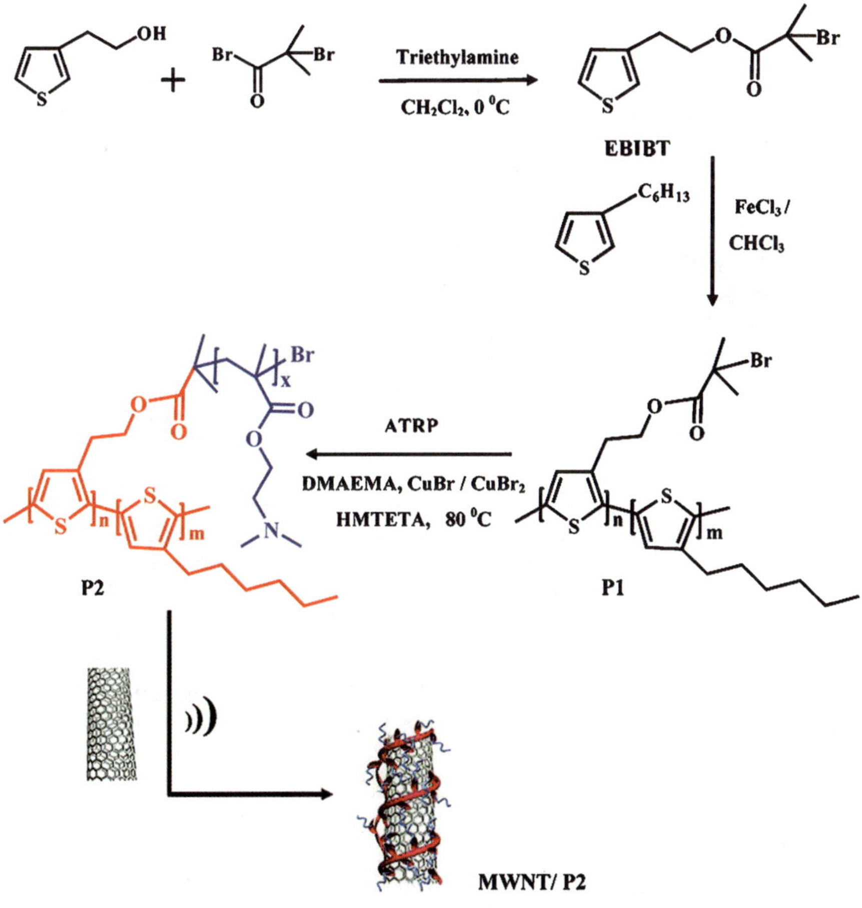

Figure 4.2 Synthesis of the compatibilizer (P2) and schematic illustration of non-covalent functionalisation of MWNTs by P2. Reproduced with permission from Mandal and Nandi.[41] EBIBT- 3-[1-ethyl-2(2-bromoisobutyrate)] thiophene; DMAEMA- 2-(dimethylamino ethyl methacrylate); HMTETA- 1,1,4,7,10,10-hexamethyltriethylenetetramine; ATRP- atom transfer radical polymerization.

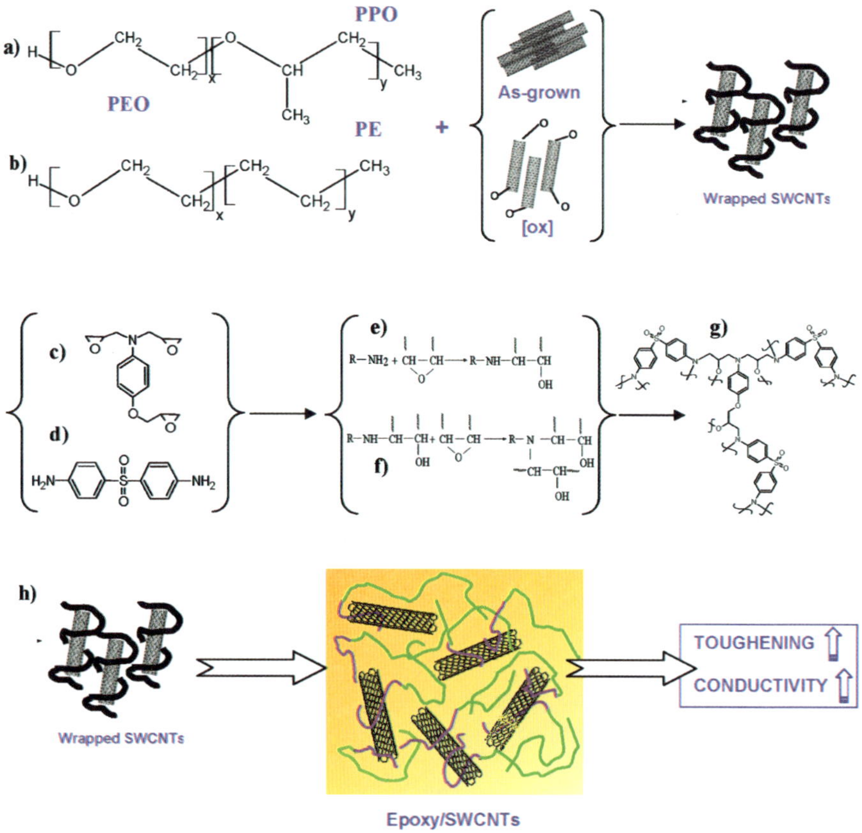

Figure 4.3 Wrapping of SWCNTs in Polyethylenoxide-Based Amphiphilic Diblock Copolymers. Chemical structures of (a) PEO-b-PPO BC; (b) PEO-b-PE BC; (c) the TGAP epoxy monomer; (d) the DDS curing agent; (e) and (f) the epoxy cross-linking mechanism, with respect to the main processes, namely, primary amine (e) and secondary amine (f) additions; (g) the TGAP/DDS cross-linked system; (h) schematic indicating the effects of wrapped SWNTs ('SWCNTs') on nanocomposites properties: green (epoxyphilic blocks) and purple lines (epoxyphobic blocks) represent the BC structures (elements are not to scale). Reproduced with permission from Gonzalez-Dominguez et al.[42] PEO- polyethylenoxide; PPO- polypropylenoxide; TGAP- riglycidyl p-aminophenol; DDS- 4,4'-diamino diphenylsulfone.

(PVDF) matrix followed by solvent casting resulted in composites that showed a Young's modulus, tensile strength and toughness in the polymer composite which were two to three times higher than that without the compatibilizer at only 0.05% of MWNTs content.[41]

Martinez and co-workers[42] reported an interesting approach which is based on non-covalent functionalisation of SWNTs with amphiphilic block copolymers (BCs) containing polyethylenoxide (Figure 4.3). They have shown that epoxy nanocomposites containing BC-wrapped SWNTs exhibit improved mechanical and electrical properties. The authors have also demonstrated that mechanical properties of composites depend on the choice of the wrapping BC and SWNT surface chemistry.[42]

Polydimethylsiloxane polymers have also been used for CNT coating *via* non-covalent CH–π interactions.[35,43,44]

In another work,[45] Kevlar-functionalised nanotubes were prepared by heating Kevlar with MWNTs in the presence of a sulphuric/nitric acid mixture under reflux. This resulted in the partial oxidation and functionalisation of CNTs with carboxylic acid groups, which formed hydrogen bonds with amido groups, as well as terminal carboxylic acid and amino groups in Kevlar. This process produced Kevlar-coated CNTs, which have been utilised for the fabrication of MWNT–poly(vinyl chloride) (PVC) composites.[45]

4.3 Covalent Functionalisation

Covalent functionalisation and the surface chemistry of CNTs have been envisaged as very important factors for the processing and applications of nanotubes. Recently, many efforts on polymer composite reinforcement have been focused on an integration of chemically modified nanotubes containing different functional groups into the polymer matrix. Covalent functionalisation of CNTs can be achieved by either direct addition reactions of reagents to the sidewalls of nanotubes or the modification of appropriate surface-bound functional (*e.g.* carboxylic acid) groups on to the nanotubes.[11,12,24]

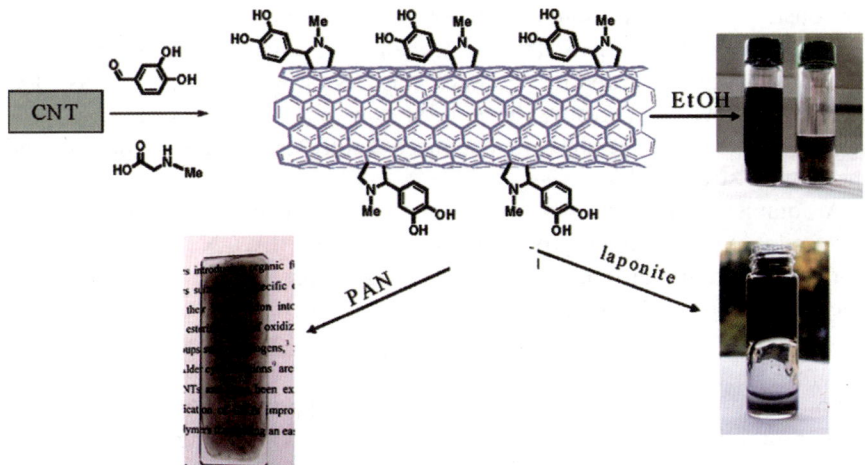

Figure 4.4 Schematic presentation of covalent functionalisation of CNTs *via* 1,3-dipolar cyclo-addition for enhancing the ability to process CNTs and facilitating the preparation of hybrid composites, which is achieved solely by mixing. EtOH, ethanol. Reproduced with permission from Georgakilas *et al.*[46]

An example of direct functionalisation of nanotubes with phenol groups *via* 1,3-dipolar cyclo-addition is shown in Figure 4.4. This functionalisation provided stable dispersions of CNTs in a range of polar solvents, including water. An advantage of the phenolic functionalities is that they allow further post-functionalisation of the MWNTs with other molecules that can be employed in preparing customised products. In fact, the functionalised CNTs were found to be compatible with polymers or layered aluminosilicate clay minerals, giving homogeneous, coherent, transparent CNT thin films and/or gels.[46]

However, the most common and facile method for the surface functionalisation of CNTs is 'nanotube oxidation', which results in the formation of a number of carboxylic acid groups (-COOH) on the surface of the nanotubes. These functionalised nanotubes are much more stable in polar solvents. For example, Feng *et al.*[47] showed that by treating MWNTs with HNO_3/H_2O_2 and HNO_3/H_2SO_4 functionalised nanotubes were formed which were stable in water at room temperature for more than 100 days. As a result, the water-stable nanotubes can be easily embedded into water-soluble polymers such as a poly(vinyl alcohol) (PVA), giving polymer–CNT composites with homogeneous nanotube dispersion.[48] Oxidised nanotubes also show excellent stability in other solvents such as caprolactam, which is used in the production of polyamide 6 (PA6).[49]

Oxidation by ozonolysis is one of generic set of techniques which is utilised to functionalise CNTs. For example, in one recent work MWNTs oxidised by ozonolysis were then modified with long-alkyl chains *via* an esterification process. These MWNTS were used to produce PVC composites, which demonstrated improved solubility in organic solvents.[50]

It has been shown that acid functionalisation significantly improves the interfacial bonding properties between the CNTs and polymer matrix. The carboxylic functional groups have been shown to give a stronger CNT–polymer interaction, leading to enhanced values in Young's modulus and mechanical strength.[51–60]

A completely different strategy for the surface functionalisation of CNTs with nitrogen-containing groups is the treatment of CNTs under atomic nitrogen flow obtained by molecular nitrogen dissociation in an $Ar + N_2$ microwave plasma. X-ray photoelectron spectroscopy of the nanotube surface demonstrated the presence of amides, oximes and mainly amine and nitrile groups.[61,62]

Some interesting work was also done on unzipped MWNTs, which have been produced by an oxidative unzipping process, involving the lengthwise cutting and opening the walls of MWNTs. These separated ribbon-like graphene layers have been used for the reinforcement of PVA-based composites. It has been demonstrated that unzipped MWNTs were more effective than pristine MWNTs in polymer reinforcement.[63]

Finally, an approach which is based on plasma polymerisation treatment enables the coating of CNTs with a very thin (~ 3 nm) polymer layer. Polymer

Table 4.1 Comparison of "grafting from" and "grafting to" approaches.

Approaches	"Grafting from"	"Grafting to"
Main differences	Initial immobilization of initiators on to the nanotube surface followed by *in situ* polymerisation.	Attaching of already preformed functionalized polymer molecules to functional groups on the nanotube surface.
Advantages	- Functionalised nanotubes with high grafting density can be produced.	- Commercially available polymers with reactive functional groups can be utilised.
	- Gives good homogeneous CNT distribution in the polymer matrix.	- Good control of polymer nature and properties.
Drawbacks	- Requires a strict control of the amounts of initiator and substrate.	- Relatively low polymer loadings.
	- Might result in low polymerisation degrees.	- Might result in cross-linking of CNTs and their aggregation.

composites based on these coated nanotubes enhanced interfacial bonding in a polystyrene polymer matrix.[64]

The presence of appropriate active functional groups such as carboxylic acids or amines on the CNT surface allows for further covalent functionalisation with polymer molecules (polymer grafting). Two main approaches for the covalent functionalisation of CNTs with polymers have been reported: "grafting from" and "grafting to".[11,12,34] The main differences, advantages and drawbacks of these approaches are summarised in Table 4.1.

4.3.1 "Grafting From" Approach

The "grafting from" approach is based on the initial immobilization of initiators on to the nanotube surface followed by *in situ* polymerisation with the formation of the polymer molecules bound to the nanotube. The benefit of this technique is that polymer-functionalised nanotubes with high grafting density can be produced. However, this method requires a strict control of the amounts of initiator and substrate.

An example of the "grafting from" strategy is a treatment of SWNTs with *sec*-butyllithium that generates carbanions on the nanotube surface. These carbanions can serve as initiators of anionic polymerisation of styrene for *in situ* preparation of polystyrene-grafted tubes.[65] This procedure allows the debundling of SWNTs and the production of a homogeneous dispersion of nanotubes in polystyrene solution.

"Grafting from" technique was widely used for the preparation poly(methyl methacrylate) (PMMA) and related polymer-grafted nanotubes. For example, Qin *et al.* reported the preparation of poly(*n*-butyl methacrylate)-grafted SWNTs by

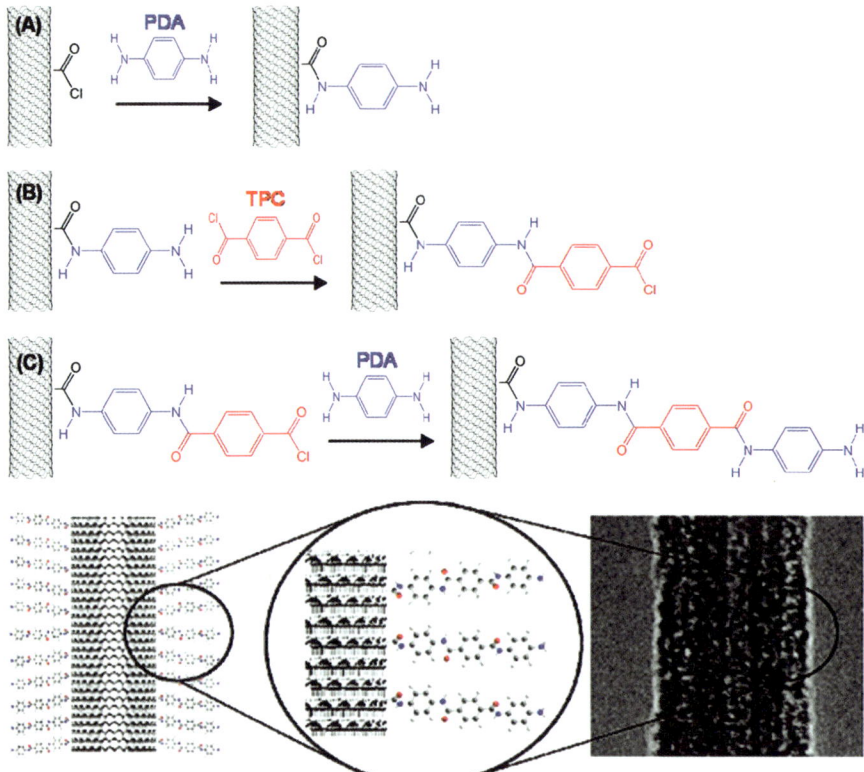

Figure 4.5 Sequential functionalisation methodology for the covalent modification of MWNTs with one and a half repeat units of Kevlar. (A) PDA-MWNTs, (B) TPC-PDA-MWNTs, and (C) PDA-TPC-PDA-MWNTs; bottom panel: the schematic presentation of functionalised MWNTs and corresponding TEM image. TPC- terephthaloyl chloride; PDA- p-phenylenediamine. Reproduced with permission from Sainsbury et al.[73]

attaching n-butyl methacrylate (nBMA) from the ends and sidewalls of SWNT via atom transfer radical polymerisation (ATRP) using methyl 2-bromopropionate as the free initiator.[66] A similar approach was reported by Hwang et al. for the synthesis of PMMA-grafted MWNTs by potassium persulfate-initiated emulsion polymerisation reactions and the use of the nanocomposites as a reinforcement for commercial PMMA by solution casting.[67] Xia et al. used an ultrasonically initiated in situ emulsion polymerisation to functionalise MWNTs with poly(butyl acrylate) (PBA) and PMMA polymers. Then these polymer-encapsulated CNTs have been used to reinforce the Nylon 6 matrix.[68]

Using the "grafting from" approach, Tong et al.[69] have modified SWNTs with polyethylene (PE) by in situ Ziegler–Natta polymerisation. In this work, the surface of the SWNTs was initially functionalised with the catalyst ($MgCl_2$/$TiCl_4$), and then the ethylene was polymerised there, giving PE-grafted SWNTs, which were then mixed with commercial PE by melt blending.[69]

Another example of the "grafting from" technique is the preparation of oligo-hydroxyamide (oHA)-functionalised MWNTs for CNT–poly(p-phenylene benzobisoxazole) (PBO) composites. In this case, pristine MWNTs were first oxidised to MWNT-COOH and then functionalised to MWNT-COCl with thionyl chloride. Then, MWNT-COCl was copolymerised with oHA to produce the corresponding grafted MWNTs (MWNT-oHA).[70,71] The "grafting from" technique was also used for the preparation of styryl-grafted nanotubes.[72] In this work,[72] the carboxylic acid groups on the surface of oxidised MWNTs were reacted with 4-vinylbenzyl chlorides *via* an esterification reaction followed by the polymerisation to produce polystyrene-grafted CNTs.

There has been a report on a very interesting approach, which involves a sequential CNT functionalisation with one and a half repeat units of the polymer poly(p-phneyleneterephthalmide) (PPTA), Kevlar (Figure 4.5). This process results in PPTA oligomer units forming an organic sheath around the nanotubes which can be readily dispersed and integrated within the polymer PPTA to form a novel composite material.[73]

Functionalisation of MWNTs *via* a microwave-induced polymerisation modification route to produce polybenzimidazole (PBI) nanocomposite films has been developed by Zhang *et al.*[74] These PBI–MWNT demonstrated an increase in the Young's modulus of approximately 43.9% at 2 wt% CNT loading, and further modulus growth was observed at higher filler loading.[74]

Another functionalisation strategy involves the preparation of CNTs with pendant self-complementary hydrogen-bonding groups. This can be achieved by functionalisation of MWNT with ureidopyrimidinone groups, which can form multiple hydrogen bonding. The functionalised CNTs were added to acrylic copolymers containing pendant ureidopyrimidinone groups, resulting in composites with significantly improved tensile performance.[75]

4.3.2 "Grafting To" Approach

The "grafting to" approach is based on the attaching of already preformed end-functionalised polymer molecules to functional groups on the nanotube surface *via* appropriate chemical reactions. An advantage of this method is that commercially available polymers containing reactive functional groups can be utilised. However, the main limitation of "grafting to" technique is that initial binding of polymer chains sterically prevents diffusion of additional macromolecules to the surface, leading to relatively low polymer loading.

The "grafting to" method was used for the preparation of composites with the polymers containing reactive functional groups. One of the first examples of the "grafting to" approach was published by Fu *et al.*[76] in 2001. In this work, carboxylic acid groups on the nanotube surface were converted into acyl chlorides by refluxing the samples in thionyl chloride. Then the acyl chloride functionalised CNTs have been reacted with hydroxyl groups of dendritic PEG polymers *via* esterification reactions.

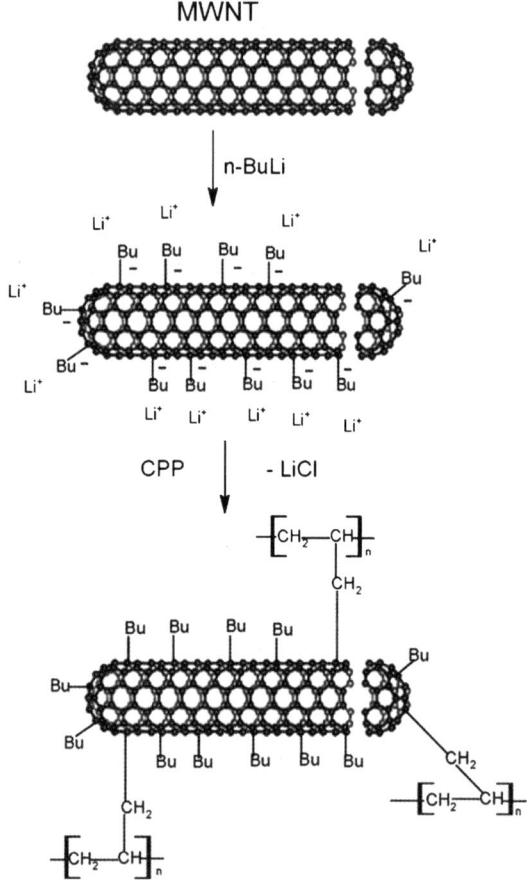

Figure 4.6 Schematic presentation for lithiation of MWNTs followed by functionalisation with CPP. Reproduced with permission from Blake et al.[78]

Another original example of the "grafting to" approach was reported by Lou et al.[77] In this work, MWNTs were grafted by polystyrene, poly(ε-caprolactone) (PCL) and their BCs by the addition reaction of the correspondent alkoxyamine-terminated precursors.

A novel organometallic modification of the "grafting to" strategy was introduced by Gun'ko and co-workers.[78] MWNTs were initially organometallically functionalised using n-butyllithium and then covalently bonded to a chlorinated polypropylene (CPP) via a coupling reaction with the elimination of LiCl. The following addition of the CPP-grafted nanotubes (Figure 4.6) to the CPP polymer matrix resulted in a significant increase in mechanical properties.

The "grafting to" method was recently applied for the preparation of epoxy-polyamidoamine SWNT composites.[79] In this case, carboxylic acid groups of oxidised nanotubes have been further functionalised using polyamido-amine generation-0 (PAMAM-0) dendrimer and then epoxy monomers. The functionalisation resulted in composites with improved dispersion of SWNTs and enhanced mechanical properties.

In another work, the functionalisation of MWNTs with a benzoxazine containing compound bis(3-furfuryl-3,4-dihydro-2H-1,3-benzoxazinyl)isopropane (BPA-FBz) and a polymer polybenzoxazine (PFBz) was achieved using the Diels–Alder reaction. These MWCNT–FBz and MWCNT–PFBz composites have then been thermally cross-linked by press molding. The nanocomposites which have been prepared from BPA–FBz and MWCNT–FBz have shown high electrical conductivities and good mechanical strength. The improvement of these properties has been attributed to the good compatibility between the polybenzoxazine matrix and MWCNT–FBz, as the benzoxazine groups of MWCNT–FBz could copolymerise with BPA–FBz nanocomposites.[80]

There is also a report on covalent grafting of a hydroxylated poly(ether ether ketone) (HPEEK) derivative to the surface of acid-treated SWNTs using two different esterification approaches. Microscopic observations revealed an increase in the bundle diameter of the SWNTs and the heterogeneous composition as result of functionalisation. It was found that the polymer-grafted SWNTs demonstrated higher decomposition temperatures and a wider range of thermal degradation than the pure HPEEK. The esterification also resulted in a decrease in the crystallinity of the HPEEK. However, dynamic mechanical studies showed an exceptional increase in the storage modulus and glass transition temperature of the polymer-grafted SWNTs which can be used as additives to prepare PEEK nanocomposites with improved performance.[81]

In another "grafting to" procedure, four polymers [PVDF, polysulfone (PSF), poly(2,6-dimethylphenylene oxide) and poly(phthalazinone ether ketone)] were chemically bonded to MWNTs using ozone-mediated process. Compared with pristine MWCNTs and PSF-modified MWCNTs, PVDF-modified MWCNTs were more efficient as additives that enhance the mechanical strength and electrical conductivity of PVDF.[82]

4.4 Combination of Non-covalent and Covalent Approaches

In some cases the combination both of non-covalent and covalent functionalisation techniques enables the production of polymer composites with significantly improved mechanical properties. Yuan and Chan-Park[83] have combined the non-covalent and covalent functionalisations of nanotubes in a single system (Figures 4.7 and 4.8) to address the dual problem of poor nanotube dispersion and the weak nanotube–matrix interface. The researchers have developed a novel dispersant which is based on hydroxyl polyimide-graft-bisphenol diglyceryl acrylate (PIOH-BDA) and used it with epoxidised SWNTs. The PIOH-BDA dispersant provided non-covalent functionalisation and good dispersion of the nanotubes due to the strong π–π interaction. In addition, PIOH-BDA also reacted with epoxide groups on the nanotubes and

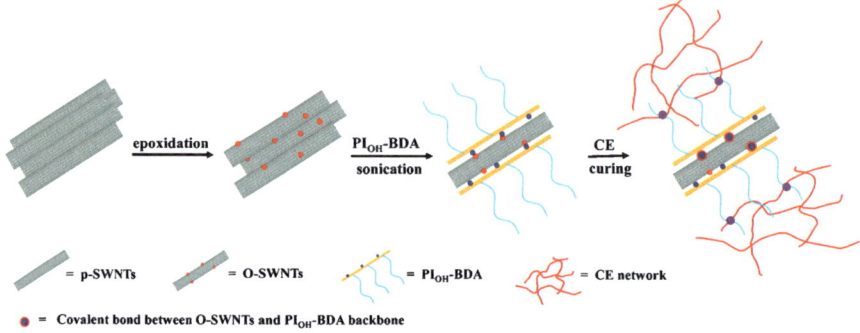

Figure 4.7 Preparation of CE composites with SWNTs functionalised by a method that combines covalent and non-covalent approaches. p-SWNTs - pristine SWNTS; O-SWNTs - epoxidized SWNTs. Reproduced with permission from Yuan and Chan-Park.[83]

functionalised SWNTs covalently in the cyanate ester (CE) polymer matrix. The resulting CE nanocomposites showed 57%, 71% and 124% increases in Young's modulus, tensile strength and toughness, respectively, compared with the neat polymer.[83]

Another paper[84] reported the development of two main strategies: (i) the covalent grafting of hydrolyzable $Si(OEt)_3$ groups on oxidised CNTs and (ii) the non-covalent adsorption of a polycation on pristine CNTs (Figure 4.9). These approaches enabled the performance of further sol-gel processing of functionalised CNTs and their incorporation into poly(methyl methacrylate)

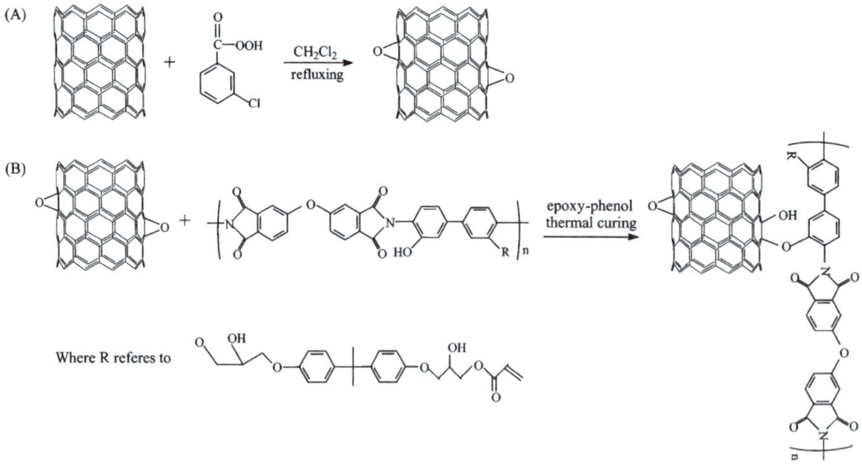

Figure 4.8 (A) Sidewall epoxidation of SWNTs. (B) Chemical reaction between epoxidised SWNTs and PIOH-BDA. Reproduced with permission from Yuan and Chan-Park.[83]

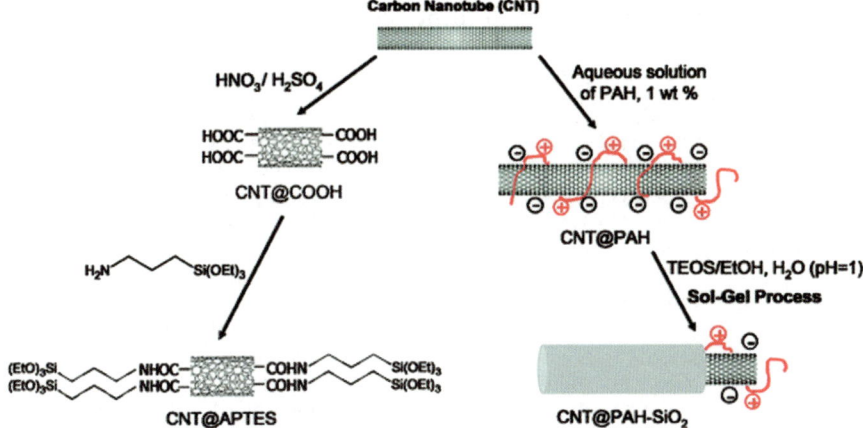

Figure 4.9 Two strategies to modify the surface of CNTs. Reproduced with permission from Mammeri et al.[84]

matrix. Nanoindentation tests of the polymer nanocomposites have demonstrated that the hardness and the elastic indentation modulus have been improved by the functionalised CNT additives.[84]

Thus, the above reports demonstrated that both non-covalent and covalent functionalisation of nanotubes can serve as effective tools for improving both the nanotube dispersion and nanotube–matrix interfacial interaction, enabling the fabrication of reinforced polymer composites.

The comparison of main principles, operation techniques, advantages and disadvantages of non-covalent and covalent approaches is summarised in Table 4.2.

4.5 Main Techniques for Fabrication of CNT–Polymer Composites

The properties of CNTs–polymer composites are strongly dependent on processing methods. These methods govern the dispersion of CNTs in the polymer matrix, which is one of the main factors driving the enhancement of mechanical properties in polymer composites. The dispersion of CNTs is defined by the degree of their aggregation and by the homogeneity of their distribution in the polymer matrix. Therefore, the careful choice of the fabrication technique is very important to prevent the aggregation of the CNTs and achieve the best stress–strain transfer in composite. Also, processing techniques frequently involve chemical functionalisation or cross-linking of nanotubes and polymer molecules at some stages. Therefore it is important to consider approaches for the fabrication and processing of CNT–polymer composites.

Table 4.2 Comparison of non-covalent and covalent approaches.

	Non-covalent	Covalent
Main principles	Binding of CNTs to polymer *via* van der Waals, π–π, CH–π or electrostatic interactions between polymer molecules and CNT surface.	Direct chemical reactions of reagents with the sidewalls of CNTs or modification of appropriate surface-bound functional groups
Operation techniques	- Surfactant-assisted dispersion. - Polymer wrapping. - Polymerisation-filling technique (PFT).	- Cyclo-addition to CNT walls. - Chemical binding to functional groups at CNT surface. - Plasma polymerisation treatment.
Advantages	Does not alter the structure of the nanotubes and therefore both initial electrical and mechanical properties can remain almost unchanged.	Provides very strong chemical bonding of CNTs to polymer matrix as result of very good load transfer.
Disadvantages	The efficiency of the load transfer is low because the forces between the wrapping molecules and the nanotube surface might be relatively weak.	- Induces some defects and damage to original CNTs. - Converts strong graphitic structure into weaker sp^3-hybridised carbon structures.

4.5.1 Solution Processing of Composites

Solution casting processing of composites is one of the most usual methods for making polymer–nanotube composites at the laboratory scale. The nanotubes and polymer are mixed in a suitable solvent before evaporating the solvent to form a composite film.[26–32] Almost all solution processing methods are variations on a general theme which can be summarised as: (i) dispersion of nanotubes in either a solvent or polymer solution by energetic agitation; (ii) mixing of nanotubes and polymer in solution by energetic agitation; (iii) controlled evaporation of solvent, leaving a composite film.

In general, agitation is provided by magnetic stirring, shear mixing, reflux or, most commonly, ultrasonication. Sonication can be provided in two forms, mild sonication in a bath or high-power sonication using a tip or horn. One of the advantages of this method is the possibility to achieve debundling and good quality dispersion of the nanotubes in the appropriate solvent. However, the solution processing technique cannot be utilised for insoluble polymers.

4.5.2 Melt Processing

Melt processing is a good alternative technique which is particularly useful for dealing with thermoplastic polymers.[27,37,85–93] This approach makes use of the fact that thermoplastic polymers soften when heated. Amorphous polymers can be processed above their glass transition temperature, whereas semi-crystalline polymers need to be heated above their melt temperature to induce sufficient softening. The advantages of melt processing are its speed and simplicity, and easy integration into standard industrial facilities (*e.g.* extruders, blow-molding machines).[7,85] Although under high temperatures, this approach can sometimes result in unexpected polymer degradation and oxidation.

Melt processing can also be used for the production of both bulk polymer composites and composite fibres. In general, melt processing involves the melting of polymer pellets to form a viscous liquid. Any additives, such as CNTs can be mixed into the melt by shear mixing. Bulk samples can then be fabricated by techniques such as compression moulding, injection moulding or extrusion. However, it is important that processing conditions are optimised, not just for different nanotube types, but for the whole range of polymer–nanotube combinations. This is because nanotubes can effect melt properties, such as viscosity, resulting in unexpected polymer degradation under conditions of high shear rates.[87]

For many applications, fibres are more suitable than bulk materials. In addition, fibre production techniques tend to be suited to the alignment of nanotubes with in the fibre. A number of studies have focused on the production of composite fibres by melt processing. Fibre processing is generally similar to melt processing, but usually involves a process such as extrusion to produce an elongated sample which can then be drawn into a fibre.[7,27,37,85,94,95]

4.5.3 *In situ* Polymerisation Processing

Over the last decade, *in situ* polymerisation in the presence of CNTs has been intensively explored for the preparation of polymer-grafted nanotubes and the processing of the corresponding polymer composite materials. The main advantage of this method is that it enables grafting of polymer macromolecules on to the convex walls of CNTs. This then provides better nanotube dispersion and the formation of a strong interface between the nanotube and polymer matrix. In addition, *in situ* polymerisation is a very convenient processing technique, which allows the preparation of composites with high nanotube loading and very good miscibility with almost any polymer type. This technique is particularly important for the preparation of insoluble and thermally unstable polymers, which cannot be processed by solution or melt processing.

Initially, *in situ* radical polymerisation was applied for the synthesis of PMMA–MWNT composites by Jia *et al.*[96] In this work, *in situ* polymerisation was performed using the radical initiator 2,2′-azobisisobutyronitrile (AIBN). The authors believed that π-bonds in CNTs were initiated by AIBN and therefore nanotubes could participate in PMMA polymerisation to form a strong interface between the MWNTs and the PMMA matrix. Then, Velasco-Santos *et al.*[97] and Putz *et al.*[98] also used AIBN as an initiator of *in situ* radical polymerisation to incorporate functionalised MWNTs and SWNTs into the PMMA matrix.

In situ polymerisation was also very useful for the preparation of polyamide–CNT polymer composites. For instance, PA6–MWNTs composites have been prepared by *in situ* hydrolytic polymerisation of ε-caprolactam in the presence of pristine and carboxylated tubes.[99] The ε-caprolactam monomer was found to form an electron-transfer complex with MWNTs, giving a homogeneous polymerisable master solution, which facilitated the formation of composites with homogeneously dispersed nanotubes. In another work, Gao *et al.*[100] reported a new improved chemical processing technology that allows the continuous spinning of SWNTs–Nylon 6 (PA6) fibres by the *in situ* ring-opening polymerisation of ε-caprolactam in the presence of nanotubes (Figure 4.10).[100] This process results in a new hybrid material with characteristics of both the fibre and the matrix, with an excellent compatibility between the SWNTs and Nylon 6.

In situ epoxidation reaction has also been used for the preparation and processing of epoxy polymer composites. For instance, carboxyl- and fluorine-functionalised SWNTs have been integrated into an epoxy polymer *via* the

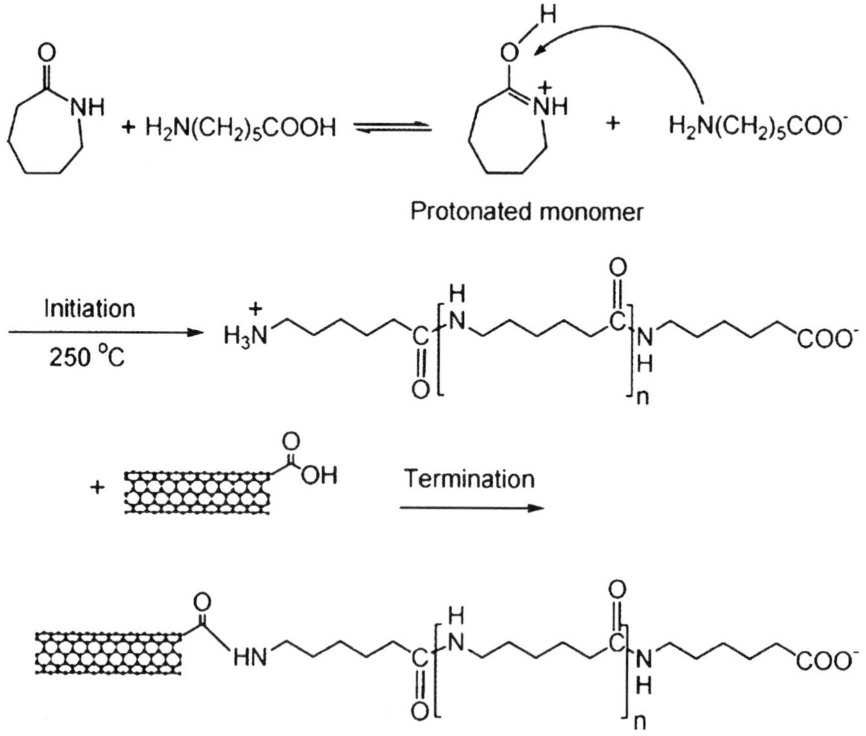

Figure 4.10 Scheme of the preparation of SWNT– Nylon 6 composites by ring-opening polymerisation of ε-caprolactam in the presence of SWNTs. Reproduced with permission from Gao *et al.*[100]

formation of covalent bonds by *in situ* epoxy ring-opening esterification and amine-curing chemical reactions.[101] The same group has further developed a fully integrated nanotube epoxy composite material with direct covalent bonding between the matrix and SWNTs using functionalised SWNTs prepared *via* diamine reactions with alkyl-carboxyl groups directly attached to the SWNTs sidewalls.[102]

In situ polymerisation reactions have also been applied to the preparation of other polymer–nanotube composites. For example, Kumar *et al.*[103] have synthesised new ultra-strong PBO composites in the presence of SWNTs in poly(phosphoric acid) (PPA) by *in situ* PBO polymerisation.[103] After the polymerisation, PBO–SWNT composite fibres have been spun from the liquid crystalline solutions using dry-jet spinning.

In general, *in situ* polymerisation can be used for the preparation of almost any polymer composites containing CNTs which can be non-covalently or covalently bound to the polymer matrix, as discussed above.

PFT also involves *in situ* co-polymerisation of olefins catalysed directly from nanotubes pre-treated by a methylaluminoxane (MAO) or highly active metallocene-based complexes (*e.g.* Cp_2ZrCl_2).[33] This approach destroys nanotube bundles and results in homogeneously coated CNTs.

In another work, using the amidation reaction of octadecylamine with nanotubes, Yang *et al.*[104] prepared functionalised soluble MWNTs and mixed them in solution with a copolymer of methyl and ethyl methacrylate [P(MMA-*co*-EMA)].[104]

Broza *et al.*[105] prepared new poly(butylene terephthalate) (PBT)–nanotube composites by introducing the oxidised SWNTs into the reaction mixture during two-stage polycondensation of buthylene terephthalate in a molten state.

In overall, the use of polymer-coated CNTs produced by *in situ* polymerisation, whether by covalent or non-covalent methods leads to the production of polymer nanocomposites displaying much better thermo-mechanical, flame retardant and electrical conductive properties, even at very low nanotube loadings. As mentioned above, the covalent approach allows the formation of a strong interface between the nanotube and polymer matrix due to strong chemical bonding of polymer molecules to the CNT surface. The *in situ* polymerisation technique also enables the preparation of composites with very high nanotube loadings.

4.5.4 Processing of Composites Based on Thermosets

A closely related method to the *in situ* polymerisation processing of composites is based on epoxy resins thermosets. In this approach, CNTs can be dispersed in a liquid epoxy precursor and then the mixtures can be cured by the addition of hardener, such as triethylene tetramine (TETA), and the application of temperature or pressure.[21,102,106,107] In most cases, the epoxy monomer exists in liquid state, facilitating nanotube dispersion. Curing is then carried out to convert the viscous liquid mixture into the final solid composite. In the simplest cases, nanotubes are dispersed by sonication in a liquid epoxy such as Shell EPON 828 epoxy.[106] This blend can then be cured by the addition of hardener such as TETA and the application of temperature or pressure.[102,107,108]

4.5.5 Coagulation Spinning and Electrospinning

Widely used industrial approaches for the production of polymer fibres and yarns, such as coagulation spinning and electrospinning, have also been utilised for the fabrication of polymer–nanotube composites. In coagulation spinning, for example, composite fibres can be produced by an injection of surfactant-stabilised nanotube dispersion in water into a rotating bath of polymer (*e.g.* PVA) dissolved in water such that nanotube and polymer dispersions flowed in the same direction at the point of injection. In this case, polymer molecules replace surfactant molecules on the nanotube surface, thus destabilising the nanotubes dispersion which collapses to form a fibre. These fibres can then be retrieved from the bath, rinsed and dried.[109–111]

This method was further developed and improved by Dalton *et al.*[110] They injected the nanotube dispersion into the centre of a co-flowing PVA/water stream in a closed pipe. The wet fibre was then allowed to flow through the pipe for approximately 1 m before being wound on a rotating mandrel. Flow in more controllable and more uniform conditions of the pipe resulted in longer ($\sim$100 m) and more stable fibres. Crucially, wet fibres were not rinsed to remove PVA, although they were dried to produce fibres with final diameters of tens of microns. Furthermore Miaudet *et al.*[111] have shown that these fibres can be drawn at temperatures above the PVA glass transition temperature, resulting in improved nanotube alignment and polymer crystallinity.

Another method used recently to form composite-based fibres from solution is electrospinning. The electrospinning technique involves electrostatically driving a jet of polymer and nanotubes dispersions in an appropriate solvent out of a nozzle on to a metallic counter-electrode. This technique has been used to produce man-made fibres since 1934[112] and involves electrostatically driving a jet of polymer solution out of a nozzle on to a metallic counter-electrode. In 2003, Ko *et al.*[113] described using electrospinning to fabricate polymer–nanotube composite fibres and yarns. Composite dispersions of SWNT and either poly(lactic acid) (PLA) or polyacrylonitrile (PAN) in *N,N*-dimethylformamide (DMF) were initially produced. The dispersion was then placed in a pipette with a 0.9 mm nozzle. A wire was placed in the pipette and connected to a steel plate, *via* a high voltage power supply (25 kV). The plate was 15 cm below the nozzle. When the power supply is turned on, the composite solution becomes charged. This forces it out of the nozzle and towards the counter electrode. Charging of the solvent causes rapid evaporation resulting in the coalescence of the composite into a fibre which can be collected from the steel plate. Fibres with diameters between 10 nm and 1 micron can be produced in this fashion. Yarns were also produced by collecting the fibres on a rotating drum and twisting them.[112,113]

In a similar study, Sen et al.[114] formed fibre-based membranes. They were spun from SWNT dispersed in either polystyrene or polyurethane with a 3 cm needle-plate gap. In this work, the solution was pumped slowly out of a needle under application of 15 kV. Spinning was continued for 1 h until the counter-electrode was covered in a membrane built up from the spun fibres.

In a more recent work, MWNTs have been incorporated into surface-modified, reactive P(St-co-GMA) nanofibres by electrospinning. Then resulting nanofibres have been functionalised with epoxide groups and added to the epoxy matrix producing reinforced epoxy resins. The polymer composites have demonstrated over a 20% increase in flexural modulus, when compared with neat epoxy, despite a very low composite fibre weight fraction (at approximately 0.2% by a single-layer fibrous mat). The increase is attributed to the combined effect of the well-dispersed MWNTs and the surface chemistry of the electrospun fibres that enabled an effective cross-linking between the polymer matrix and the nanofibres.[115]

4.5.6 Buckypaper-based Approaches

Buckypaper is a thin porous assembly of CNTs usually formed by filtration from their dispersion in a solvent. Over last few years, the electrical and mechanical properties of buckypaper have been studied extensively.[116–118] In order to prepare buckypaper CNTs must be dispersed in an appropriate solvent. One of the solvents used to prepare buckypaper is N-methylpyrrolidone (NMP), which routinely fabricates good quality buckypaper without the need for centrifugation.[119–122] DMF,[123] 3-aminopropyltriethoxy silane[124] and γ-butyrolactone, also known as 'liquid ecstasy', well known for its narcotic properties, have all been shown to be good solvents for CNTs.[125] Using an appropriate surfactant in water can be cheaper than using DMF and NMP, which have also the added disadvantage of having high boiling points. The most common surfactants are sodium dodecyl benzene sulfonate (SDBS) and SDS. Aqueous dispersion of SWNTs in the presence of the water-soluble perylene derivatives has also been reported.[126] After the nanotubes are dispersed in a solvent, the buckypaper can usually be fabricated by a simple Buchner filtration.

Another way to produce buckypaper without the dispersion and filtration method is the 'domino pushing effect'. Ding et al.[127] present a simple and effective macroscopic manipulation of aligned CNT arrays. The domino pushing of the CNT arrays can efficiently ensure that most of the CNTs are well aligned tightly in the buckypaper structure.

Buckypaper normally has a laminar structure with the orientations of the bundles of tubes being random in the plane of the sheet.[128] This paper is extremely porous and contains up to 70% free volume. The pore size is theoretically tunable depending on the nanotube size and shape and the casting solvent.[129] Normally, the composite preparation involves the infiltration of polymer from solution into pre-fabricated buckypaper sheets. This concept was first demonstrated by Coleman *et al.*[130] in 2003. They first produced thin buckypaper sheets of SWNTs by Buchner filtration. These porous sheets were then soaked in polymer solutions for various times before rinsing. In this way, polymer mass fractions of up to 30 wt% could be incorporated.

A similar technique was demonstrated by Wang *et al.*[131] They also fabricated buckypaper but incorporated an epoxy-hardener blend by Buchner filtration. To reduce the viscosity, the blend was mixed with acetone. Very good infiltration was observed throughout the paper. The epoxy was cured by hot pressing 3–5 stacked sheets together at 177 °C to form a thick composite film.

In overall, infusing the porous buckypaper with polymer is a facile way of improving the buckypaper mechanical properties and create polymer composites with high loadings of nanotubes (>60%). To increase mechanical properties, sheets of buckypaper can also be inserted between laminates.[132,133] Polymers or epoxies have been layered on top of buckypaper to develop a ply material that has electrochemical actuation properties.[134] In addition, nanotubes in the buckypaper may also be chemically cross-linked to create much stronger composites.[135] For example, using the nitrene reaction, functionalised nanotubes can be linked within a bundle as well as between bundles. By filtering and drying, a buckypaper of linked nanotubes was obtained.[136] This linked buckypaper has the potential to have higher mechanical properties than pristine buckypaper.

4.5.7 Layer-by-layer (LBL) Technique

The LBL approach involves building up a layered composite film by alternate dipping of a substrate into dispersions of CNTs and polyelectrolyte solutions.[137,138] Additionally, to improve the structural integrity of the film, cross-linking can be induced. LBL assembly is a simple, versatile and relatively inexpensive approach which provides multifunctional molecular assemblies of tailored architectures and material properties for various reaction and sensing materials of nanometre thickness, and will enable large-scale, reproducible production of membrane-based, highly integrated microsensors.[139,140] This method has significant advantages as thickness and polymer/nanotube ratio can be easily controlled and very high nanotube loading levels can be obtained. Therefore, this has led to recent exceptional growth in the use of LBL-generated nanocomposites. This method has been used extensively to incorporate inorganic nanoparticles, nanowires and nanotubes into organic polymers.[141] For example, in the process of the LBL assembly of active ester-modified MWNTs and poly(allylamine hydrochloride) (PAH) on an activated surface of a quartz slide, a reaction occurred between the active ester on the surface of MWNTs and the amine groups of polyallylamine, yielding amide bonds. This resulted in a mechanically stable free-standing thin film.[142]

Kotov and co-workers have published several very interesting papers[137,138,140,141,143–145] on LBL assembly of CNTs and polymers. This approach enabled the production PVA/[SWNT + poly(sodium 4-styrenesulfonate]-layered coatings and free-standing films, which displayed high electrical conductivities and excellent mechanical strength.[143] These new materials demonstrated high chemical/mechanical durability, electrical conductivity and wearability, as well as interesting chemo- and biosensing opportunities. The high strength and conductivities of these composites are attributed to the unique homogeneity of the LBL-assembled composites.[144] In overall, the LBL technique in combination with chemical cross-linking allows the fabrication of CNT–polymer composites with outstanding mechanical properties.[145]

4.5.8 Swelling Under Ultrasound Technique

This technique was initially reported by Gun'ko and co-workers[146] in 2009. High-strength, high-toughness Kevlar–nanotube composite fibres were produced by the swelling of commercially available Kevlar fibres in a suspension of MWNTs in NMP under ultrasonication. This process allowed the nanotubes to diffuse into the interior of the fibre. The resulting composites were stronger and tougher than the original Kevlar fibres at only 1–2 wt% of nanotube loadings. There is also another recent report on new conductive CNT–PE composites, which have been prepared similarly by swelling a thin PE film in MWNT dispersions in tetrahydrofuran (THF) under ultrasonication.[147] This new approach is a very promising post-processing technique, which allows us to incorporate CNTs into already formed polymer products, including insoluble or temperature-sensitive polymers, such as Kevlar. In addition, this technology enables the inclusion of nanotubes into a very thin (several hundred nm) top polymer layer. Therefore, only a very small percentage ($\sim$1–2%) of nanotubes is needed to produce polymer composites with potentially high electrical surface conductivity and improved mechanical parameters.

4.6 Influence of Nanotube Functionalisation on Mechanical Properties of CNT–Polymer Composites

In order to estimate the role of nanotube functionalisation and compare the mechanical parameters of corresponding composites, we normally use the rate of increase (dY/dV_f) of the Young's modulus (Y) to the volume fraction of nanotubes to polymer (V_f).[148] These values enable the quantitative estimate of the reinforcement of both modulus and strength at low nanotube volume fractions. For a thin film composite, in which the nanotubes are aligned in the plane of the film, the maximum theoretically expected value of dY/dV_f is $\sim$400

GPa.[11,149] In our calculations, densities of nanotubes used were assumed to be $\rho = 1350$ kg m^{-3} for SWNTs and $\rho = 2150$ kg m^{-3} for MWNTs.[150] There are a huge amount of variables that effect the mechanical properties of CNT–polymer composites including: type of CNTs used; choice of solvent; nanotube functionality; polymer used; composite preparation *etc.*[151] As this procedure potentially has some errors, these values are to be taken as a guideline only.

In this Section we will discuss and compare mechanical properties of selected reported nanotube–polymer composites. We are going to focus on functionalisation of nanotubes and polymer composite fabrication techniques as the two main factors which have the strongest influence on the mechanical properties of CNT–polymer composites.

Functionalisation of nanotubes is extremely important for their processing and has a direct impact on mechanical characteristics of CNT–polymer composites. Chemical functionalisation of the nanotube surface is expected to maximize composite interfacial shear strength and facilitate SWNT debundling. While it has been shown that covalent chemical attachments may decrease the maximum buckling force of nanotubes by approximately 15%,[152] the fact that composites based on functionalised nanotubes are expected to have large interfacial shear strengths should more than compensate. Covalently grafted long-chain molecules entangle with the polymer matrix, ensuring very good stress transfer from matrix to nanotube. In addition, the functional groups make the nanotubes more compatible both with polymer hosts and solvents. This tends to dramatically improve the nanotube dispersion and hence further improve composite properties.

The Young's moduli and dY/dV_f values of some polymer composites that contain various functionalised CNTs are summarised in Table 4.3.

As we can see, the highest dY/dV_f values are observed for polymers loaded with alkyl-, amine- or ferritin protein-functionalised nanotubes at very low nanotube contents (<1 wt%). Normally, nanotube loadings of greater than 2 wt% decrease the mechanical properties due to the aggregation of nanotubes and the reduction in the nanotube–polymer interaction. In general, there are greater increases in mechanical parameters and dY/dV_f values for covalently functionalised CNTs due to more efficient interfacial stress–strain transfer between nanotubes and polymer matrix. However, sometimes CNTs can act as nucleation centres, enhancing polymer crystallinity, as it has been reported by Coleman *et al.*[28] It has been shown that in nanotube–PVA composites the polymer crystallinity increases linearly with the increasing volume fraction of nanotube, indicating a crystalline polymer coating at the nanotube surface.[148] In these cases, very high values of Young's moduli (*e.g.* 7.04 GPa) and dY/dV_f (*e.g.* 754 GPa) can be achieved, even for non-functionalised nanotubes.[28,148]

Hwang *et al.*[67] reinforced PMMA by the addition of PMMA-grafted arc-MWNTs. They observed an increase in the modulus as measured by dynamic mechanical analysis (DMA) from 2.9 to 29 GPa on the addition of 20 wt% nanotubes ($dY/dV_f \sim 116$ GPa). This is significant for two reasons. Firstly, they managed to successfully functionalise arc-MWNT, which is challenging compared with functionalisation of Chemical Vapour Deposition (CVD)–MWNT. This means that the fillers have a very high modulus. In addition, good dispersion and a continued increase in modulus enhancement was observed up to 20 wt%, which is unprecedented. Failure of the nanotubes by the "sword and sheath" method was observed, suggesting excellent interfacial stress transfer (Figure 4.11).

Melt spun fibres from PA6-grafted SWNTs in a PA6 matrix were produced by Gao et al.[100] The fibres were stiffened from 440 MPa to 1200 MPa on the addition of 1.5 wt% tubes ($dY/dV_f \sim 120$ GPa). The strength was doubled from 41 to 86 MPa. Both strength and toughness were maximised at mass fractions of 0.2%.

Liu et al.[153] reinforced PVA with hydroxyl-functionalised SWNT. The idea was that the hydroxy groups would be linked by hydrogen bonding with the hydroxyl functionalities of the PVA. A reasonably good modulus enhancement was observed with an increase from 2.4 to 4.3 GPa on the addition of 0.8 wt% nanotubes, corresponding to a reinforcement value of ~ 305 GPa, whereas the strength increases from 74 to 107 MPa was also observed. These results were explained by the observation of good load transfer by Raman spectroscopy.

In another work, fluoronanotubes have been treated with asymmetric diamine molecules and N-Boc-1,6-diaminohexane, which were used to replace fluorines. The Boc deprotection resulted in amino groups, which were employed to create strong covalent bonds with the polymer matrix via epoxy ring-opening etherification and curing chemical reactions. The composites loaded with 0.5 wt% of these functionalised nanotubes demonstrated improvement in the Young's modulus, ultimate tensile strength and storage modulus by 30%, 25% and 10%, respectively, compared with the neat epoxy matrix.[154]

In summary, nanotubes with carboxylic acid functionalities have demonstrated significant increases in Young's moduli for various polymer composites, in particular for those which are produced by solution casting techniques.[57] For example, acid- or alkali-treated CNTs in PVA increased the tensile strength and toughness of the PVA/CNT-coated thread by 117% and 560%, respectively.[48] Acid-treated CNTs in PCL increased the strength and modulus of the composite 12.1% and 164.3% respectively at 1.2 wt% of MWNT content.[155] Acid-treated CNTs in polyurethane composites gave a remarkable 740% increase in the Young's modulus from 0.05 to 0.42 GPa at 20 vol% with dY/dV_f of 4.5 GPa.[59]

Similar trends are also observed for other functional groups. For instance, triethylene-tetramine (TETA)-functionalised MWNT–epoxy composites showed an increase in the Young's modulus of 38% and about 30% in the tensile strength at very low nanotube loadings, which corresponds to dY/dV_f of 355 GPa.[156] In another work, MWNTs coated with silica and then functionalised with 3-methacryloxypropyltrimethoxysilane (3-MPTS) were added to polypropylene (PP). As expected, the 3-MPTS-functionalised MWNT–PP composite has a higher tensile strength than the pristine MWNT–PP composite.[157]

Table 4.3 Mechanical properties of selected polymer composites containing functionalised CNTs.

Nanotube/polymer composite	Nanotube functionality	Preparation techniques	Y_{Poly} (GPa)	Y_{Max} (GPa)	Nanotube content (wt%)	dY/dV_f (GPa)	Reference and year
CVD-MWNT/methyl and ethyl methacrylate P(MMA-co-EMA)	Octadecylamine	Solution casting	1.64	2.62	10	15	104 2004
SWNT/epoxy	Large organic groups	Solution casting and curing	2.02	3.4	4	95	102 2004
SWNT/polyamide 6 (PA6)	PA6	Solution casting and melt spinning	0.44	1.2	1.5	120	100 2005
SWNT/poly(vinyl alcohol) (PVA)	Hydroxyl	Solution casting	2.4	4.3	0.8	305	153 2005
CVD-MWNT/PVA	Ferritin protein	Solution casting	3.4	7.2	1.5	380	158 2005
MWNT/polycarbonate (PC)	Epoxide	Solution casting and injection moulding	2.0	3.8	5	95.5	162 2006
MWNT/PMMA	PMMA	Melt mixing and extrusion	2.7	2.9	3	18.15	163 2006
CVD-MWNT/PVC	CPP	Solution casting	0.56	0.9	1	115	160 2006
CVD-MWNT/PS	CPP	Solution casting	1.48	2.63	1	304	160 2006
SWNT/copolymers of styrene and vinyl phenol (PSVPh)	Carboxylic acid	Solution casting	1.5	2.1	5	24.25	57 2006
CVD-MWNT/polyurethane (PU)	Carboxylic acid	Solution casting	0.05	0.42	20	4.5	59 2006

Table 4.3 (Continued)

Nanotube/polymer composite	Nanotube functionality	Preparation techniques	Y_{Poly} (GPa)	Y_{Max} (GPa)	Nanotube content (wt%)	dY/dV_f (GPa)	Reference and year
SWNT/biosteel (synthetic spider silk)	Octadecylamine	Solution casting	1.6	1.9	0.125	381	[149] 2007
SWNT/PAMAM	Epoxy	Solution casting and curing	2.76	3.49	1	153	[79] 2008
CVD-MWNT/Kevlar	PVC	Solution casting	1.5	2.5	2	300	[45] 2008
CVD-MWNT/epoxy	TETA	Cast moulding	1.56	2.4	0.6	355	[156] 2008
CVD-MWNT/PS	Butyl	Solution casting	1.29	1.63	0.25	433	[161] 2008

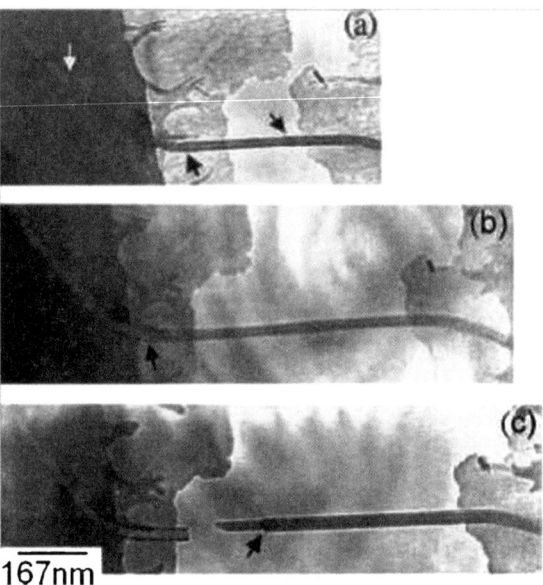

Figure 4.11 TEM images of a MWNT-containing PMMA thin film taken at different times: (a) $t = 0$, (b) $t = 4$ and (c) $t = 10$ min. Reproduced with permission from Hwang et al.[67]

The protein, ferritin, was successfully grafted on to CVD–MWNT.[158] This material was mixed with PVA. The authors claimed that the ferritin was expected to form bonds with the PVA. Indeed, a significant modulus enhancement from 3.4 to 7.2 GPa was observed corresponding to a reinforcement of 380 GPa.

Very considerable increases in mechanical properties of polymer composites have also been demonstrated by CNTs functionalised *via* organometallic approaches, *e.g.* using butyl lithium.[159] This method allowed the preparation of MWNTs covalently functionalised with CPP and then utilised as additives in both polystyrene and PVC composites. This gave a two-fold increase in both modulus and strength in polymer composite films at nanotube loading levels of less than 1 wt%.[160] Also, polystyrene composites based on butyl lithium - functionalised MWNTs demonstrated an increase of up to 25% in their Young's moduli and up to 50% in their tensile strength over pure polystyrene at low nanotube loadings of 0.25 wt%, which corresponds to dY/dV_f of 433 GPa. This value corresponds to the theoretical maximum (~ 400 GPa).[161] Thus, organometallic chemistry approaches enable the achievement of very strong binding between nanotube and polymer matrix, and as result demonstrate an excellent interfacial stress transfer in corresponding polymer composites.

Non-covalent functionalisation approaches also enable the preparation of polymer composites with significantly improved mechanical properties. For example, in the recent work of Yuan *et al.*[164] three different polymers such as polyimide without side-chain (PI), polyimide-graft-glyceryl 4-nonylphenyl

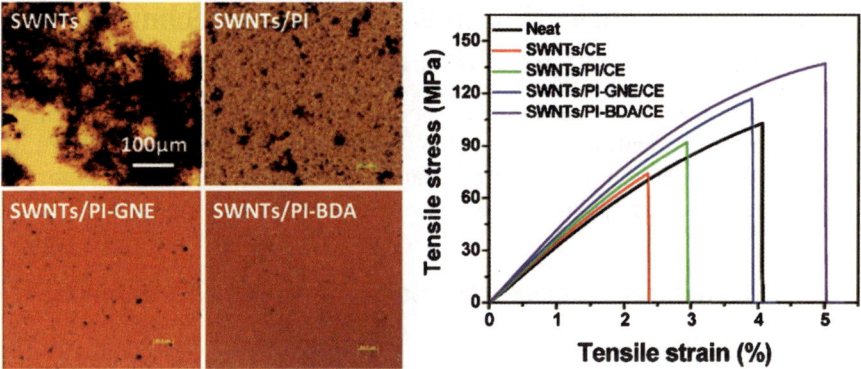

Figure 4.12 Left: AFM images of polymer-coated SWNTs; right: stress–strain curves of the corresponding SWNT polymer composites. Reproduced with permission from Yuan et al.[164]

ether (PI-GNE) and polyimide-graft-bisphenol A diglyceryl acrylate (PI-BDA) have been used to disperse SWNTs and to improve the interfacial bonding between SWNTs and CE matrix. Atomic force microscopy (AFM) images (Figure 4.12) showed that both PI-GNE and PI-BDA are highly effective at dispersing and debundling of SWNTs in DMF, whereas PI was less effective. A series of CE-based composite films reinforced with different loadings of SWNTs, SWNTs/PI, SWNTs/PI-GNE and SWNTs/PI-BDA have been prepared. It was found that SWNTs/PI-BDA/CE composites have the greatest improvement in mechanical properties of the composites (Figure 4.12), because of better dispersion of nanotubes.[164]

Another group has synthesised a new reactive comb-like polymer, polyimide-graft-bisphenol diglyceryl acrylate (PI-BDA), which is shown to be highly effective in non-covalent functionalisation and dispersing SWNTs into individual nanotubes or small bundles. These PI-BDA-functionalised SWNTs have been added to a thermosetting resin blend of cyanate ester and epoxy (CE-EP) and composite fibres with SWNT have been produced. It was shown that the addition of only 1 wt% SWNTs results in an increase in the tensile modulus, strength and toughness by 80% (to 4.70 ± 0.24 GPa), 70% (to 142.3 ± 6.9 MPa) and 58% (4.1 ± 0.4 MJ m^{-3}), respectively, compared with the neat resin blend. The authors believe that the reinforcement can be explained by the strong π–π interactions between SWNTs and the PI-BDA backbone.[165]

La Saponara and co-workers[166] have demonstrated that the non-covalent surface treatment of the CNTs using hexamethylene diamine and one of two surfactants, (Triton X-100) or cetyl pyridinium chloride (CPC), enabled the production of reinforced fibreglass-reinforced panels, which were prepared with treated CNT/epoxy by hand lay-up. The surface treatment with diamine/CPC resulted in the most superior composites with improved mechanical performance, higher resistance to fatigue and impact damage (over 30 J).[166]

4.7 Role of Fabrication and Processing Techniques in Reinforcement of Polymers by CNTs

As we mentioned above, mechanical properties of polymer composites strongly depend on the fabrication and processing approaches used. In this Section, we will discuss and compare mechanical properties of polymer–nanotube composites, which are produced by various processing techniques.

4.7.1 Mechanical Properties of Solution-processed Composites

Solution casting is one of the most frequently used techniques for the fabrication of nanotube–polymer composites. This method allows for the preparation of quite strong composites at the low nanotube loadings. Although a number of reports appeared in the literature before 2002,[30,167–170] the first significant increase in stiffness was observed by in a study by Cadek et al.,[26] in which PVA was used as a matrix material. They carried out nanohardness tests on spun cast films of arc-discharge MWNTs in both PVA and poly(vinyl carbazole) (PVK). Increases in the Young's modulus from 7 GPa to 12.6 GPa with 0.6 vol% MWNT in PVA and from 2 GPa to 5.6 GPa with 4.8 vol% in PVK were observed. These increases are equivalent to reinforcement values of $dY/dV_f = 990$ GPa and $dY/dV_f = 75$ GPa for PVA and PVK, respectively. However the value for PVA is probably artificially inflated as the modulus values for PVA are known to be closer to 2 GPa.[171] This suggests that the reinforcement value for PVA should be scaled down to $dY/dV_f \sim 280$ GPa. Crucially, Cadek et al.[26] studied the morphology of the polymer matrices using differential scanning calorimetry (DSC). They observed that the PVA crystallinity increased linearly with the nanotube content. This suggests the nucleation of crystallinity by the nanotubes. No such effect was observed for the PVK samples. This suggests that the difference in reinforcement may be related to the presence of a crystalline interface for PVA composites, but an amorphous one for PVK materials. This indicates the possibility that stress transfer may be maximised by the presence of an ordered interface.

In another work, Velasco-Santos et al.[58] measured reasonably large increases in modulus from 0.71 GPa for a methyl-ethyl-methacrylate co-polymer to 2.34 GPa at 1 wt% arc-MWNT. This corresponds to a reinforcement of $dY/dV_f \sim 272$ GPa which is on a par with the value (scaled) for PVA composites from Cadek et al.[26] However, in this work no nucleation of crystallinity was observed. This suggests that good stress transfer can be obtained at an amorphous interface, depending on the polymer.

Coleman et al.[28] also fabricated PVA-based composites using commercial low diameter ($D \sim 15$ nm) MWNT as a filler. They observed significant reinforcement with modulus enhancement from 1.92 GPa to 7.04 GPa at 0.6% nanotube volume fraction. This represents a reinforcement of $dY/dV_f = 754$ GPa. Also, strength enhancement from 81 MPa to 348 MPa was observed. In addition, a microscopy study of composite fracture surfaces showed nanotube pullout. However, the pullout diameter showed that a layer of polymer had remained attached to the tube after fracture. The thickness of this layer was very similar to the measured thickness of a crystalline polymer coating nucleated by the MWNT. This strongly suggests that the presence of interfacial crystallinity has a large bearing on the mechanical properties of composites.

Another example is the use of the solution casting technique to produce PVC composites with MWNTs, which have values of dY/dV_f of 204 GPa. With Kevlar-functionalised MWNT–PVC composites, however, there is a substantial increase in dY/dV_f to 300 GPa.[45] These values are quite close to the maximum expected value of dY/dV_f.[11,149] This indicates on very good stress–strain transfer between PVC and the Kevlar-coated nanotubes. Most importantly, the reinforcement was achieved at less than 2 wt% nanotube content, which a great advantage if we take into account the nanotube costs.

In another work, SWNT–epoxy composites gave dY/dV_f of 107.3 GPa. However, PAMAM-0-functionalised SWNT–epoxy composites had a higher dY/dV_f of 153.6 GPa.[79] In this paper, the authors used the Halpin–Tsai equation to predict the modulus of fibre reinforced composites.[172,173] The experimental values were only half of their model prediction. The reason for this was that most of the SWNTs in epoxy showed significant curvature. If the experimental values of their work were scaled up, their theoretical maximum values would be $dY/dV_f \sim 300$ GPa, which is in excellent agreement with previous theoretical predictions.[11,45,149]

4.7.2 Mechanical Properties of Melt-processed Composites

In this Section we discuss mechanical properties of CNT–polymer composites produced by melt processing. Although a number of papers have reported this approach, we will only discuss those which reported highest values of dY/dV_f.

Gorga and Cohen[174] increased the stiffness of PMMA from 2.7 to 3.7 GPa on addition of 10 wt% of CVD–MWNT ($dY/dV_f \sim 17$ GPa). In addition, increases in strength from 64 to 80 MPa and a 170% increase in toughness were observed.[174]

However, it was in 2004 that a number of researchers began to report significant improvements. Meincke et al.[86] fabricated composites from CVD–MWNT in PA6. This almost doubled the modulus from 2.6 to 4.2 GPa at 12.5 wt%, corresponding to a reinforcement value of $dY/dV_f \sim 34$ GPa. This was, however, accompanied by a significant reduction in ductility, from 40% to approximately 4%. In addition, they made blend composites from PA6, acrylonitrile-butadiene-styrene (ABS) and nanotubes. Here, the results were more modest with increases from 1.97 to 2.51 GPa at 7.5 wt%. This gives a slightly lower reinforcement value of 11 GPa. However, this can be explained partly by the fact that the nanotubes were observed by transmission electron microscopy (TEM) to reside only in the PA6 phase. In the three-phase blends, the ductility fell from 105% to 40%, a fall that was accompanied by a decrease in impact strength by a factor of two.

Liu and co-workers[175,176] made composites from CVD–MWNT in PA6. They observed a three-fold increase in modulus from 0.4 GPa to 1.24 GPa on the addition of only 2 wt% nanotubes corresponding to an impressive reinforcement value of 64 GPa. In addition, a significant increase in yield strength from 18 to 47 MPa was observed with similar increases in ultimate tensile strength. Furthermore, no decrease in toughness was observed as the ductility only fell slightly from 150% to 110%. These impressive results were attributed to very good dispersion and adhesion, as observed by microscopy measurements.

In summary, the melt-processing technique allows the preparation of composites with higher nanotube loadings (over 10 wt%). Young's moduli increases of 100%,[86] 150%[85] and even 200%[175] are obtained by this method. However, the higher content of nanotubes normally results in much lower dY/dV_f values. In this case, polymer composites with lower nanotube loadings should be considered. If we look at dY/dV_f values more closely and calculate percentage increases for lower optimal nanotube contents, we can see that the melt mixing technique is quite efficient. For example, Potschke et al.[177] reported SWNT–PC composites which had increases in Young's modulus up to 7.5 wt % nanotube loadings. If we calculate the dY/dV_f values at 7.5 wt% (at which the Young's modulus reaches its maximum), we obtain $dY/dV_f = 7.8$, if we calculate the dY/dV_f value at 1 wt% (the most linear increase), we obtain a significantly higher number of 52.6. The same applies to MWNT–PA6[86] and MWNT–PC[87] composites if dY/dV_f are calculated at lower nanotube contents (4–6 wt%). Therefore for composites formed by melt mixing the dY/dV_f values corresponding to maximal Young's modulus increases in Table 4.3 should be used as a standard reference only.

There is also a great difference in mechanical properties between samples produced by shear-controlled orientation in injection moulding (dynamic samples) and conventional injection moulding (static samples). For example, the morphological observation of MWNT-filled polymer composites based on polycarbonate (PC) and PE indicated that in the dynamic samples the MWNTs predominantly localized in the PC microfibrils without obvious migration to the PE matrix. With such unique morphology, the tensile properties of the dynamic samples are considerably higher than for the static samples.[178] In addition, injection moulding can give a high degree of CNT orientation, which can affect significantly the stress–strain curves, as shown by Weidisch and co-workers.[179] They took stress–strain curves for nanocomposites in the injection-moulded direction and perpendicular to the injection-moulded direction. Not only were the shape of the stress–strain curves different, but the strain at breaking was considerably higher when the composites were measured in the direction of the direction of injection moulding.

4.7.3 Mechanical Properties of Composites Based on Thermosetting Polymers

Due to their wide range of industrial uses, thermosetting polymers, in general, and epoxy resins, in particular, have been widely studied as a potential matrix for nanotube-based composites.[180–182] In the following Sections, all matrices were epoxy resins unless otherwise stated. As in previous Sections, only those papers with values of dY/dV_f above the median (18 GPa) are discussed.

The first record of the formation of epoxy–nanotube composites comes from a paper by Ajayan et al.[107] in 1994. In this work, nanotubes were aligned within the epoxy matrix by the shear forces induced by cutting with a diamond knife. However, no quantitative mechanical measurements were made.

The first true mechanical study was made by Schadler et al.[183] in 1998. They measured the stress–strain properties of a MWNT–epoxy composite during both tension and compression. In tension, the modulus increased from 3.1 GPa to 3.71 GPa on the addition of 5 wt% nanotubes, a reinforcement of $dY/dV_f =$ 18 GPa. However, better results were seen in compression, with an increase in the modulus from 3.63 to 4.5 GPa, which corresponds to a reinforcement of 26 GPa. No significant increases in the strength of toughness were observed. The difference between tension and compression was explained by Raman studies which showed significantly better stress transfer to the nanotubes in compression than in tension. This can be explained by the fact that load transfer in compression can be thought of as a hydrostatic pressure effect, whereas load transfer in tension relies on the matrix–nanotube bond. However, it should be pointed out that later studies showed the reverse to be true, *i.e.* load transfer in tension but none in compression.[106] In further contrast, work by Wood et al.[184] has shown that the mechanical response of SWNTs in tension and compression are identical.

Further work was carried out on CVD-MWNT–epoxy composites by Breton et al.[185] Significant increases in the modulus from 2.75 GPa to 4.13 GPa were observed on the addition of 6 wt% nanotubes, corresponding to $dY/dV_f \sim 45$ GPa. They attributed the large reinforcement values to residual oxygen-containing groups that were covalently bonded to the nanotubes during purification. The presence of these groups may have improved interfacial bonding. Bai et al.[186] observed a doubling of the Young's modulus from 1.2 to 2.4 GPa on the addition of 1 wt% CVD–MWNT. This corresponds to a reinforcement value of $dY/dVf \sim 330$ GPa, which is the largest observed for an epoxy-based composite as measured by tensile testing. In addition, significant increases in strength from 30 to 41 MPa were recorded. Excellent matrix–nanotube adhesion was confirmed by the observation of nanotube breakage during fracture surface studies.[186]

Finally, Li et al.[181] studied SWNT–epoxy composites by nano-hardness testing. They observed an increase in modulus from 4 to 7 GPa with 5 wt% SWNT ($dY/dV_f \sim 94$ GPa). Increases in hardness from 0.35 to 0.45 GPa were also observed.

4.7.4 Mechanical Properties of Composites Prepared by *In Situ* Polymerisation

In situ polymerisation methods are expected to achieve large interfacial shear strength because they can provide strong, frequently covalent, polymer–nanotube interactions.

While early reports were promising,[187] a breakthrough came from a study by Kumar *et al.*[103] They polymerised PBO in the presence of SWNT before performing fibre wet-spinning. The PBO modulus of 138 GPa was increased to 167 GPa on the addition of 10 wt% SWNT, a reinforcement value of ~550 GPa. In addition, the fibre strength was improved from 2.6 GPa to 4.2 GPa. Although this does not match the strength of commercial PBO fibres (~6 GPa), it is certainly an impressive start. Furthermore, incorporation of SWNTs actually increased the ductility, thereby improving the toughness which is an important consideration for fibre materials.

In another study by Velasco-Santos *et al.*,[97] PMMA was polymerised in the presence of arc-MWNT. This is important as they have higher intrinsic strength or stiffness than CVD-MWNT or bundles of SWNT. Stiffness was increased from 1.5 to 2.5 GPa by 1 wt% MWNT, corresponding to dY/dV_f ~ 150 GPa. In addition, the strength increased from 30 to 50 MPa, with a proportional increase in toughness.

Finally, Putz *et al.*[98] in another study on PMMA and SWNT observed an increase in stiffness as measured by DMA from 0.3 to 0.38 GPa. However, this increase occurred at the extremely low volume fraction of 8×10^{-5}. This corresponds to a reinforcement value of dY/dV_f ~ 960 GPa. This is close to the theoretical rule of mixtures maximum for long, well-aligned, well-graphitised nanotubes. However, it should be pointed out that this study was carried out at 100 rad s^{-1}, which is a reasonably high frequency. Time–temperature equivalence of mechanical properties for visco-elastic materials suggests that this may be an over-estimate as compared with values obtained by pseudo-static measurements.[188]

Acid-treated functionalised nanotubes are normally very good precursors for the preparation of nanotube–polymer composites by *in situ* polymerisation. For example, composite films of CNTs with polyaniline (PANI),[189] PPy or poly(3,4-ethylenedioxythiophene) (PEDOT) were prepared *via* electrochemical co-deposition from solutions containing acid-treated CNTs and the corresponding monomer.[190] All of the composites showed improved mechanical integrity, and higher electronic and ionic conductivities. In another work,[49] PA6–MWNT composites produced *in situ* have shown large dY/dV_f values of 143. In fact, Ma *et al.*[191] have found that the polymerisation process was 4500 times faster when a self-doped PANI nanocomposite was fabricated using *in situ* polymerisation in the presence of single-stranded DNA-dispersed and functionalised SWNTs. The mechanical properties of the MWNT–PMMA nanocomposites were studied as a function of both nanotube concentration and surface treatment. The Young's modulus of non-functionalised MWNT–PMMA composites was comparable with the pristine PMMA, both gave

Young's modulus values of ~2.7 GPa. Plasma-enhanced chemical vapour deposition (PECVD) was successfully used to produce a PMMA conformal coating (using methyl methacrylate monomers) on MWNTs. This increased the Young's modulus to 2.85 GPa at 3 wt% which corresponds to $\mathrm{d}Y/\mathrm{d}V_\mathrm{f}$ of 28.4 GPa.[163] The overall set of mechanical properties indicates that the polymer coating had a significant effect on the mechanical properties at a 1 wt% concentration of tubes, suggesting improved interfacial adhesion between the filler and the matrix material.[163]

Direct shear testing of poly(butyl methacrylate) (PBMA)–CNT composites prepared by *in situ* polymerisation showed strong mechanical behavior, with up to a 200% increase in the Young's modulus over PBMA. The PBMA–CNTs hybrid composite was sandwiched between two opposing steel plates, which also serve as the grips for the shear testing. The steel plates and the composite were heated to 200 °C, and then bonded together under pressure (1 MPa) for 10 min. The load applied and the strain induced are recorded at frequent intervals to determine a stress–strain curve for the confining stress. From thermo-gravimetric analysis (TGA) analysis there is ~40% of nanotubes in these composites.[192] For PMMA-functionalised MWNT–PMMA composites prepared *in situ*, all of the mechanical parameters increased. These increases are approximately 94% for the Young's modulus, 360% for ultimate tensile strength, 373% for breaking strength, 1282% for toughness, and 526% for the elongation at break. This reaches a maximum at 0.5 wt%.[193]

In another work,[194] it was demonstrated that *in situ* polymerisation with the addition of ultra-small amounts of SWNTs results in cross-linked poly(urethane urea) elastomers. Even at a very low SWNT concentration of 0.002 wt%, the maximum values of the modulus and strength increased by factors of 2.5 and 1.5, respectively, compared with the corresponding values for the neat polymer.[194]

It should be mentioned that nucleation of crystallinity [28] in the presence of nanotubes may occur in solution casting, melt processing and *in situ* polymerisation processing of some polymer–nanotube composites.

4.7.5 Mechanical Properties of Composites Fibres Prepared by Spinning

In 2000, Vigolo *et al.*[109] demonstrated that nanotube composite ribbons and fibres could be spun by coagulation spinning. These fibres displayed reasonable moduli and strength of 9–15 GPa and ~150 MPa respectively.[109] They were extremely flexible and could easily be tied into knots. The mechanical properties of these fibres could be significantly improved by stretching when wet. This tended to align the nanotubes resulting in moduli of ~40 GPa and strengths of up to 230 MPa.[195]

However, Dalton *et al.*[110] demonstrated, in 2003, that the properties could be markedly improved by leaving the polymer in the fibre. This resulted in large increases in the Young's modulus and strength to 80 GPa and 1.8 GPa respectively. In addition, these fibres could be stretched until strains of ∼100%, resulting in massive toughness values of up to 570 J g^{-1}, which is an order of magnitude larger than Kevlar. Miaudet et al.[111] have recently shown that toughness values as high as 870 J g^{-1} and 690 J g^{-1} can be achieved for similar fibres spun from SWNTs and MWNTs, respectively. In addition, by drawing the fibres at elevated temperature, they have shown that SWNT-based fibres with excellent all-round properties can be obtained. These fibres had moduli, strength and toughness values of 45 GPa, 1.8 GPa and 55 J g^{-1}. Crucially, this toughness value was obtained at the low strain value of 11%.

Fibres made from SWNTs and PVA have also been spun using DNA as a dispersant. These materials demonstrated moduli in the range 12–18 GPa and strengths of 90–110 MPa.[196]

CNT–polymer composite fibres have also been spun by electrospinning. Ko et al.[113] demonstrated stabilised PAN–SWNT fibres with a significantly increased Young's modulus. The fibre modulus was 140 GPa at 4 wt% nanotubes which compares well with the PAN fibre modulus of 60 GPa. However, it should be pointed that these measurements were carried out in compression using AFM. Sen at al.[114] built up membranes from electrospun fibres of SWNT and polyurethane. These could then by characterised by tensile testing. Polyurethane porous membranes had tangent moduli and strengths of 7 MPa and 7 MPa, respectively. Nevertheless, the membranes fabricated from composite fibres demonstrated moduli and strengths of up to 25 and 15 MPa.

4.7.6 Mechanical Properties of Composites Prepared Using Buckypaper

The reported mechanical properties of buckypaper composed of SWNTs show a wide variation. The reported moduli range from 0.5 MPa to 2 GPa,[121,197] and the reported breaking strengths range from 6.3 to 33 MPa.[128,198–200] The mechanical properties of buckypaper are dominated by the nature of the junctions between individual tubes,[201] although tube entanglement is also important.[202] Composites with high nanotube loadings based on buckypapers use mostly polymer intercalation to reinforce bulk nanotube materials. Binding agents such as organic polymers can be intercalated into the porous internal structure of the buckypaper. The mechanical properties vary so vastly with buckypaper type and preparation, thus any increases will be presented as a percentage increase over the control sample given in the paper. In some cases the control sample given was buckypaper (without polymer), denoted as $Y_{Buckypaper}$. In other cases, the control samples given was the polymer without any nanotubes, denoted as $Y_{Polymer}$.

Yang et al.[104] used a low viscosity resin solution which was allowed to infiltrate the buckypapers to impregnate the SWNT rope networks. A hot-press moulding process was used for moulding and curing to produce the final composites. Nanocomposites with preformed tube networks and high SWNT loading (up to 39% by weight) were manufactured.[104] The permeability of the produced buckypapers was measured to predict the resin infiltration time. The storage modulus of the resulting composites was found to be 15 GPa, which is 429% higher than the neat resin modulus. In this way it is possible to transform a brittle polymer into a conductive flexible film.

CNT–PANI composites have important potential applications as the electrodes in energy storage devices due to their attractive electrochemical properties. Meng *et al.*[203] very recently reported a novel method to prepare the interesting paper-like CNT–PANI composites by using the CNT network as the template. Compared with the conventional brittle CNT–PANI composites, these paper-like composites were much thinner and more flexible.[203] Oxidised multiwalled CNT (oxidised MWNT)–PVA composite sheets show that the best compromise of mechanical and electrical properties was obtained for a PVA weight fraction of approximately 30 wt%.[204] The same group shows the critical importance of covalent or non-covalent surface functionalisation for the conductive properties of nanocomposites.[205] Coleman *et al.*[206] intercalated PVA into SWNT buckypaper. These composites displayed improvements in the Young's modulus and tensile strength by factors of 3 and 9, respectively.[206]

Intercalating both high-molecular weight and low-molecular weight poly(vinyl pyrrolidone) (PVP) into buckypaper from solution gave increases in mechanical properties. There was a 3.5 times increase in the Young's modulus and a 6 times increase in tensile strength in these new composites at low polymer loadings (<50% of the total composite mass) compared with original buckypaper. A large increase in toughness (25 times) occurred for the low-molecular weight polymer.[207] SWNT buckypaper can also be impregnated with PC solution to make thermoplastic nanocomposites.[208,209] Dynamic mechanical property tests indicated that the storage modulus of the resulting nanocomposites at 20 wt% nanotubes loading was improved by a factor of 3.4 compared with neat PC material.[209] Composites based on polyurethane, with mass fractions of up to 80% PEG-functionalised nanotubes gave increases in the Young's modulus by up to 800 times compared with polyurethane films.[210]

4.7.7 Mechanical Properties of Composites Prepared Using LBL Approach

The LBL technique enables the fabrication of nanotube–polymer composite films consisting of alternating layers of nanotubes and polyelectrolyte, which have shown great promise as high-strength light-weight materials. Mamedov *et al.*[137] demonstrated such materials from SWNT and poly(ethyleneimine) (PEI). Compared with PEI, significant increases in modulus, from 0.3 GPa to ∼11 GPa, were observed for composites containing ∼50 wt% SWNT. Increases in strength from ∼9 MPa to up to 325 MPa were observed. Although the composites were less ductile than the neat polymer (1% *versus* 4%), they still showed a large increase in toughness by a factor of five to ten.

In a similar study, Olek *et al.*[138] fabricated LBL composites using MWNT and PEI. Two types of MWNT were used: "hollow" and "bamboo". The bamboo composites displayed moduli and strengths of ~4.5 GPa and ~150 MPa respectively. The hollow MWNT composites showed slightly lower values of ~2 GPa and ~110 MPa, respectively. To further emphasise the importance of tube type and quality, hollow nanotubes were boiled in nitric acid before composite fabrication. This resulted in much lower modulus and strength values of ~0.2 GPa and 35 MPa, respectively.

Another example, PVA–SWNT–poly(sodium 4-styrene-sulfonate) free-standing films, produced by LBL, were highly flexible and possessed tensile strength of 160 MPa at the volume fraction of SWNTs of approximately 10%.[143] Unfortunately, Young's moduli and corresponding dY/dV_f values have not been reported, therefore it is difficult to compare this samples with those made by other methods. The same group has also used LBL to reinforce cotton yarns, which resulted in more than two-fold increase in the ultimate yield strength (from 41.6 to 87.8 MPa) and the Young's modulus (from 140 to 342 MPa). Finally, the same group also reported LBL fabrication of SWNT–PVA composites, which were cross-linked together by thermal ester bond formation between carboxylic acid groups in SWNTs and hydroxyl functionalities in PVA or by glutaraldehyde treatment. The resulting SWNT–PVA stiffest non-fibrous SWNT composites made to date demonstrated maximum values of tensile strength up to 600.1 MPa, stiffness of 20.6 GPa and toughness of 152.1 J g^{-1}. These values are 2–10 times higher than those observed for other bulk composites (even exceeding the toughness of Kevlar by three-fold). Such high performance was attributed to both high nanotube content and efficient stress–strain transfer in the composites.[145]

Thus several researchers have produced nanotube–polymer composites with mechanical increases close to the theoretical maximum using a variety of preparation and functional procedures. However, it could be argued that in some cases where nucleation of crystallinity occurs, nanotubes should only be considered as a pseudo-reinforcement agent in polymer composites.

4.8 Conclusions and Future Outlook

In conclusion, there has been significant progress in the utilisation of CNTs for polymer composite reinforcement over the last decade. A range of new composites have shown astonishing mechanical parameters. It has been clearly demonstrated that composites based on chemically covalently functionalised nanotubes show the best results. This is not surprising, as functionalisation significantly improves both nanotube dispersion and stress–strain transfer between the nanotubes and polymer matrix. Both the rule of mixtures and the Halpin–Tsai equations (taking $Y_m \sim 1$ GPa and $l/D \sim 1000$) predict maximum dY/dV_f values of approximately 1 TPa. A number of the systems studied report values in this range.[28,95,98] been performed in this area. We expect that this issue must be addressed CNT–polymer composites are to be widely used in the future.

Finally, we need to identify and optimise processing techniques for CNT–polymer composite fabrication. Among the sophisticated approaches, the LBL technique seems to be an excellent choice for the production of mechanically strong and highly conductive nanocomposites, because it minimises the damage to the nanotubes during processing. Also, new swelling under ultrasound, post-processing techniques have a great potential for the preparation of both mechanically strong and potentially conductive polymer composites.

In summary, there is a large interest in the further development of polymer–nanotube composite materials, which would have a broad range of important applications, including new ultra-strong polymer–nanotube materials for bullet-proof vests, protective clothing, high-performance composites for aircraft and automotive industries (*e.g.* seat belts, cables, reinforcement of tires, brake-linings, bumpers, *etc.*). These large sectors will require huge quantities of CNTs. For these reasons, the development of new cost-effective CNT–polymer composites at a large scale will be necessary to meet these needs.

References

1. M.-F. Yu, O. Lourie, M. J. Dyer, K. Moloni, T. F. Kelly and R. S. Ruoff, *Science*, 2000, **287**, 637–640.
2. M.-F. Yu, B. S. Files, S. Arepalli and R. S. Ruoff, *Phys. Rev. Lett.*, 2000, **84**, 5552.
3. G. O. Shonaike and S. G. Advani, ed, *Advanced Polymeric Materials: Structure Property Relationships*, C.R.C. Press, Boca Raton, FL, USA, 2003.
4. E. T. Thostenson, Z. F. Ren and T. W. Chou, *Compos. Sci. Technol.*, 2001, **61**, 1899–1912.
5. R. H. Baughman, A. A. Zakhidov and W. A. de Heer, *Science*, 2002, **297**, 787–792.
6. R. Andrews and M. C. Weisenberger, *Curr. Opin. Solid State Mater. Sci.*, 2004, **8**, 31–37.
7. O. Breuer and U. Sundararaj, *Polym. Compos.*, 2004, **25**, 630–645.
8. P. J. F. Harris, *Int. Mater. Rev.*, 2004, **49**, 31–43.
9. C. C. Wang, Z. X. Guo, S. K. Fu, W. Wu and D. B. Zhu, *Prog. Polym. Sci.*, 2004, **29**, 1079–1141.
10. X. L. Xie, Y. W. Mai and X. P. Zhou, *Mater. Sci. Eng.. R: Rep.*, 2005, **49**, 89–112.
11. J. N. Coleman, U. Khan, W. J. Blau and Y. K. Gun'ko, *Carbon*, 2006, **44**, 1624–1652.
12. J. N. Coleman, U. Khan and Y. K. Gun'ko, *Adv. Mater.*, 2006, **18**, 689–706.
13. M. Moniruzzaman and K. I. Winey, *Macromolecules*, 2006, **39**, 5194–5205.

However, there are several inadequate aspects where significant improvement is urgently needed. The problems with mechanical reinforcement of melt-processed composites must be addressed. It is very important as melt processing is the most promising approach for scaling up and the industrial manufacturing of CNT-reinforced composites. It is likely that the problems are based on dispersion or interfacial shear strength issues. If this is the case, these may be addressed by appropriate functionalisation of the nanotubes. This will result in them being more compatible with the polymer melt and so improve dispersion. Another significant problem is the disappointing values for CNT–polymer composite strength. One possibility to address this issue is to optimise the nanotube type (SWNT or MWNT) and functionalisation. While much higher volume fractions have been attained, all SWNT exist in bundles under these conditions. Under these circumstances, SWNTs lose their intrinsic advantages. In the case of MWNTs with diameters of ~ 10 nm, however, even at $V_f = 10\%$, nanotubes are on average separated by 20 nm. This gives us much more scope to attain high volume fractions with MWNTs. The volume fraction at which every polymer strand is within 5 nm of a nanotube is a massive 25 vol%. Thus, in the ideal case, arc-MWNTs would be the obvious choice as the ultimate filler. The length and diameter of nanotubes can also be optimised. Once we have identified the optimum nanotube type we must maximise nanotube solubility, dispersion (and hence volume fraction) and stress transfer. All of these factors can again be addressed as one by functionalisation of the nanotubes. The choice of the right functional group will optimise interactions with either the solvent or the melt and aid stress transfer. In order to perfect the functionalisation scheme, there are several aspects to consider. The molecular structure of the group must be chosen. This may, for example, be a short-chain analogue to the host polymer. In addition, the group length and surface density would be expected to affect dispersion and stress transfer. Once the nanotubes have been fully optimized, we can consider what mechanical properties are attainable. We can calculate these properties for a typical thermoplastic polymer with low intrinsic modulus and strength. Assuming that the optimum tube length is a few times the critical length,[211] we can use the simple rules of mixtures for modulus and strength.[171] We can assume that the upper limit for SWNT loading is 1 vol% (as discussed) and that $Y_{SWNT} = 1$ TPa and $\sigma_{SWNT} = 50$ GPa. Then for an aligned composite Y_C is ~ 10 GPa and σ_C is ~ 0.5 GPa.

It is important to notice that the fatigue behaviour, durability and lifetime of polymer composites may all be substantially improved by the addition of CNTs.[135,212,213] For example, it has been demonstrated that incorporating CNTs into the glass fibre composites inhibits the formation of cracks, because a large density of nucleation sites is provided by the CNTs. Also, the increase in energy absorption from the fracture of nanotubes bridging across nanoscale cracks and nanotube pull-out from the matrix contributes to the higher fatigue life of composites containing CNTs.[212] However, very little research has sti

14. L. Y. Jiang, H. L. Tan, J. Wu, Y. G. Huang and K. C. Hwang, *Nano*, 2007, **2**, 139–148.
15. C. Li, E. T. Thostenson and T. W. Chou, *Compos. Sci. Technol.*, 2008, **68**, 1227–1249.
16. M. Endo, M. S. Strano and P. M. Ajayan, in *Carbon Nanotubes*, 2008, pp. 13–61.
17. M. T. Byrne and Y. K. Gun'ko, *Adv. Mater.*, 2010, **22**, 1672–1688.
18. A. L. Martinez-Hernandez, C. Velasco-Santos and V. M. Castano, *Curr. Nanosci.*, 2010, **6**, 12–39.
19. P. C. Ma, N. A. Siddiqui, G. Marom and J. K. Kim, *Compos. A: Appl. Sci. Manuf.*, 2010, **41**, 1345–1367.
20. J. L. Bahr and J. M. Tour, *J. Mater. Chem.*, 2002, **12**, 1952–1958.
21. T. Lin, V. Bajpai, T. Ji and L. M. Dai, *Aust. J. Chem.*, 2003, **56**, 635–651.
22. S. Banerjee, T. Hemraj-Benny and S. S. Wong, *Adv. Mater.*, 2005, **17**, 17–29.
23. D. Tasis, N. Tagmatarchis, A. Bianco and M. Prato, *Chem. Rev.*, 2006, **106**, 1105–1136.
24. P. Singh, S. Campidelli, S. Giordani, D. Bonifazi, A. Bianco and M. Prato, *Chem. Soc. Rev.*, 2009, **38**, 2214–2230.
25. C. Velasco-Santos, A. L. Martinez-Hernandez and V. M. Castano, *Compos. Interfac.*, 2005, **11**, 567–586.
26. M. Cadek, J. N. Coleman, V. Barron, K. Hedicke and W. J. Blau, *Appl. Phys. Lett.*, 2002, **81**, 5123–5125.
27. J. K. W. Sandler, S. Pegel, M. Cadek, F. Gojny, M. van Es, J. Lohmar, W. J. Blau, K. Schulte, A. H. Windle and M. S. P. Shaffer, *Polymer*, 2004, 45, 2001–2015.
28. J. N. Coleman, M. Cadek, R. Blake, V. Nicolosi, K. P. Ryan, C. Belton, A. Fonseca, J. B. Nagy, Y. K. Gun'ko and W. J. Blau, *Adv. Funct. Mater.*, 2004, **14**, 791–798.
29. F. Dalmas, L. Chazeau, C. Gauthier, K. Masenelli-Varlot, R. Dendievel, J. Y. Cavaillé and L. Forró, *J. Polym. Sci. B: Polym. Phys.*, 2005, **43**, 1186–1197.
30. A. Dufresne, M. Paillet, J. L. Putaux, R. Canet, F. Carmona, P. Delhaes and S. Cui, *J. Mater. Sci.*, 2002, **37**, 3915–3923.
31. B. E. Kilbride, J. N. Coleman, J. Fraysse, P. Fournet, M. Cadek, A. Drury, S. Hutzler, S. Roth and W. J. Blau, *J. Appl. Phys.*, 2002, **92**, 4024–4030.
32. B. McCarthy, J. N. Coleman, R. Czerw, A. B. Dalton, M. in het Panhuis, A. Maiti, A. Drury, P. Bernier, J. B. Nagy, B. Lahr, H. J. Byrne, D. L. Carroll and W. J. Blau, *J. Phys. Chem. B*, 2002, **106**, 2210–2216.
33. S. Bredeau, S. Peeterbroeck, D. Bonduel, M. Alexandre and P. Dubois, *Polym. Int.*, 2008, **57**, 547–553.
34. P. Liu, *Euro. Polym. J.*, 2005, **41**, 2693–2703.
35. D. Baskaran, J. W. Mays and M. S. Bratcher, *Chem. Mater.*, 2005, **17**, 3389–3397.

36. A. Hirsch, *Angew. Chem. Int. Ed.*, 2002, 41, 1853–1859.
37. R. Andrews, D. Jacques, D. L. Qian and T. Rantell, *Acc. Chem. Res.*, 2002, **35**, 1008–1017.
38. J. Liu, O. Bibari, P. Mailley, J. Dijon, E. Rouviere, F. Sauter-Starace, P. Caillat, F. Vinet and G. Marchand, *New J. Chem.*, 2009, **33**, 1017–1024.
39. J. N. Coleman and M. S. Ferreira, *Appl. Phys. Lett.*, 2004, 84, 798–800.
40. M. Foroutan and A. T. Nasrabadi, *J. Phys. Chem. B*, 2010, **114**, 5320–5326.
41. A. Mandal and A. K. Nandi, *J. Phys. Chem. C*, 2012, **116**, 9360–9371.
42. J. M. Gonzalez-Dominguez, M. A. Tesa-Serrate, A. Anson-Casaos, A. M. Diez-Pascual, M. A. Gomez-Fatou and M. T. Martinez, *J. Phys. Chem. C*, 2012, **116**, 7399–7408.
43. S. H. Liao, C. Y. Yen, C. H. Hung, C. C. Weng, M. C. Tsai, Y. F. Lin, C. C. M. Ma, C. Pan and A. Su, *J. Mater. Chem.*, 2008, **18**, 3993–4002.
44. R. Verdejo, F. Barroso-Bujans, M. A. Rodriguez-Perez, J. A. d. Saja, M. Arroyo and M. A. Lopez-Manchado, *J. Mater. Chem.*, 2008, **18**, 3933–3939.
45. I. O'Connor, H. Hayden, S. O'Connor, J. N. Coleman and Y. K. Gun'ko, *J. Mater. Chem.*, 2008, **18**, 5585–5588.
46. V. Georgakilas, A. Bourlinos, D. Gournis, T. Tsoufis, C. Trapalis, A. Mateo-Alonso and M. Prato, *J. Am. Chem. Soc.*, 2008, **130**, 8733–8740.
47. J. T. Feng, J. H. Sui, W. Cai and Z. Y. Gao, *J. Compos. Mater.*, 2008, **42**, 1587–1595.
48. B. Zhao, J. Wang, Z. J. Li, P. Liu, D. Chen and Y. F. Zhang, *Mater. Lett.*, 2008, **62**, 4380–4382.
49. J. Gao, M. E. Itkis, A. Yu, E. Bekyarova, B. Zhao and R. C. Haddon, *J. Am. Chem. Soc.*, 2005, **127**, 3847–3854.
50. H. J. Salavagione, G. Martinez and C. Marco, *J. Mater. Chem.*, 2012, **22**, 7020–7027.
51. J. Gao, B. Zhao, M. E. Itkis, E. Bekyarova, H. Hu, V. Kranak, A. Yu and R. C. Haddon, *J. Am. Chem. Soc.*, 2006, **128**, 7492–7496.
52. G. Sui, W. H. Zhong, X. P. Yang, Y. H. Yu and S. H. Zhao, *Polym. Adv. Technol.*, 2008, **19**, 1543–1549.
53. S. M. Yuen, C. C. M. Ma, C. C. Teng, H. H. Wu, H. C. Kuan and C. L. Chiang, *J. Polym. Sci. B: Polym. Phys.*, 2008, **46**, 472–482.
54. S. M. Yuen and C. C. M. Ma, *J. Appl. Polym. Sci.*, 2008, **109**, 2000–2007.
55. J. T. Luo, H. C. Wen, W. F. Wu and C. P. Chou, *Polym. Compos.*, 2008, **29**, 1285–1290.
56. S. M. Yuen, C. C. M. Ma, C. Y. Chuang, K. C. Yu, S. Y. Wu, C. C. Yang and M. H. Wei, *Compos. Sci. Technol.*, 2008, **68**, 963–968.
57. A. Rasheed, H. G. Chae, S. Kumar and M. D. Dadmun, *Polymer*, 2006, **47**, 4734–4741.
58. C. Velasco-Santos, A. L. Martinez-Hernandez, F. Fisher, R. Ruoff and V. M. Castano, *J. Phys. D.–Appl. Phys*, 2003, **36**, 1423–1428.
59. N. G. Sahoo, Y. C. Jung, H. J. Yoo and J. W. Cho, *Macromol. Chem. Phys.*, 2006, **207**, 1773–1780.

60. K. K. Wong, S. Q. Shi and K. T. Lau, *Key Eng. Mater.*, 2008, **334-335**, 705-709.
61. B. Ruelle, A. Felten, J. Ghijsen, W. Drube, R. L. Johnson, D. Liang, R. Erni, G. Van Tendeloo, P. Sophie, P. Dubois, T. Godfroid, M. Hecq and C. Bittencourt, *Micron*, 2009, **40**, 85-88.
62. B. Ruelle, S. Peeterbroeck, R. Gouttebaron, T. Godfroid, F. Monteverde, J.-P. Dauchot, M. Alexandre, M. Hecq and P. Dubois, *J. Mater. Chem.*, 2007, **17**, 157-159.
63. Y. Wang, Z. X. Shi and J. Yin, *J. Phys. Chem. C*, 2010, **114**, 19621-19628.
64. D. Shi, J. Lian, P. He, L. M. Wang, F. Xiao, L. Yang, M. J. Schulz and D. B. Mast, *Appl. Phys. Lett.*, 2003, **83**, 5301-5303.
65. G. Viswanathan, N. Chakrapani, H. Yang, B. Wei, H. Chung, K. Cho, C. Y. Ryu and P. M. Ajayan, *J. Am. Chem. Soc.*, 2003, **125**, 9258-9259.
66. S. Qin, D. Qin, W. T. Ford, D. E. Resasco and J. E. Herrera, *Macromolecules*, 2004, **37**, 752-757.
67. G. L. Hwang, Y.-T. Shieh and K. C. Hwang, *Adv. Funct. Mater.*, 2004, **14**, 487-491.
68. H. Xia, Q. Wang and G. Qiu, *Chem. Mater.*, 2003, **15**, 3879-3886.
69. X. Tong, C. Liu, H.-M. Cheng, H. Zhao, F. Yang and X. Zhang, *J. Appl. Polym. Sci.*, 2004, **92**, 3697-3700.
70. C. J. Zhou, S. F. Wang, Y. Zhang, Q. X. Zhuang and Z. W. Han, *Polymer*, 2008, **49**, 2520-2530.
71. C. J. Zhou, S. F. Wang, Q. X. Zhuang and Z. W. Han, *Carbon*, 2008, **46**, 1232-1240.
72. X. Chen, F. Tao, J. Wang, H. Yang, J. Zou, X. Chen and X. Feng, *Mater. Sci. and Eng.: A*, 2009, **499**, 469-475.
73. T. Sainsbury, K. Erickson, D. Okawa, C. S. Zonte, J. M. J. Frechet and A. Zettl, *Chem. Mater.*, 2010, **22**, 2164-2171.
74. L. Zhang, Q. Q. Ni, A. Shiga, T. Natsuki and Y. Q. Fu, *Polym. Eng. Sci.*, 2011, **51**, 1525-1532.
75. A. Kokil, T. Saito, W. Depolo, C. L. Elkins, G. L. Wilkes and T. E. Long, *J. Macromol. Sci. A: Pure Appl. Chem.*, 2011, **48**, 1016-1021.
76. K. Fu, W. Huang, Y. Lin, L. A. Riddle, D. L. Carroll and Y. P. Sun, *Nano Lett.*, 2001, **1**, 439-441.
77. X. Lou, C. Detrembleur, V. Sciannamea, C. Pagnoulle and R. Jerome, *Polymer*, 2004, **45**, 6097-6102.
78. R. Blake, Y. K. Gun'ko, J. Coleman, M. Cadek, A. Fonseca, J. B. Nagy and W. J. Blau, *J. Am. Chem. Soc.*, 2004, **126**, 10226-10227.
79. L. Sun, G. L. Warren, J. Y. O'Reilly, W. N. Everett, S. M. Lee, D. Davis, D. Lagoudas and H. J. Sue, *Carbon*, 2008, **46**, 320-328.
80. Y. H. Wang, C. M. Chang and Y. L. Liu, *Polymer*, 2012, **53**, 106-112.
81. A. M. Diez-Pascual, G. Martinez, J. M. Gonzalez-Dominguez, A. Anson, M. T. Martinez and M. A. Gomez, *J. Mater. Chem.*, 2010, **20**, 8285-8296.
82. C. M. Chang and Y. L. Liu, *Carbon*, 2010, **48**, 1289-1297.

83. W. Yuan and M. B. Chan-Park, *ACS Appl. Mater. Interfaces*, 2012, **4**, 2065–2073.
84. F. Mammeri, J. Teyssandier, C. Connan, E. Le Bourhis and M. M. Chehimi, *RSC Adv.*, 2012, **2**, 2462–2468.
85. R. Andrews, D. Jacques, M. Minot and T. Rantell, *Macromol. Mater. Eng.*, 2002, **287**, 395–403.
86. O. Meincke, D. Kaempfer, H. Weickmann, C. Friedrich, M. Vathauer and H. Warth, *Polymer*, 2004, **45**, 739–748.
87. P. Potschke, A. R. Bhattacharyya, A. Janke and H. Goering, *Compos. Interfac.*, 2003, **10**, 389–404.
88. P. Potschke, T. D. Fornes and D. R. Paul, *Polymer*, 2002, **43**, 3247–3255.
89. J. Sandler, G. Broza, M. Nolte, K. Schulte, Y. M. Lam and M. S. P. Shaffer, *J. Macromol. Sci. – Phys.*, 2003, **B42**, 479–488.
90. P. Potschke, A. R. Bhattacharyya, A. Janke, S. Pegel, A. Leonhardt, C. Taschner, M. Ritschel, S. Roth, B. Hornbostel and J. Cech, *Fullerenes Nanotubes Carbon Nanostruct.*, 2005, **13**, 211 – 224.
91. P. Potschke, M. Abdel-Goad, I. Alig, S. Dudkin and D. Lellinger, *Polymer*, 2004, **45**, 8863–8870.
92. P. Potschke, A. R. Bhattacharyya and A. Janke, *Euro. Polym. J.*, 2004, **40**, 137–148.
93. P. Potschke, H. Brunig, A. Janke, D. Fischer and D. Jehnichen, *Polymer*, 2005, **46**, 10355–10363.
94. R. Haggenmueller, H. H. Gommans, A. G. Rinzler, J. E. Fischer and K. I. Winey, *Chem. Phys. Lett.*, 2000, **330**, 219–225.
95. J. C. Kearns and R. L. Shambaugh, *J. Appl. Polym. Sci.*, 2002, **86**, 2079–2084.
96. Z. Jia, Z. Wang, C. Xu, J. Liang, B. Wei, D. Wu and S. Zhu, *Mater. Sci. Eng. A*, 1999, **271**, 395–400.
97. C. Velasco-Santos, A. L. Martinez-Hernandez, F. T. Fisher, R. Ruoff and V. M. Castano, *Chem. Mater.*, 2003, **15**, 4470–4475.
98. K. W. Putz, C. A. Mitchell, R. Krishnamoorti and P. F. Green, *J. Polym. Sci. B: Polym. Phys.*, 2004, **42**, 2286–2293.
99. C. Zhao, G. Hu, R. Justice, D. W. Schaefer, S. Zhang, M. Yang and C. C. Han, *Polymer*, 2005, **46**, 5125–5132.
100. J. Gao, M. E. Itkis, A. Yu, E. Bekyarova, B. Zhao and R. C. Haddon, *J. Am. Chem. Soc.*, 2005, **127**, 3847–3854.
101. J. Zhu, J. Kim, H. Peng, J. L. Margrave, V. N. Khabashesku and E. V. Barrera, *Nano Lett.*, 2003, **3**, 1107–1113.
102. J. Zhu, H. Peng, F. Rodriguez-Macias, J. L. Margrave, V. N. Khabashesku, A. M. Imam, K. Lozano and E. V. Barrera, *Adv. Funct. Mater.*, 2004, **14**, 643–648.
103. S. Kumar, T. D. Dang, F. E. Arnold, A. R. Bhattacharyya, B. G. Min, X. Zhang, R. A. Vaia, C. Park, W. W. Adams, R. H. Hauge, R. E. Smalley, S. Ramesh and P. A. Willis, *Macromolecules*, 2002, **35**, 9039–9043.
104. J. Yang, J. Hu, C. Wang, Y. Qin and Z. Guo, *Macromol. Mater. Eng.*, 2004, **289**, 828–832.

105. G. Broza, M. Kwiatkowska, Z. Roslaniec and K. Schulte, *Polymer*, 2005, **46**, 5860–5867.
106. P. M. Ajayan, L. S. Schadler, C. Giannaris and A. Rubio, *Adv. Mater.*, 2000, **12**, 750–753.
107. P. M. Ajayan, O. Stephan, C. Colliex and D. Trauth, *Science*, 1994, **265**, 1212–1214.
108. Q. Q. Li, M. Zaiser and V. Koutsos, *Phys. Status Solidi A: Appl. Res.*, 2004, **201**, R89–R91.
109. B. Vigolo, A. Penicaud, C. Coulon, C. Sauder, R. Pailler, C. Journet, P. Bernier and P. Poulin, *Science*, 2000, **290**, 1331–1334.
110. A. B. Dalton, S. Collins, E. Munoz, J. M. Razal, V. H. Ebron, J. P. Ferraris, J. N. Coleman, B. G. Kim and R. H. Baughman, *Nature*, 2003, **423**, 703.
111. P. Miaudet, S. Badaire, M. Maugey, A. Derre, V. Pichot, P. Launois, P. Poulin and C. Zakri, *Nano Lett.*, 2005, **5**, 2212.
112. A. Formhals, ed. U. S. P. Office, United States, 1934.
113. F. Ko, Y. Gogotsi, A. Ali, N. Naguib, H. H. Ye, G. L. Yang, C. Li and P. Willis, *Adv. Mater.*, 2003, **15**, 1161.
114. R. Sen, B. Zhao, D. Perea, M. E. Itkis, H. Hu, J. Love, E. Bekyarova and R. C. Haddon, *Nano Lett.*, 2004, **4**, 459–464.
115. E. Ozden-Yenigun, Y. Z. Menceloglu and M. Papila, *ACS Appl. Mater. Interfaces*, 2012, **4**, 777–784.
116. A. G. Rinzler, J. Liu, H. Dai, P. Nikolaev, C. B. Huffman, F. J. Rodríguez-Macías, P. J. Boul, A. H. Lu, D. Heymann, D. T. Colbert, R. S. Lee, J. E. Fischer, A. M. Rao, P. C. Eklund and R. E. Smalley, *Appl. Phys. A: Mater. Sci. Process.*, 1998, **67**, 29–37.
117. J. Liu, A. G. Rinzler, H. Dai, J. H. Hafner, R. K. Bradley, P. J. Boul, A. Lu, T. Iverson, K. Shelimov, C. B. Huffman, F. Rodriguez-Macias, Y.-S. Shon, T. R. Lee, D. T. Colbert and R. E. Smalley, *Science*, 1998, **280**, 1253–1256.
118. M. Endo, H. Muramatsu, T. Hayashi, Y. A. Kim, M. Terrones and N. S. Dresselhaus, *Nature*, 2005, **433**, 476–476.
119. S. D. Bergin, V. Nicolosi, P. V. Streich, S. Giordani, Z. Y. Sun, A. H. Windle, P. Ryan, N. P. P. Niraj, Z. T. T. Wang, L. Carpenter, W. J. Blau, J. J. Boland, J. P. Hamilton and J. N. Coleman, *Adv. Mater.*, 2008, **20**, 1876–1881.
120. K. D. Ausman, R. Piner, O. Lourie, R. S. Ruoff and M. Korobov, *J. Phys. Chem. B*, 2000, **104**, 8911–8915.
121. F. M. Blighe, P. E. Lyons, S. De, W. J. Blau and J. N. Coleman, *Carbon*, 2008, **46**, 41–47.
122. S. Giordani, S. D. Bergin, V. Nicolosi, S. Lebedkin, M. M. Kappes, W. J. Blau and J. N. Coleman, *J. Phys. Chem. B*, 2006, **110**, 15708–15718.
123. J. Liu, M. J. Casavant, M. Cox, D. A. Walters, P. Boul, W. Lu, A. J. Rimberg, K. A. Smith, D. T. Colbert and R. E. Smalley, *Chem. Phys. Lett.*, 1999, **303**, 125.

124. Z. Sun, V. Nicolosi, S. D. Bergin and J. N. Coleman, *Nanotechnology*, 2008, **19**, 485702.
125. S. D. Bergin, V. Nicolosi, S. Giordani, A. de Gromard, L. Carpenter, W. J. Blau and J. N. Coleman, *Nanotechnology*, 2007, **18**, 455705.
126. C. Backes, C. D. Schmidt, F. Hauke, C. Boîttcher and A. Hirsch, *J. Am. Chem. Soc.*, 2009, **131**, 2172–2184.
127. W. Ding, S. Pengcheng, L. Changhong, W. Wei and F. Shoushan, *Nanotechnology*, 2008, **19**, 075609.
128. L. Berhan, Y. B. Yi, A. M. Sastry, E. Munoz, M. Selvidge and R. Baughman, *J. Appl. Phys.*, 2004, **95**, 4335–4345.
129. R. L. D. Whitby, T. Fukuda, T. Maekawa, S. L. James and S. V. Mikhalovsky, *Carbon*, 2008, **46**, 949–956.
130. J. N. Coleman, W. J. Blau, A. B. Dalton, E. Munoz, S. Collins, B. G. Kim, J. Razal, M. Selvidge, G. Vieiro and R. H. Baughman, *Appl. Phys. Lett.*, 2003, **82**, 1682–1684.
131. Z. Wang, Z. Liang, B. Wang, C. Zhang and L. Kramer, *Compos. A: Appl. Sci. Manuf.*, 2004, **35**, 1225–1232.
132. X. Zhao, J. Gou, G. Song and J. Ou, *Compos. B: Eng.*, 2009, **40**, 134–140.
133. J. L. Abot, Y. Song, M. J. Schulz and V. N. Shanov, *Compos. Sci. Technol.*, 2008, **68**, 2755–2760.
134. Y. H. Yun, V. Shanov, M. J. Schulz, S. Narasimhadevara, S. Subramaniam, D. Hurd and F. J. Boerio, *Smart Mater. Struct.*, 2005, **14**, 1526–1532.
135. C. S. Grimmer and C. K. H. Dharan, *J. Wuhan Univ. Technol. – Mater. Sci. Ed.*, 2009, **24**, 167–173.
136. M. Holzinger, J. Steinmetz, D. Samaille, M. Glerup, M. Paillet, P. Bernier, L. Ley and R. Graupner, *Carbon*, 2004, **42**, 941–947.
137. A. A. Mamedov, N. A. Kotov, M. Prato, D. M. Guldi, J. P. Wicksted and A. Hirsch, *Nat. Mater.*, 2002, **1**, 190–194.
138. M. Olek, J. Ostrander, S. Jurga, H. Mohwald, N. Kotov, K. Kempa and M. Giersig, *Nano Lett.*, 2004, **4**, 1889–1895.
139. T. J. Kang, M. Cha, E. Y. Jang, J. Shin, H. U. Im, Y. Kim, J. Lee and Y. H. Kim, *Adv. Mater.*, 2008, **20**, 3131–3137.
140. B. S. Shim, P. Podsiadlo, D. G. Lilly, A. Agarwal, J. Leet, Z. Tang, S. Ho, P. Ingle, D. Paterson, W. Lu and N. A. Kotov, *Nano Lett.*, 2007, **7**, 3266–3273.
141. S. Srivastava and N. A. Kotov, *Acc. Chem. Res.*, 2008, **41**, 1831–1841.
142. H. J. Park, J. Kim, J. Y. Chang and P. Theato, *Langmuir*, 2008, **24**, 10467–10473.
143. B. S. Shim, Z. Y. Tang, M. P. Morabito, A. Agarwal, H. P. Hong and N. A. Kotov, *Chem. Mater.*, 2007, **19**, 5467–5474.
144. B. S. Shim, W. Chen, C. Doty, C. L. Xu and N. A. Kotov, *Nano Lett.*, 2008, **8**, 4151–4157.
145. B. S. Shim, J. Zhu, E. Jan, K. Critchley, S. S. Ho, P. Podsiadlo, K. Sun and N. A. Kotov, *ACS Nano*, 2009, **3**, 1711–1722.

146. I. O'Connor, H. Hayden, J. N. Coleman and Y. K. Gun'ko, *Small*, 2009, **5**, 466–469.
147. I. O' Connor, S. De, J. N. Coleman and Y. K. Gun'ko, Carbon, 2009, 47, 1983.
148. M. Cadek, J. N. Coleman, K. P. Ryan, V. Nicolosi, G. Bister, A. Fonseca, J. B. Nagy, K. Szostak, F. Beguin and W. J. Blau, *Nano Lett.*, 2004, **4**, 353–356.
149. D. Blond, D. N. McCarthy, W. J. Blau and J. N. Coleman, *Biomacromolecules*, 2007, **8**, 3973–3976.
150. J. N. Coleman, M. Cadek, K. P. Ryan, A. Fonseca, J. B. Nagy, W. J. Blau and M. S. Ferreira, *Polymer*, 2006, **47**, 8556–8561.
151. U. Khan, K. Ryan, W. J. Blau and J. N. Coleman, *Compos. Sci. Technol.*, 2007, **67**, 3158–3167.
152. A. Garg and S. B. Sinnott, *Chem. Phys. Lett.*, 1998, **295**, 273–278.
153. L. Liu, A. H. Barber, S. Nuriel and H. D. Wagner, *Adv. Funct. Mater.*, 2005, **15**, 975–980.
154. Y. Zhao and E. V. Barrera, *Adv. Funct. Mater.*, 2010, **20**, 3039–3044.
155. H. S. Kim, Y. S. Chae, J. H. Choi, J. S. Yoon and H. J. Jin, *Adv. Compos. Mater.*, 2008, **17**, 157–166.
156. S. Q. Li, F. Wang, Y. Wang, J. W. Wang, J. Ma and J. Xiao, *J. Mater. Sci.*, 2008, **43**, 2653–2658.
157. Z. Zhou, S. Wang, L. Lu, Y. Zhang and Y. Zhang, *Compos. Sci. Technol.*, 2008, **68**, 1727–1733.
158. S. Bhattacharyya, C. Sinturel, J. P. Salvetat and M.-L. Saboungi, *Appl. Phys. Lett.*, 2005, **86**, 113104.
159. R. Blake, Y. K. Gun'ko, J. Coleman, M. Cadek, A. Fonseca, J. B. Nagy and W. J. Blau, *J. Am. Chem. Soc.*, 2004, **126**, 10226–10227.
160. R. Blake, J. N. Coleman, M. T. Byrne, J. E. McCarthy, T. S. Perova, W. J. Blau, A. Fonseca, J. B. Nagy and Y. K. Gun'ko, *J. Mater. Chem.*, 2006, **16**, 4206–4213.
161. M. T. Byrne, W. P. McNamee and Y. K. Gun'ko, Nanotechnology, 2008, 19.
162. A. Eitan, F. T. Fisher, R. Andrews, L. C. Brinson and L. S. Schadler, *Compos. Sci. Technol.*, 2006, **66**, 1162–1173.
163. R. E. Gorga, K. K. S. Lau, K. K. Gleason and R. E. Cohen, *J. Appl. Polym. Sci.*, 2006, **102**, 1413–1418.
164. W. Yuan, W. F. Li, Y. G. Mu and M. B. Chan-Park, *ACS Appl. Mater. Interfaces*, 2011, **3**, 1702–1712.
165. W. Yuan, J. L. Feng, Z. Judeh, J. Dai and M. B. Chan-Park, *Chem. Mater.*, 2010, **22**, 6542–6554.
166. S. Yesil, C. Winkelrnann, G. Bayram and V. La Saponara, *Mater. Sci. Eng. A: Struct. Mater. Prop. Microstruct. Process.*, 2010, **527**, 7340–7352.
167. D. Qian, E. C. Dickey, R. Andrews and T. Rantell, *Appl. Phys. Lett.*, 2000, **76**, 2868–2870.
168. S. L. Ruan, P. Gao, X. G. Yang and T. X. Yu, *Polymer*, 2003, **44**, 5643–5654.

169. B. Safadi, R. Andrews and E. A. Grulke, *J. Appl. Polym. Sci.*, 2002, **84**, 2660–2669.
170. M. S. P. Shaffer and A. H. Windle, *Adv. Mater.*, 1999, 11, 937–941.
171. W. D. Callister, *Materials Science and Engineering: An Introduction*, Wiley, New York, 2003.
172. J. C. Halpin, *J. Compos. Mater.*, 1969, **3**, 732.
173. J. C. Halpin and J. L. Kardos, *Polym. Eng. Sci.*, 1976, **16**, 344–352.
174. R. E. Gorga and R. E. Cohen, *J. Polym. Sci. B: Polym. Phys.*, 2004, **42**, 2690–2702.
175. T. X. Liu, I. Y. Phang, L. Shen, S. Y. Chow and W.-D. Zhang, *Macromolecules*, 2004, **37**, 7214–7222.
176. W. De Zhang, L. Shen, I. Y. Phang and T. X. Liu, *Macromolecules*, 2004, **37**, 256–259.
177. P. Potschke, S. Pegel, A. Janke, I. Alig and S. Dudkin, *AIP Conf. Proc.*, 2004, **723**, 478.
178. S. N. Li, B. Li, Z. M. Li, Q. Fu and K. Z. Shen, *Polymer*, 2006, **47**, 4497–4500.
179. M. Ganss, B. K. Satapathy, M. Thunga, R. Weidisch, P. Potschke and D. Jehnichen, *Acta Mater.*, 2008, **56**, 2247–2261.
180. A. Allaoui, S. Bai, H. M. Cheng and J. B. Bai, *Compos. Sci. Technol.*, 2002, **62**, 1993–1998.
181. X. D. Li, H. S. Gao, W. A. Scrivens, D. L. Fei, X. Y. Xu, M. A. Sutton, A. P. Reynolds and M. L. Myrick, *Nanotechnology*, 2004, **15**, 1416–1423.
182. T. Ogasawara, Y. Ishida, T. Ishikawa and R. Yokota, *Compos. A: Appl. Sci. Manuf.*, 2004, **35**, 67–74.
183. L. S. Schadler, S. C. Giannaris and P. M. Ajayan, *Appl. Phys. Lett.*, 1998, **73**, 3842–3844.
184. J. R. Wood, Q. Zhao, M. D. Frogley, E. R. Meurs, A. D. Prins, T. Peijs, D. J. Dunstan and H. D. Wagner, *Phys. Rev. B*, 2000, **62**, 7571–7575.
185. Y. Breton, G. Desarmot, J. P. Salvetat, S. Delpeux, C. Sinturel, F. Beguin and S. Bonnamy, *Carbon*, 2004, **42**, 1027–1030.
186. J. Bai, *Carbon*, 2003, **41**, 1325–1328.
187. C. Park, Z. Ounaies, K. A. Watson, R. E. Crooks, J. Smith, Joseph, S. E. Lowther, J. W. Connell, E. J. Siochi, J. S. Harrison and T. L. S. Clair, *Chem. Phys. Lett.*, 2002, **364**, 303–308.
188. I. M. Ward and J. Sweeney, An Introduction to the Mechanical Properties of Solid Polymers, J. Wiley & Sons, 2nd edition, Chichester, 2004.
189. E. Lafuente, M. A. Callejas, R. Sainz, A. M. Benito, W. K. Maser, M. L. Sanjuan, D. Saurel, J. M. de Teresa and M. T. Martinez, *Carbon*, 2008, **46**, 1909–1917.
190. C. Peng, J. Jin and G. Z. Chen, *Electrochim. Acta*, 2007, **53**, 525–537.
191. Y. F. Ma, P. L. Chiu, A. Serrano, S. R. Ali, A. M. Chen and H. X. He, *J. Am. Chem. Soc.*, 2008, **130**, 7921–7928.
192. W. H. Li, X. H. Chen, C. S. Chen, L. S. Xu, Z. Yang and Y. G. Wang, *Polym. Compos.*, 2008, **29**, 972–977.

193. D. Blond, V. Barron, M. Ruether, K. P. Ryan, V. Nicolosi, W. J. Blau and J. N. Coleman, *Adv. Funct. Mater.*, 2006, **16**, 1608–1614.
194. Y. I. Estrin, E. R. Badamshina, A. A. Grishchuk, G. S. Kulagina, V. A. Lesnichaya, Y. A. Ol'khov, A. G. Ryabenko and S. N. Sul'yanov, *Polym. Sci. Ser. A*, 2012, **54**, 290–298.
195. B. Vigolo, P. Poulin, M. Lucas, P. Launois and P. Bernier, *Appl. Phys. Lett.*, 2002, **81**, 1210–1212.
196. J. N. Barisci, M. Tahhan, G. G. Wallace, S. Badaire, T. Vaugien, M. Maugey and P. Poulin, *Adv. Funct. Mater.*, 2004, **14**, 133–138.
197. J. Fraysse, A. I. Minett, G. Gu, S. Roth, A. G. Rinzler and R. H. Baughman, *Curr. Appl. Phys.*, 2001, **1**, 407–411.
198. R. H. Baughman, C. Cui, A. A. Zakhidov, Z. Iqbal, J. N. Barisci, G. M. Spinks, G. G. Wallace, A. Mazzoldi, D. De Rossi, A. G. Rinzler, O. Jaschinski, S. Roth and M. Kertesz, *Science*, 1999, **284**, 1340–1344.
199. J. Coleman, N. W. Blau, J. A. Dalton, B. E. Munoz, S. Collins, B. Kim, G. J. Razal, M. Selvidge, G. Vieiro and R. Baughman, H., *Appl. Phys. Lett.*, 2003, **82**, 1682–1684.
200. U. Dettlaff-Weglikowska, V. Skakalova, R. Graupner, S. H. Jhang, B. H. Kim, H. J. Lee, L. Ley, Y. W. Park, S. Berber, D. Tomanek and S. Roth, *J. Am. Chem. Soc.*, 2005, **127**, 5125–5131.
201. M. Motta, Li, I. Kinloch and A. Windle, *Nano Lett.*, 2005, **5**, 1529–1533.
202. X. Zhang, T. V. Sreekumar, T. Liu and S. Kumar, *J. Phys. Chem. B*, 2004, **108**, 16435–16440.
203. C. Meng, C. Liu and S. Fan, *Electrochem. Commun.*, 2009, **11**, 186–189.
204. C. Bartholome, A. Derre, O. Roubeau, C. Zakri and P. Poulin, *Nanotechnology*, 2008, **19**, 325501.
205. C. Bartholome, P. Miaudet, A. Derre, M. Maugey, O. Roubeau, C. Zakri and P. Poulin, *Compos. Sci. Technol.*, 2008, **68**, 2568–2573.
206. J. N. Coleman, W. J. Blau, A. B. Dalton, E. Munoz, S. Collins, B. G. Kim, J. Razal, M. Selvidge, G. Vieiro and R. H. Baughman, *Appl. Phys. Lett.*, 2003, **82**, 1682.
207. C. J. Frizzell, M. in het Panhuis, D. H. Coutinho, K. J. Balkus, A. I. Minett, W. J. Blau and J. N. Coleman, *Phys. Rev. B*, 2005, **72**, 245420.
208. G. T. Pham, Y. B. Park, S. R. Wang, Z. Y. Liang, B. Wang, C. Zhang, P. Funchess and L. Kramer, *Nanotechnology*, 2008, **19**, 325705.
209. S. Wang, Z. Liang, G. Pham, Y.-B. Park, B. Wang, C. Zhang, L. Kramer and P. Funchess, *Nanotechnology*, 2007, **18**, 095708.
210. F. M. Blighe, W. J. Blau and J. N. Coleman, Nanotechnology, 2008, **19**, 415709.
211. A. Kelly and W. R. Tyson, *J. Mech. Phys. Solids*, 1965, **13**, 329.
212. C. S. Grimmer and C. K. H. Dharan, *J. Mater. Sci.*, 2008, **43**, 4487–4492.
213. D. R. Bortz, C. Merino and I. Martin-Gullon, *Compos. Sci. Technol.*, 2012, **72**, 446–452.

CHAPTER 5
Polymer-grafted Carbon Nanotubes via "Grafting From" Approach

CHAO GAO*, ZHENG LIU, LIANG KOU AND
XIAOLI ZHAO

MOE Key Laboratory of Macromolecular Synthesis and Functionalization, Department of Polymer Science and Engineering, Zhejiang University, 38 Zheda Road, Hangzhou 310027, P. R. China
*E-mail: chaogao@zju.edu.cn

5.1 Linear Polymer-functionalized Carbon Nanotubes (CNTs)

5.1.1 Atom Transfer Radical Polymerization (ATRP) Approach to Polymer-grafted CNTs

CNTs can be functionalized with polymers by three grafting strategies: "grafting from", "grafting to", and "random grafting".[1-8] Among the various methods of controlled/living radical polymerization (CRP), ATRP is one of the most powerful synthetic techniques,[9-20] and is also a convenient manner for the preparation of polymeric materials and surface modifications, because the corresponding initiators (*e.g.*, ethyl 2-bromoisobutyrate) or reactive initiating functional compounds (*e.g.*, 2-bromopropionyl bromide and 2-bromoisobutyryl bromide), catalysts of metal halides (*e.g.*, CuCl and CuBr, *etc.*) and ligands of multiamines (*e.g.*, 2,2′-Bipyridine (BPy) and *N,N,N',N'',N''*-pentamethyldiethylenetriamine,

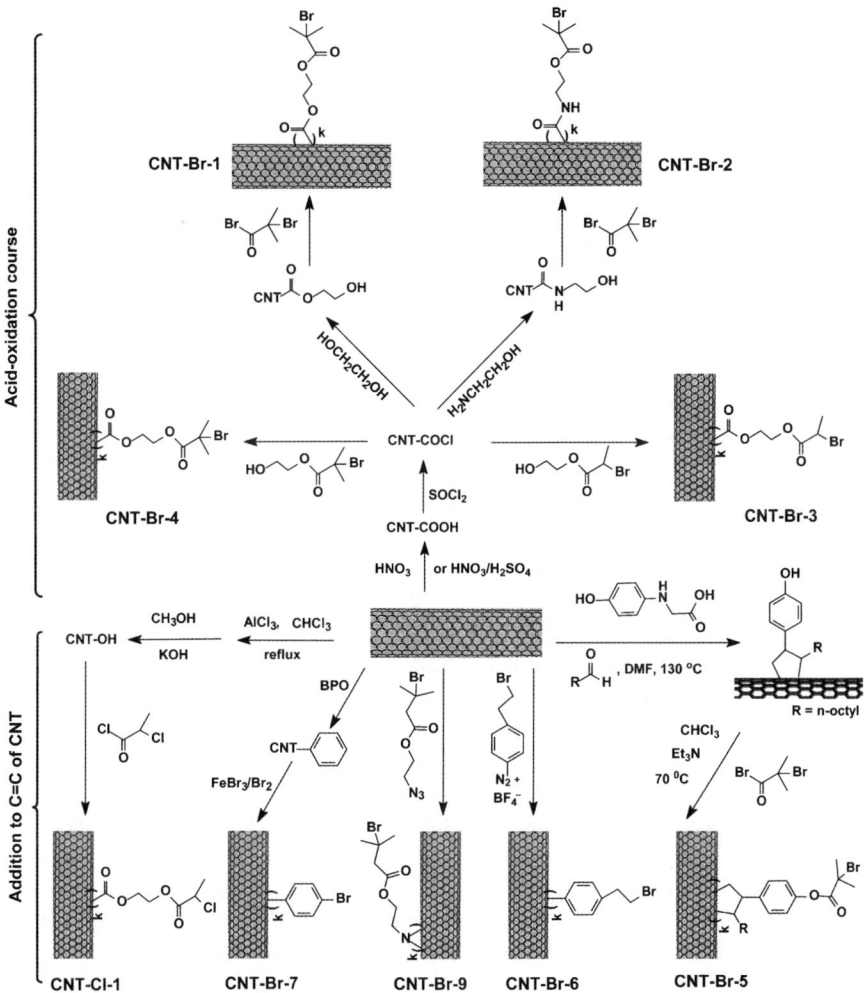

Scheme 5.1 CNT-based macroinitiators for surface-initiating atom transfer radical polymerization (ATRP).[6,7]

pentamethyldiethylenetriamine (PMDETA), *etc.*) are commercially available. Accordingly, ATRP was the first method used to functionalize CNTs. Three common classes of monomers, including (meth)acrylates, styrenics and acrylamides, have been employed to grow polymer/copolymer brushes from CNT surfaces by the *in situ* ATRP approach.

5.1.1.1 Synthesis of CNT-based Macroinitiators

CNT-linked ATRP macroinitiators were first prepared to functionalize CNTs, as depicted in Scheme 5.1. Two manners were adopted to synthesize macroinitiators on the basis of carboxylic acid groups of oxidized CNTs

(CNT-COOH) and the carbon–carbon double bonds of CNTs. Four macroinitiators (CNT-Br) with similar structures were derived from CNT-COOH. Gao and co-workers[21–23] synthesized CNT-Br-1 using pristine multi-walled carbon nanotubes (MWCNTs) *via* a four-step protocol: (1) oxidation of MWCNTs with concentrated HNO_3 for 24 h or a mixture of concentrated HNO_3 and H_2SO_4 (1:3 by volume) for 1–2 h to obtain MWCNT-COOH; (2) the reaction of MWCNT-COOH and thionyl chloride to obtain acyl chloride-functionalized MWCNTs (MWCNT-COCl); (3) a subsequent reaction with glycol to produce hydroxyl-functionalized CNTs (MWCNT-OH); and (4) a final reaction with 2-bromo-2-methylpropionyl bromide to give rise to CNT-Br-1.[21–23] Such a macroinitiator was widely used to successfully synthesize a series of polymer brushes, which is described in the following sections. Likewise, Narain *et al.*[24] prepared 2-bromo-2-methylpropionyl-functionalized single-walled carbon nanotubes (SWCNTs) (CNT-Br-2) by sequentially reacting acyl chloride-functionalized SWCNTs (SWCNT-COCl) with 2-aminoethanol and 2-bromo-2-methylpropionyl bromide. Ford and co-workers[25] obtained CNT-Br-3 through a reaction between SWCNT-COCl and pre-synthesized 2-hydroxyethyl 2'-bromopropionate in anhydrous tetrahydrofuran (THF). Similarly, CNT-Br-4 was obtained *via* the reaction of MWCNT-COCl and hydroxyethyl-2-bromoisobutyrate in toluene in the presence of triethylamine at 100 °C for 24 h.[26]

Alternatively, functional groups can be introduced on to CNTs by coupling carbon–carbon double bonds. Adronov and co-workers[27] synthesized CNT-Br-5, 2-bromo-2-methylpropionyl-functionalized SWCNTs, by 1,3-dipolar cycloaddition[28,29] of 4-hydroxyphenyl glycine and octyl aldehyde in *N,N*-dimethylformamide (DMF) at 130 °C for 5 days, followed by an esterification with 2-bromoisobutyryl bromide in DMF at 70 °C. Paik and co-workers[30] introduced hydroxyl groups on to the surface of SWCNTs by direct electrophilic addition of chloroform using $AlCl_3$ as the catalyst and subsequent hydrolysis in the presence of KOH, and the macroinitiator of CNT-Cl-1 was obtained by subsequent esterification with 2-chrolopropyl chloride. Chehimi and co-workers[31] introduced brominated aryl initiating groups on to aligned MWCNTs (CNT-Br-6) by electrochemical reduction of the diazonium salt BF_4-$^+N_2$-C_6H_4-CH_2CH_2-Br. By reaction of N-doped MWCNTs with benzoyl peroxide (BPO) in toluene at 105 °C for 5 h, followed by bromination at 55 °C with Br_2 in CCl_4 using $FeBr_3$ as the catalyst, Terrones and co-workers[32] also obtained the nanotube-supported ATRP macroinitiator (CNT-Br-7). Liu *et al.*[53] introduced ATRP initiator groups on to SWCNTs (CNT-Cl-2 and CNT-Br-8) by treatment of purified SWCNTs with peroxy organic acids containing Cl or Br substituents, *m*-chloroperbenzoic acid (MCPBA) and 2-bromo-2-methylperpropionic acid (BMPPA). As a comparison, Zhi *et al.*[33] introduced ATRP initiating groups on boron nitride nanotubes (BNNTs) by the reaction of an amino-containing BNNT with chloroacetyl chloride at 120 °C.

Recently, Gao and co-workers[34] developed a facile approach to the functionalization of MWCNTs and SWCNTs *via* nitrene chemistry, and functional groups such as hydroxyl, carboxyl and amino could be immobilized

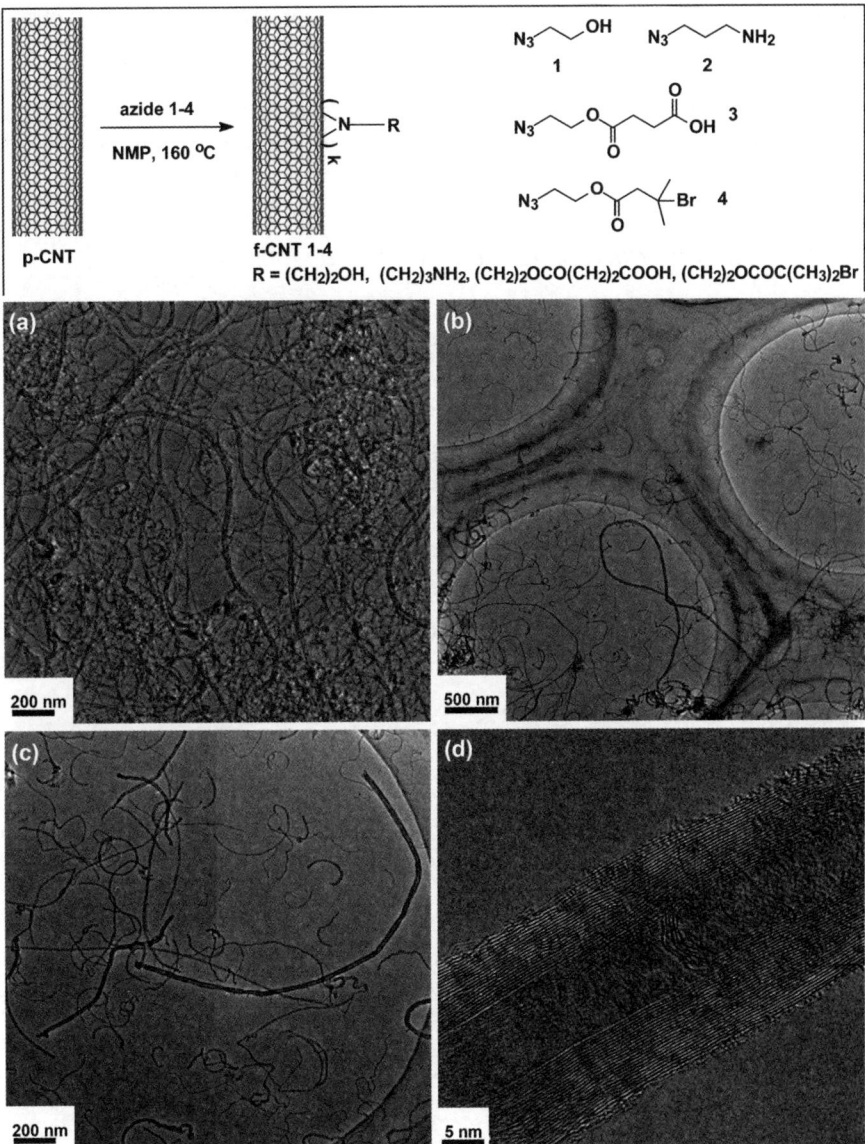

Figure 5.1 Nitrene chemistry approach to functionalize CNTs (top), and representative TEM images (bottom) of pristine (p-) MWCNTs (a) and hydroxyl-functionalized MWCNTs (MWCNT-OH) (b–d). Other functionalized CNTs (f-CNTs) showed similar TEM images to the MWCNT-OH. Reprinted with permission from Gao et al.[34]

on to CNTs in one-step electrophilic [2 + 1] cycloaddition reaction. The functionalized CNTs became individually dispersed (Figure 5.1). The CNT macroinitiator (CNT-Br-9) was synthesized by a one-step reaction between pristine CNTs and 2-azidoethyl-2-bromo- 2-methylpropanoate. The Br density achieved 0.6 mmol g^{-1} of functionalized CNTs or 8.2 groups per 1000 carbons, which is close to the value of CNT-Br-1. Subsequent ATRP of styrene, methyl methyacrylate (MMA) and 3-azido-2-hydroxy propyl methacrylate (GMA-N$_3$) produced various polymer-grafted CNTs. This versatile approach has been also extended to functionalize two-dimensional (2D) carbon such as graphene.[35]

5.1.1.2 In Situ *Polymerization of (Meth)Acrylates*

Polymer brushes can be grown from CNTs in the presence of CNT-based macroinitiators by an *in situ* ATRP technique with or without a sacrificial initiator. The monomers used in such a protocol are shown in Scheme 5.2, and the corresponding polymerization conditions and results are listed in

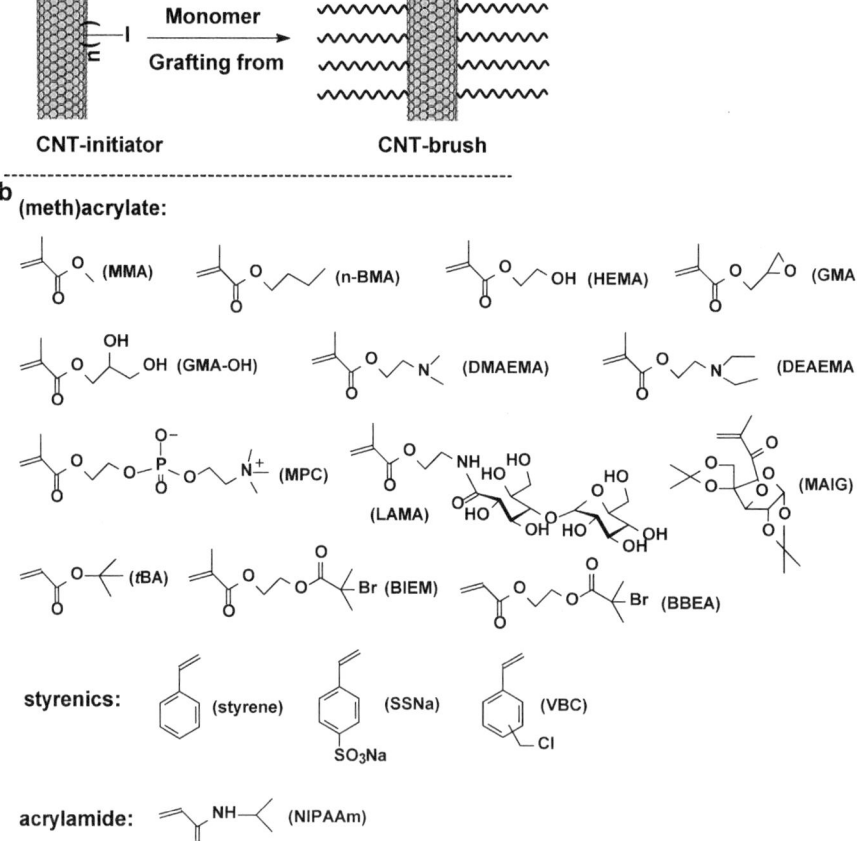

Scheme 5.2 A cartoon for polymer brushes grown from the CNT surface (a), and the monomers used in the functionalization of CNTs *via* the ATRP "grafting from" approach (b).[6,7]

Table 5.1. Various polymers from polar to apolar, from basic to acidic, and from water-soluble to oil-soluble have been successfully grafted from CNTs.

Among three common classes of monomers including (meth)acrylates, styrenics and acrylamides, (meth)acrylates were the most widely used because various functional polymers with hydroxyl, carboxyl, amino and other groups can be facilely obtained from the corresponding functional (meth)acrylate monomers, except the commonly used monomers of methyl methacrylate (MMA) and n-butyl methacrylate (n-BMA).

Almost at the same time, Gao and co-workers,[21] Ford and co-workers[25] and Adronov and co-workers [27] independently developed the *in situ* ATRP protocol to grow polymer brushes on MWCNTs and SWCNTs using a CNT-supported macroinitiator, which opened the door for the controlled functionalization of CNTs. Using CNT-Br-1 as the initiator and CuBr/PMDETA as the catalyst system, Gao and co-workers[21] prepared poly(methyl methacrylate) (PMMA)-grafted MWCNTs. Clear core-shell structures were observed under high-resolution transmission electron microscopy (HRTEM) with CNT as the core and the grafted polymer layer as the shell, strongly and directly demonstrating the high density and even grafting efficiency. The thickness of the grafted PMMA layer on the nanotube surface can be controlled from ~ 3.8 to ~ 14 nm by the feed ratio of MMA to CNT-Br-1 (Figure 5.2). This result also answered well the long-term debate of the location of chemical functionalization for CNTs (at the tube ends or sidewalls): oxidation and the subsequent reactions occurred at the whole tube surface, including the ends and the body. Thermogravimetric analysis (TGA) measurements indicated that the grafted polymer content varied from 31.9 to 82.0 wt% when the feed ratio of MMA to CNT-Br-1 was changed from 1:1 to 10:1. The as-prepared PMMA-grafted MWNTs were soluble in weak polar solvents such as chloroform and THF.[21]

This approach to functionalize CNTs has been widely extended to other monomer systems including 2-(dimethylamino)ethyl methacrylate (DMAEMA),[36] 2-(diethyl amino)ethyl methacrylate (DEAEMA),[37,38] glycerol monomethacrylate (GMA-OH),[23] glycidyl methacrylate (GMA),[39] 3-*O*-methacryloyl-1,2:5,6-di-*O*-isopropylidene-*D*- glucofuranose (MAIG)[40] and *tert*-butyl acrylate (*t*BA)[41] with either CuBr/PMDETA or CuBr/1,1,4,7,10,10-hexamethyl-triethylenetetramine (HMTETA) as the catalyst/ligand, grafting various functional polymers from the CNT surfaces successfully. The same conclusions can be drawn for other polymers as in the case of PMMA-grafted MWCNTs: (1) core-shell nanostructures can be clearly observed with HRTEM, especially when the grafted polymer amount is higher than 50 wt%; (2) the whole nanotube surface, including both the ends and body, is covered by polymer brushes; (3) the grafted polymer content can be readily adjusted by the feed ratio of monomer to CNT-Br-1 and the reaction time; and (4) the solubility and dispersibility of CNTs are highly improved due to the polymer grafting. These experiments all manifest the good reproducibility and high generality of this approach.

Interestingly, the THF dispersion of PMMA-grafted MWCNTs brushes (MWCNT-g-PMMA, 85 wt% polymer) could self-assemble into suprastructures

Table 5.1 Selected conditions and results of ATRP in the presence of CNT-based macroinitiator.[6,7]

Macroinitiator	Monomer	Catalyst/ligand	Solvent/temperature/time	Highest f_{wt}%	Ref.
CNT-Br-1	MMA	CuBr/PMDETA	DMF/60 °C/30 h	82	21
	MMA + HEMA	CuBr/PMDETA	DMF/60 °C/20 h	54.5	21
	DMAEMA	CuBr/PMDETA	THF/60 °C/48 h	80	36
	DEAEMA	CuBr/PMDETA	MeOH/60 °C/48 h	67	38
	GMA-OH	CuBr/PMDETA	MeOH/40 °C/48 h	90	23,37
	GMA	CuBr/PMDETA	Diphenyl ether/50 °C/48 h	82	39
	MAIG	CuBr/HMTETA	Ethyl acetate/60 °C/29 h	71	40
	BIEM	CuBr/(PPh$_3$)$_2$NiBr$_2$	Ethyl acetate/100 °C/4.5 h	38	40
	MAIG + BIEM	CuBr/(PPh$_3$)$_2$NiBr$_2$	Ethyl acetate/100 °C/4.5 –29.5 h	40–53	40
	tBA	CuBr/PMDETA	DMF/60 °C/48 h	75	41
	Styrene	CuBr/PMDETA	Diphenyl ether/100 °C/50 h	77.9	22
	Styrene + tBA	CuBr/PMDETA	—	70	48
	SSNa	CuBr/PMDETA	DMF/130 °C/30 h	68	41
	VBC	CuBr/PMDETA	Toluene/130 °C/ 48 h	50	182
	NIPAAm	CuBr/PMDETA	Water/RT/48 h	84	51
CNT-Br-2	MPC	CuBr/bpy	MeOH/25 °C/24 h	54	24
	MPC	CuBr/bpy	H$_2$O/25 °C/12 h	59	24
	LAMA	CuBr/bpy	H$_2$O/25 °C/12 h	67	24
	LAMA	CuBr/bpy	NMP/25 °C/12 h	70	24
CNT-Br-3	n-BMA	CuCl/bpy	DCB/60 °C/19 h	69	25
	Styrene	CuBr/bpy	DCB/110 °C/14 h	75	49
CNT-Br-4	MMA	CuBr/PMDETA	Toluene/90 °C/24 h	70.9	26
	Styrene	CuBr/PMDETA	Toluene/100 °C/24 h	33.0	26
	MMA + styrene	CuBr/PMDETA	Toluene/100 °C/24 h	24.6	26
	Styrene	CuBr/PMDETA	Bulk/80 °C/24 h	60	50
	CH$_2$CHCN	CuBr/PMDETA	Bulk/70 °C/24 h	40	50
	Styrene + CH$_2$CHCN	CuBr/PMDETA	Bulk/80 °C/24 h	70	50
	BBEA	CuBr/PMDETA	Toluene/100 °C/24 h	80	161

Table 5.1 (Continued)

Macroinitiator	Monomer	Catalyst/ligand	Solvent/temperature/time	Highest $f_{wt}\%$	Ref.
CNT-Br-5	MMA	CuBr/bpy	DMF + H$_2$O/RT/24 h	—	27
	tBA	CuBr/bpy	DMF + H$_2$O/RT/24 h	—	27
CNT-Br-6	MMA	CuCl + CuCl$_2$/PMDETA	Bulk/90 °C/6 h	—	31
	Styrene	CuBr/PMDETA	Bulk/110 °C/16 h	—	32
CNT-Br-7	Styrene	CuBr/dNbpy	Toluene/110 °C/18 h	40	32
CNT-Br-8	MMA	CuBr/PMDETA	DCB/60 °C/24 h	30	47
CNT-Cl-1	Styrene	CuCl/PMDETA	Acetone/60 °C/96 h	6 nm	30

RT, room temperature.

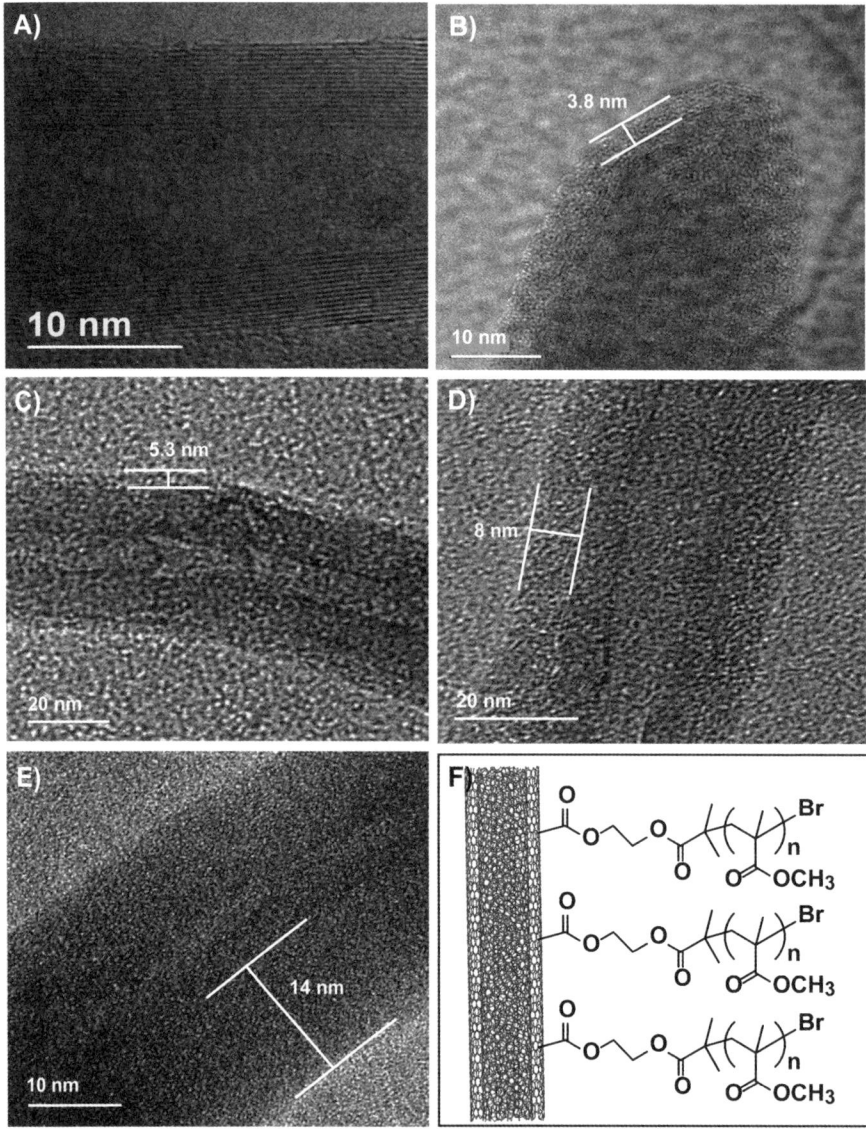

Figure 5.2 TEM images of pristine MWCNT (A) and PMMA-grafted MWCNT (B–E) prepared by the *in situ* ATRP technique, and a schematic illustration of PMMA-grafted MWCNTs (F). Reprinted permission from Kong *et al.*[21]

on solid surfaces such as gold, mica, silicon, quartz or carbon films.[42] The combination of scanning electron microscopy (SEM), atomic force microscopy (AFM) and TEM measurements confirmed the morphology of the assembled structures, and revealed the assembly mechanism. With decreasing the concentration of MWCNT-g-PMMA from 3 to 0.1 mg mL^{-1}, the assembled structures changed from cellular and basketwork-like forms to multi-layer

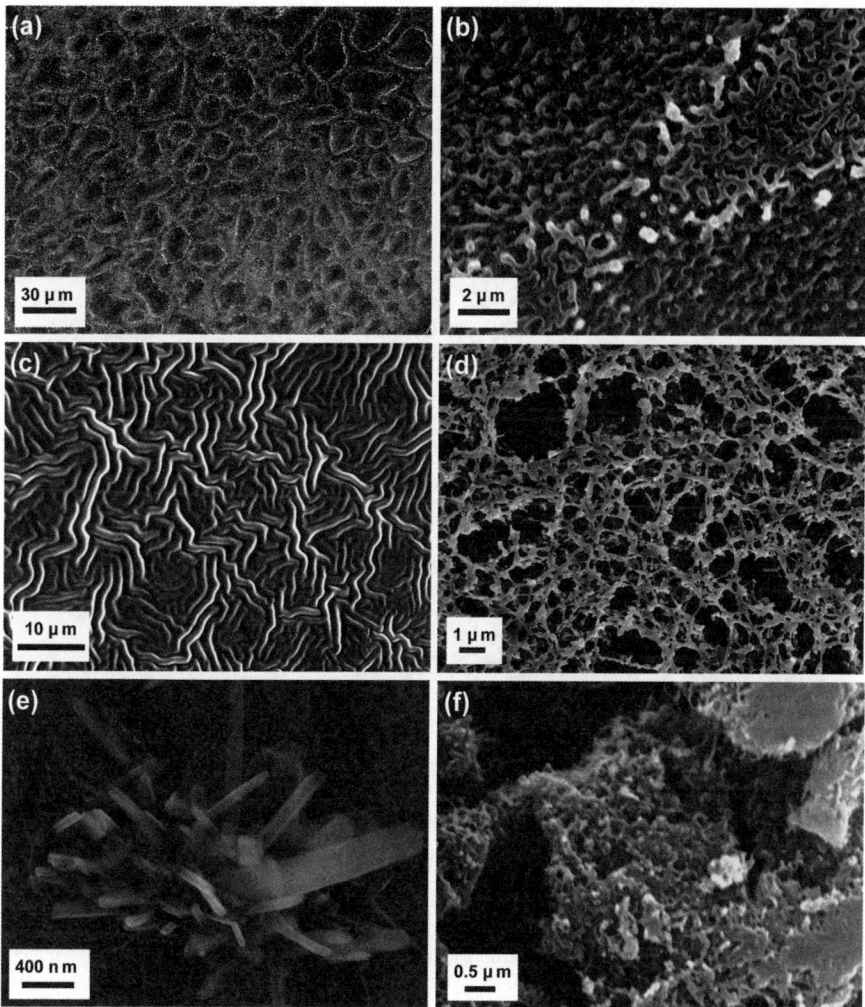

Figure 5.3 Representative SEM images of the self-assembled patterns or structures of the MWNT-*g*-PMMA on a gold surface with concentrations of 3 (a, b), 2 (c), 1 (d), and 0.1 (e) mg mL^{-1}, and an SEM image of the as-prepared MWNT-*g*-PMMA bulk material (f). Reprinted with permission from Gao.[42]

cellular networks and individual needles (Figure 5.3). The phase separation during evaporation of the solvent probably drove the MWCNT-*g*-PMMA nanohybrids to assemble and form the suprastructured objects, and the rigid MWCNTs stabilize the formed structures, as demonstrated by TEM and SEM measurements (see Figure 5.4). The self-assembly behavior of polymer-grafted CNTs is possibly a general phenomenon for highly soluble core-shell nanostructures.

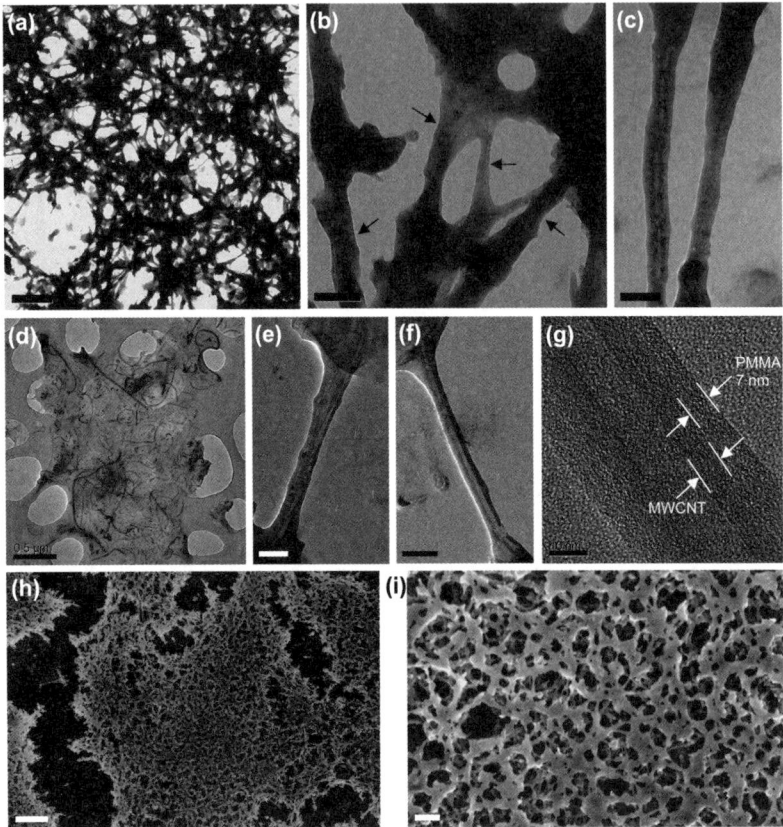

Figure 5.4 Low-voltage (100 kV) TEM images of MWNT-*g*-PMMA on a carbon-coated TEM sample grid (a, b, c), high-voltage (200 kV) TEM image of MWNT-*g*-PMMA on a multi-pore lacey carbon TEM sample grid (d), and high-resolution high-voltage TEM images of MWNT-*g*-PMMA on the carbon-coated TEM sample grid (e, f, g), and SEM images of the same sample measured by TEM (shown in a) of assembling on the TEM grid (h, i). The concentration of solution is 1 mg of MWNT-*g*-PMMA per 1 mL of THF. The scale bars are 3 μm (a), 200 nm (b), 50 nm (c), 0.5 μm (d), 50 nm (e), 50 nm (f), 10 nm (g), 5 μm (h) and 1 μm (i). Reprinted with permission from Gao.[42]

Poly[2-(diethylamino)ethyl methacrylate] (PDEAEMA)-grafted MWCNTs showed an interesting pH-responsive solubility/dispersibility because of the pH-responsive PDEAEMA shell.[37] Polyelectrolyte-grafted CNTs were used as nanoplatforms to fabricate hybrid nanostructures and nanomaterials by layer-by-layer (LbL) electrostatic attraction (Figure 5.5). Gao and co-workers[36,38] grafted PDEAEMA or the dimethylamino analogue (PDMAEMA) from MWCNTs by ATRP and used as templates to load negative iron oxide (Fe_3O_4) or CdTe quantum dots (QDs), obtaining supraparamagnetic or fluorescent CNTs (Figure 5.6). The magnetic CNTs were adsorbed on to sheep red blood

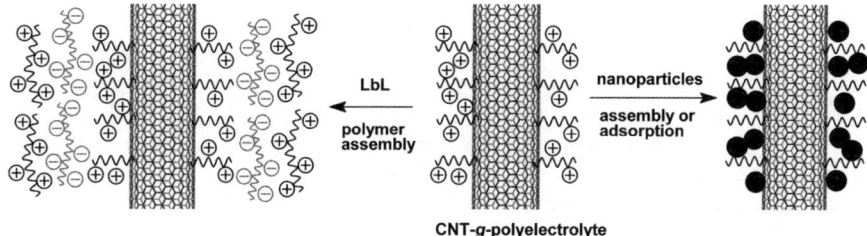

Figure 5.5 Polyelectrolyte-grafted CNTs used as nanosubstrate for multi-layer polymer assembly or nanoparticle-loading.[6]

cells, so that individual cells could be manipulated in a magnetic field, opening the door for biomanipulation with magnetic nanotubes.[38]

Poly(acrylic acid) (PAA)-grafted CNTs, derived from the hydrolysis of *t*BA-grafted CNTs, could be applied as a charged nanosubstrate to coat polymer

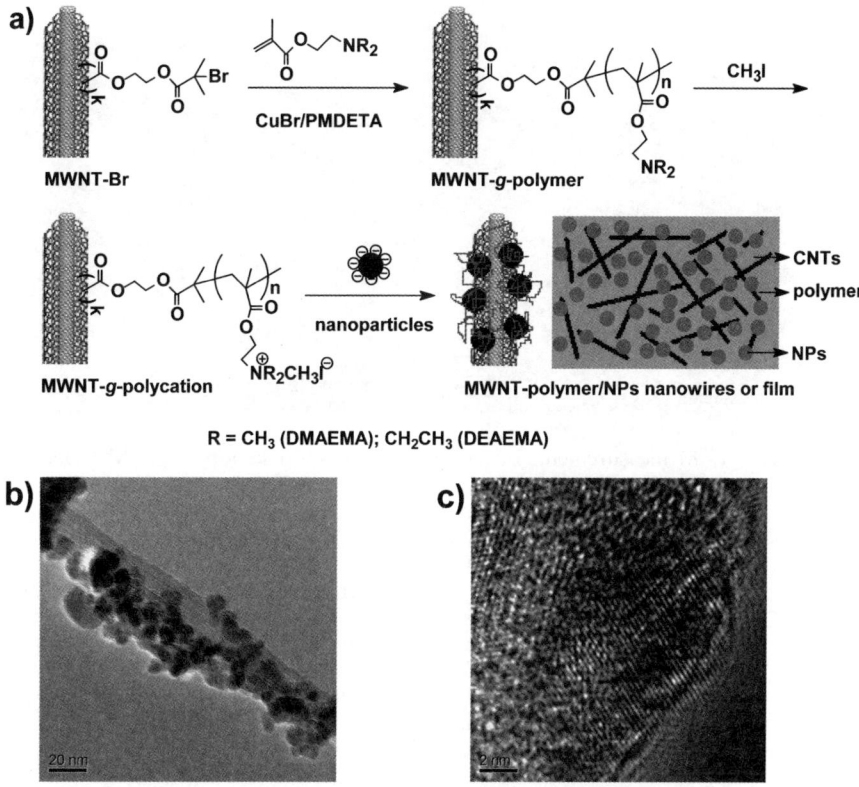

Figure 5.6 (a) Synthesis of cationic polymer-grafted MWCNTs, and assembly of nanoparticles on the polycation-functionalized MWCNTs; (b, c) TEM images of PDEAEMA-grafted MWCNT/Fe_3O_4 hybrid at different magnifications.[38]

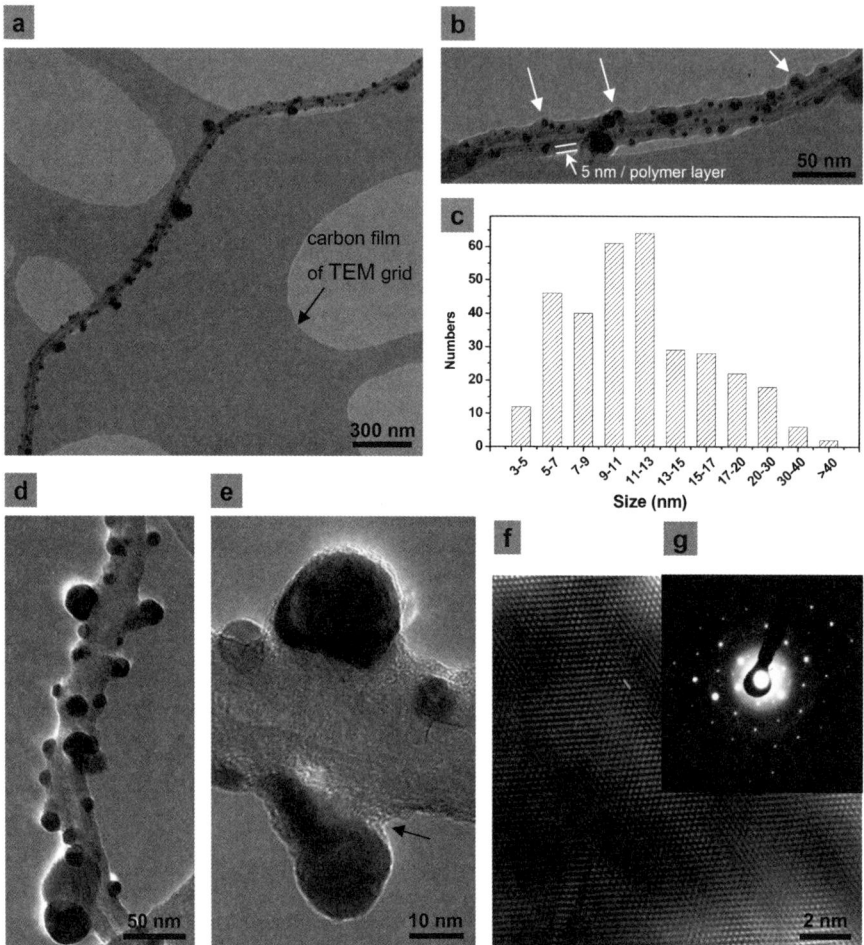

Figure 5.7 TEM measurements of nanohybrids of silver nanoparticles/MWCNT-*g*-PAA (60 wt% of PAA, 2 h reaction). (a) TEM image at 100 kV. (b) Higher magnification of (a) displaying the respective phases of polymer layer, silver nanoparticles and CNTs. (c) Size distribution of the produced silver nanoparticles. (d) TEM image at 200 kV. (e) Higher magnification of (d) showing the polymer interlayer between silver nanoparticle and nanotube wall. (f) High-resolution and high-magnification TEM image of a silver nanoparticle, presenting the silver crystal lattice. (g) Selected area electron diffraction (SAED) patterns showing the face-centred cubic (fcc) crystal structure of the silver single nanocrystals. Reprinted with permission from Chao et al.[43]

with opposite charges *via* LbL assembly approach (Figure 5.5).[41] Moreover, PAA-grafted CNTs are good templates for loading silver nanocrystals. As shown in Figure 5.7, Ag nanoparticles with a mean diameter of ∼10 nm could be evenly grown on CNT surfaces in water without any additional chemicals and

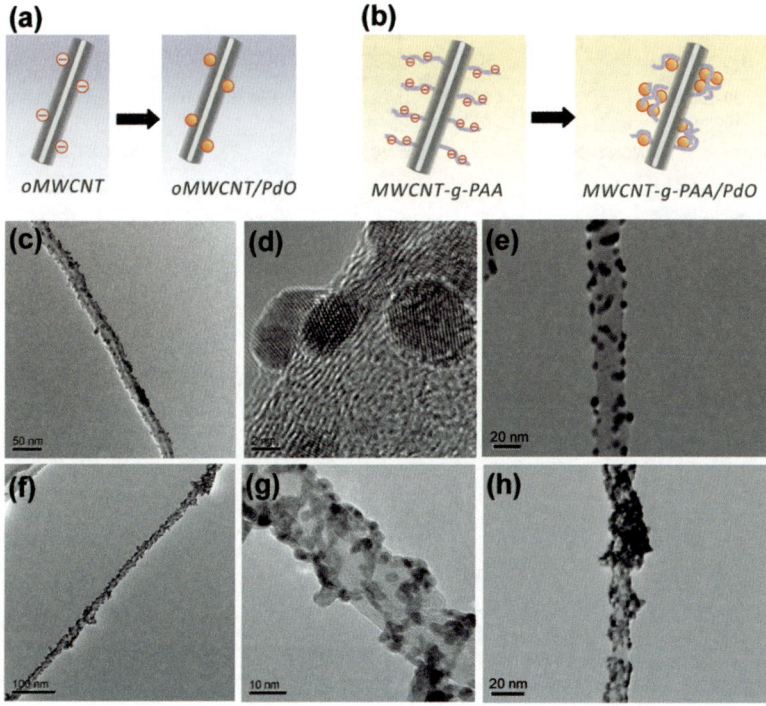

Figure 5.8 (a) Direct decoration of PdO nanoparticles on oxidized MWCNTs (oMWCNT) from aqueous solution of $Pd(NO_3)_2$. (b) Decoration of PdO nanoparticles on PAA-grafted MWCNTs (MWCNT-g-PAA) from $Pd(NO_3)_2$ aqueous solution. (c–e) TEM images of oMWCNT/PdO. (f–h) TEM images of MWCNT-g-PAA/PdO.[44]

physical treatments.[43] Similarly, uniform PdO nanoparticles with diameters of 2.3–2.6 nm were loaded on CNT surfaces by simple decomposition of $Pd(NO_3)_2$ in water at room temperature, and the PAA grafting made the deposited PdO nanoparticles highly dense (Figure 5.8).[44]

Alternatively, hydrophilic multi-hydroxyl poly(GMA-OH)-grafted MWCNTs could be converted into multicarboxyl polymer-functionalized CNTs by reaction with succinic anhydride and then used as templates to efficiently sequestrate metal ions such as Ag^+, Co^{2+}, Ni^{2+}, Au^{3+}, La^{3+} and Y^{3+} (Scheme 5.3), generating MWCNT-polymer/metal hybrid nanocomposites, nanowires or necklace-like nanostructures, depending on the grafted polymer content and the nature of the captured metals.[23] The combination of SEM, TEM and energy dispersive spectroscopy (EDS) characterizations demonstrated the structure and elements of the hybrid nano-objects.

Both carboxylic polymer-grafted CNTs and oxidized CNTs (CNT-COOH) could induce the formation and stabilization of spherical vaterite crystals in the biomineralization of $CaCO_3$.[45] CNT-COOH could stabilize the formed crystals up to 1 week in water, whereas in the case of carboxylic polymer-functionalized

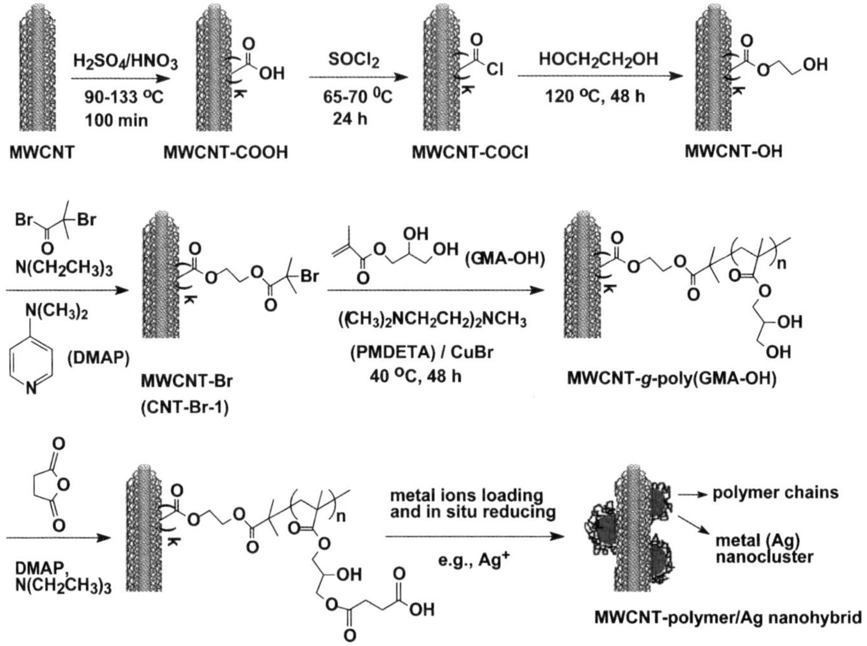

Scheme 5.3 Functionalization of MWCNTs with poly(GMA-OH) by ATRP, conversion of the hydroxyl groups of poly(GMA-OH) into carboxylic groups, and metal sequestration/reduction by the grafted polyacid chains.[23]

CNTs, the formed vaterite crystals completely transformed into thermodynamically stable calcite crystals in water within 10 h. "Offline" TEM observations of the mineralization process of $CaCO_3$ in the presence of CNT-COOH or pristine CNTs revealed that the crystals nucleated at the carboxyl groups of CNT-COOH, grew around the CNTs, and finally formed spherical vaterite crystals impenetrated by the CNTs. The strong interaction between CNT-COOH and crystals together with the strong mechanical strength of CNTs stabilized the formed vaterite crystals and made them difficult to dissolve in water. In the cases of polymer-functionalized MWCNTs, the nucleation sites were located on the polymer chains instead of on the CNTs. Thus, it is the polymer chains that have strong interactions with crystals, resulting in the significant influence of polymer chains on the crystallization or mineralization of $CaCO_3$. These findings demonstrate that nanomaterials could strongly influence the mineralization of biomatters, which may help us prepare novel biomaterials and bionanomaterials. Recently, liquid crystalline polymer-grafted CNTs was prepared by *in situ* ATRP technique.[46]

Using CNT-Br-4 as the initiator, Baskaran *et al.*[26] also successfully grafted PMMA from the MWCNT surfaces by ATRP with CuBr/PMDETA as the catalyst/ligand in toluene at 90 °C. The amount of grafted PMMA approached 70.9 wt%.[26] CNT-Br-5 and CuBr/bpy were employed by Adronov and co-workers[27] to prepare PMMA-grafted SWCNTs in DMF/H$_2$O solvent at room temperature for 1–48 h.[27] It was found that the polydispersity index (PDI) of the recovered PMMA is greater than 1.6 and chain-extension of the PMMA-functionalized SWCNTs did not result in any significant mass increase, indicating that the polymerization initiated by the CNT-Br-5 was not living anymore and the polymer chains grafted on CNTs had been terminated irreversibly.[27] The authors explained that the CNTs might act as radical scavengers, thereby removing radical species from solution and resulting in the uncontrollability of the polymerization.[27] On the contrary, Ford and co-workers[25] obtained poly(n-butyl methacrylate) (PnBMA)-grafted SWCNTs with good controllability in dichlorobenzene (DCB) by the addition of methyl 2-bromopropionate as the free or sacrificial initiator using CNT-Br-3 as the initiator and CuCl/BPy as the catalyst/ligand. Chehimi and co-workers[31] functionalized MWCNT-coated silicon wafer with PMMA at 90 °C for 6 h using CNT-Br-6 as the initiator and CuCl + CuCl$_2$/PMDETA as the catalyst system. The grafted polymer layer was observed with HRTEM. X-ray photoelectron spectroscopy (XPS) analysis confirmed the presence of PMMA by its characteristic C1s and valence band features. Liu et al.[47] prepared PMMA-grafted SWCNTs at 60 °C for 24 h using CNT-Br-8 as the initiator and CuBr/PMDETA as the catalyst system, and they found that the grafted polymer content increased from 17 to 30 wt% by the addition of ethyl 2-bromoisobutyrate as a coinitiator.

Functionalization of CNTs with biocompatible and biodegradable polymers is one of the most interesting topics in this field due to the great potential in the applications of bionanotechnology. Narain et al.[24] grafted poly(2-methacryloyloxyethyl phosphorylcholine) (polyMPC) and poly(lactobionamidoethyl methacrylate) (polyLAMA) from the SWCNT surfaces by ATRP of MPC and LAMA at 25 °C for 12–24 h with CNT-Br-2 as the initiator and CuBr/BPy as the catalyst, and the grafted polyMPC and polyLAMA contents reached 59 and 70 wt%, respectively. Gao et al.[40] functionalized MWCNTs with glycopolymer by ATRP of MAIG using CNT-Br-1 as the initiator, followed by hydrolysis of the polyMAIG chains in 80% formic acid for 48 h, and the grafted polymer content achieved 71 wt%. Figure 5.9 shows the grafting process and the representative TEM images of polyMAIG and glycopolymer-functionalized MWCNTs, revealing the high content of grafted polymer and core-shell structure for the individual functionalized nanotubes. Studies on the polymerization kinetics showed that the grafted polymer content increased linearly with the reaction time, revealing the living/controlled nature of CNT surface-initiated ATRP (Figure 5.10). Such a controlled polymerization character was also confirmed in the case of poly(GMA-OH)-grafted MWCNTs.[23]

The "living" nature of CNT-initiated ATRP promised block copolymerization from CNT surfaces. Kong et al.[21] first conducted ATRP of hydroxyethyl methacrylate (HEMA) using PMMA-grafted MWCNTs as the initiator and CuBr/PMDETA as the catalyst to prepare PHEMA-b-PMMA-grafted MWCNTs in DMF at 60 °C for 20 h. A representative TEM image is shown

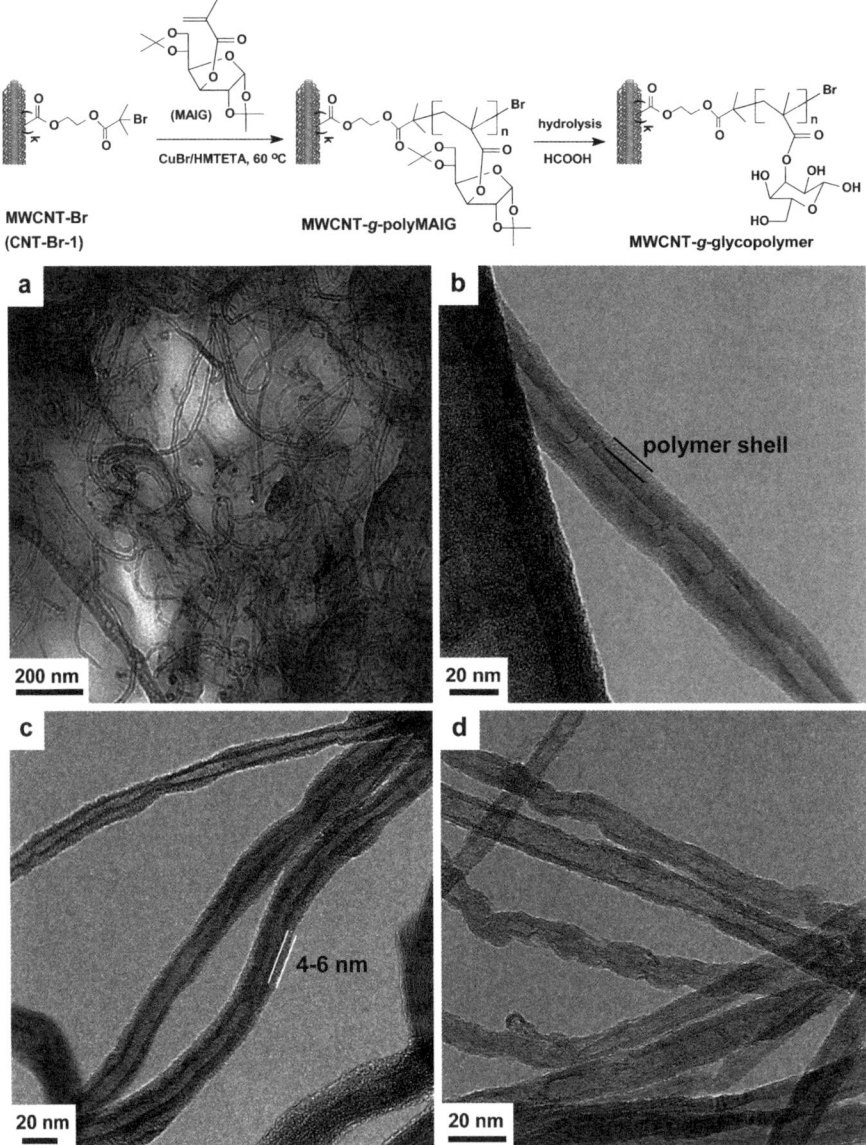

Figure 5.9 Synthesis of glycopolymer-functionalized MWCNTs (top). Representative TEM images of linear poly(MAIG)-functionalized MWCNTs at 29 (a) and 10.5 h (b), deprotected glycopolymer-functionalized MWCNTs at 29 h (c), and the macroinitiator CNT-Br-1 (d). Reprinted with permission from Gao et al.[40]

in Figure 5.11, displaying the clear core-shell structure of polymer-grafted MWCNT. The resulting nanohybrid is coated with coaxial non-polar and polar amphiphilic polymer brushes. The realization of block copolymerization from CNT surfaces not only offers a good method to fabricate tailor-made

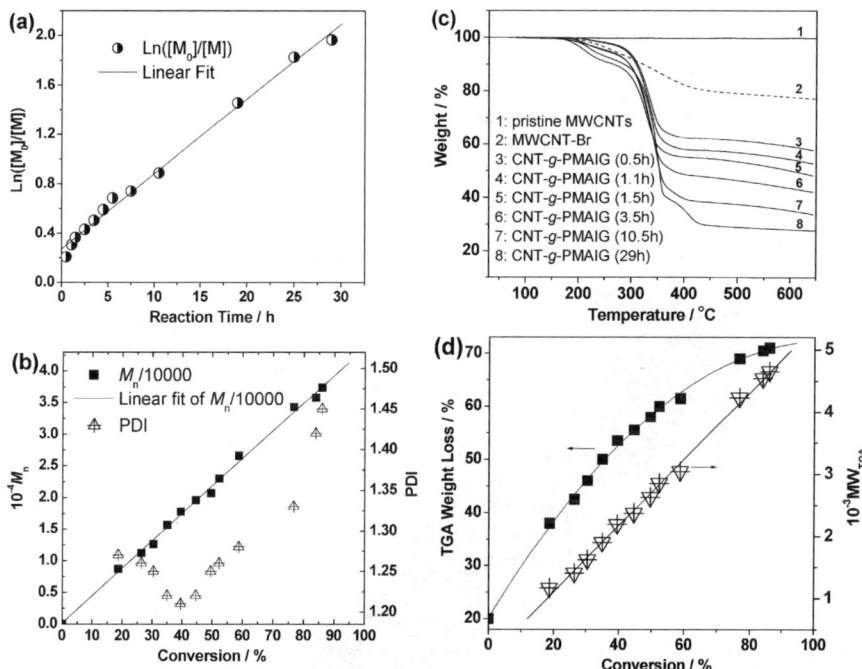

Figure 5.10 The first-order time-conversion plot (a), and apparent number-average molecular weight (M_n) and polydispersity index (PDI) of the free polymer as a function of monomer conversion (b). TGA weight loss curves for the pristine MWCNTs, MWCNT-Br, and linear glycopolymer (PMAIG)-grafted MWCNTs obtained at different reaction times (c), and the content and average molecular weight of the polymer grafted on to MWCNTs, calculated from corresponding TGA data, as a function of monomer conversion (d). Reprinted with permission from Gao et al.[40]

nanostructures, but also further proves the living polymerization characteristics of ATRP.

Another kind of amphiphilic block copolymer-grafted MWCNTs was prepared by sequential ATRP of styrene and *t*BA using CNT-Br-1 as the initiator, followed by hydrolysis of the poly(tert-butyl acrylate) (P*t*BA) block into PAA,[48] as shown in Figure 5.12. The polymer amount of the second block can also be adjusted by the feed ratio of *t*BA to polystyrene (PS)-grafted MWCNTs, demonstrating the living characteristics of the block copolymerization. The core-shell amphiphilic nano-object can self-organize into a film at the interface of chloroform and water.

5.1.1.3 In Situ *Polymerization of Styrenics*

Besides the methacrylates, styrenics is another class of monomers used in ATRP. Ford and co-workers[49] functionalized SWCNTs with PS in DCB at

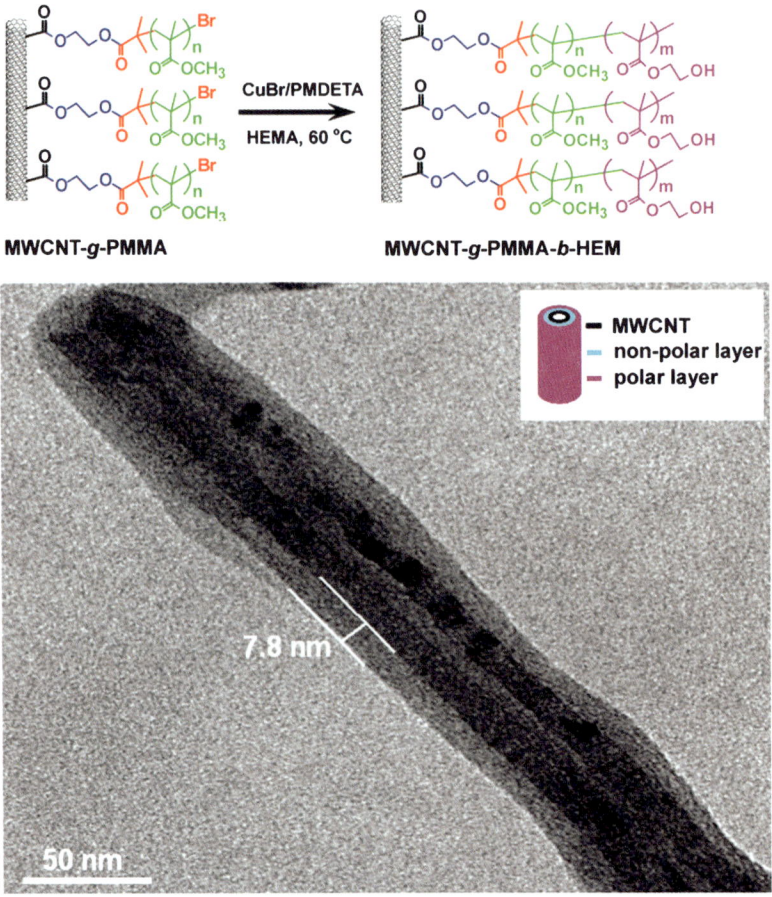

Figure 5.11 Synthesis of block copolymer-grafted MWCNTs (top), and the representative TEM image of PHEMA-*b*-PMMA-grafted MWCNT. The insert shows the cartoon of the bipolar polymer layer-coated CNT. Reprinted with permission from Kong *et al.*[21]

110 °C for 14 h using CNT-Br-3 as the initiator and CuBr/BPy as the catalyst. The grafted PS content increased with increasing the reaction time, and approached 75 wt% at 14 h. Using CNT-Br-1 as the initiator, Gao and co-workers[22] prepared PS-grafted MWCNTs in diphenyl ether at 100 °C for 50 h with a catalyst of CuBr/PMDETA. The grafted PS content rose from 28.6 to 77.9 wt% when the weight feed ratio of monomer to CNT-Br-1 increased from 1:1 to 10:1, and the M_n of the detached PS from the nanotube surfaces increased from 5000 to 11 000. The thermal decomposition temperature (T_d) of the PS chains can be dramatically improved by approximately 50–80 °C due to the covalent linkage with CNTs, and the T_g of grafted PS is also higher than that of detached PS by approximately 10–20 °C. The difference can be attributed to the chemical tethering of polymers. The detaching

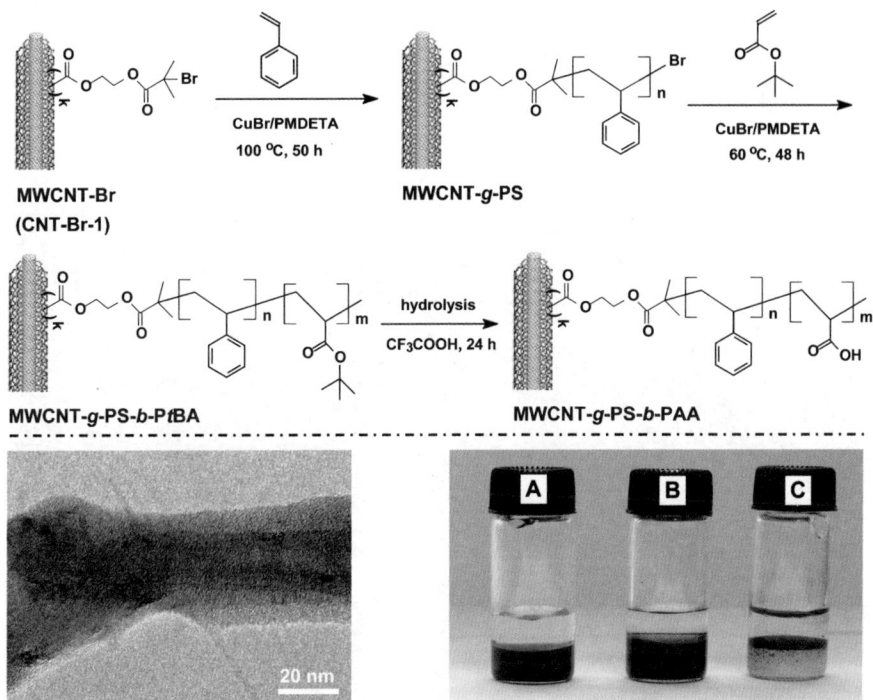

Figure 5.12 Synthesis of PS-block-PAA-grafted MWCNTs by sequential ATRP of styrene and *t*BA, followed by hydrolysis of P*t*BA block (top), a typical TEM image of PS-block-P*t*BA-grafted MWCNTs (bottom, left), and the samples of PS-grafted MWCNTs (A), PS-*b*-P*t*BA-grafted MWCNTs (B) and PS-*b*-PAA-grafted MWCNTs (C) were placed in a mixed solvent of water (upper layer) and chloroform (bottom layer). Reprinted with permission from Kong et al.[48]

of polymers confirmed the covalent grafting effect (Figure 5.13). With CNT-Br-4 as the initiator, Baskaran et al.[26] and Ryu and co-workers[50] independently functionalized MWCNTs with PS by the ATRP technique under different reaction conditions (see Table 5.1), and they also successfully carried out surface-initiated copolymerization of MMA/styrene[26] or styrene/acrylonitrile.[50] Paik and co-workers[30] developed the macroinitiator of CNT-Cl-1, by which a PS layer of 6 nm thickness was attached on to SWCNTs with the catalyst of CuCl/PMDETA in acetone at 60 °C for 96 h. N-doped CNTs can also be functionalized with PS using as initiator CNT-Br-7 and catalyst CuBr/4,4'-dinonyl-2,2'-bipyridine (dNbpy), yielding 3–5 nm thickness of polymer layer.[32] Besides, water soluble poly(sodium 4-styrenesulfonate) (PSS) was successfully grafted from MWCNT surfaces in DMF at 130 °C for 30 h with the initiator CNT-Br-1 and the catalyst CuBr/PMDETA. The content of grafted PSS varied from 25 to 68 wt% when the molar feed ratio of SSNa to CNT-Br-1 increased from 20:1 to 100:1. The PSS-grafted MWCNTs are highly soluble in water, as demonstrated by UV/Vis spectra.

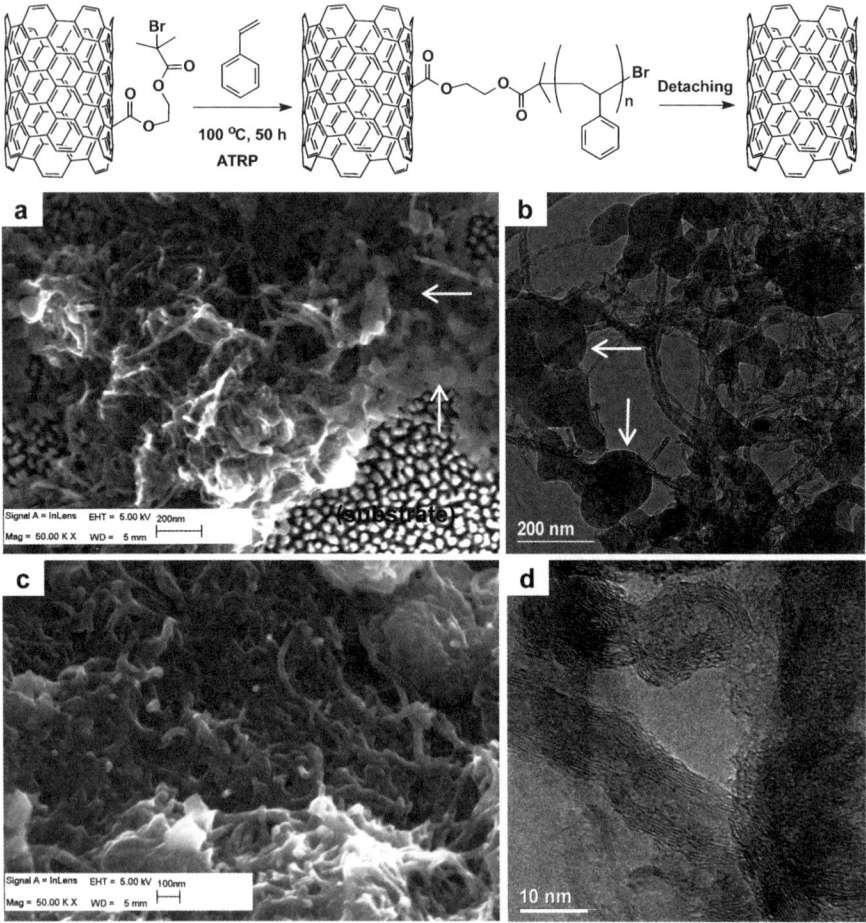

Figure 5.13 Synthesis and detaching of PS-grafted CNTs (top). SEM (a) and TEM (b) images of detached sample of MWCNT-g-PS mixed with PS (the arrows denote the mixed PS spheres), SEM image of MWCNT-g-PS in which free PS was removed completely (c), and TEM image of thermally decomposed MWCNT-g-PS at 500 °C, indicating the destruction of tubular structures (d). Reprinted with permission from Qin et al.[49]

5.1.1.4 In Situ *Polymerization of Acrylamides*

Poly(*N*-isopropylacrylamide) (PNIPAAm), a temperature-sensitive material, was grafted from CNTs by Gao and co-workers[51] through ATRP in water using CNT-Br-1 as the macroinitiator and CuBr/PMDETA as the catalyst. Similar to the cases of methacrylates and styrenics mentioned above, the content of grafted PNIPAAm can be adjusted by the feed ratio of monomer to CNT-Br-1, and improved from 51 to 84 wt% when the molar feed ratio increased from 40:1 to 200:1. HRTEM measurements confirmed the high and even grafting efficiency, as shown in Figure 5.14. The resulting nanohybrids were still thermo-sensitive, which were demonstrated by AFM, NMR, UV/Vis and differential scanning calorimetry (DSC) analyses. The transmittance measurements with a UV/Vis spectrometer indicated that the lower critical solution temperature (LCST) of the nanohybrids is 34 °C, 3 °C lower than that of the pure PNIPAAm obtained at the same condition. The thermoresponsive behavior of hydrophilic/hydrophobic phase transition is highly reversible and very fast, as detected by multi-cycle scanning by DSC (Figure 5.15).

Alternatively, PNIPAAm was grafted on to aligned MWCNTs, generating a smart surface with thermal-responsive suprahydrophilic and suprahydrophobic transition behavior.[52] CNT-initiated ATRP of another acrylamide class of monomers has rarely reported yet, leaving space for the future work. Preparation of smart nanomaterials with both temperature and pH responsibility and exploration of the application of the smart nanowires are especially expected.

In addition, Liu and Chen[53] prepared a MWCNT-based macroinitiator through an addition reaction between the ATRP initiator 1-bromoethylbenzene or the macroinitiator of bromine-terminated PS, and CNTs with CuBr/PMDETA as the catalyst system at 80 °C for 24 h. Further ATRP of styrene or NIPAAm on the functionalized CNTs afforded PS-grafted CNTs or V-shaped amphiphilic polymer PS-PNIPAAm-grafted CNTs which showed Janus

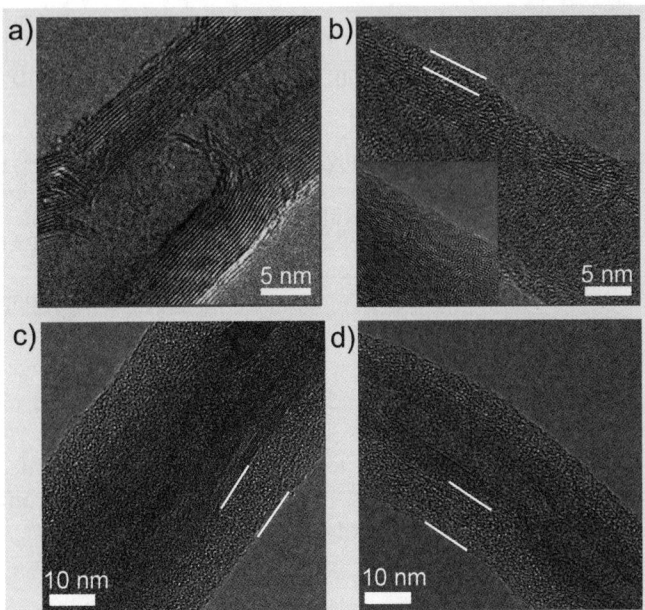

Figure 5.14 HRTEM images of pristine MWCNT (a), PNIPAAm-grafted MWCNT with polymer contents of 51 wt% (b), 68 wt% (c), and 84 wt% (d). Inset of (b): local amplification of (b). The marked polymer shells in images (b)–(d) have thicknesses of 3, 11 and 14 nm, respectively. Reprinted with permission from Kong et al.[51]

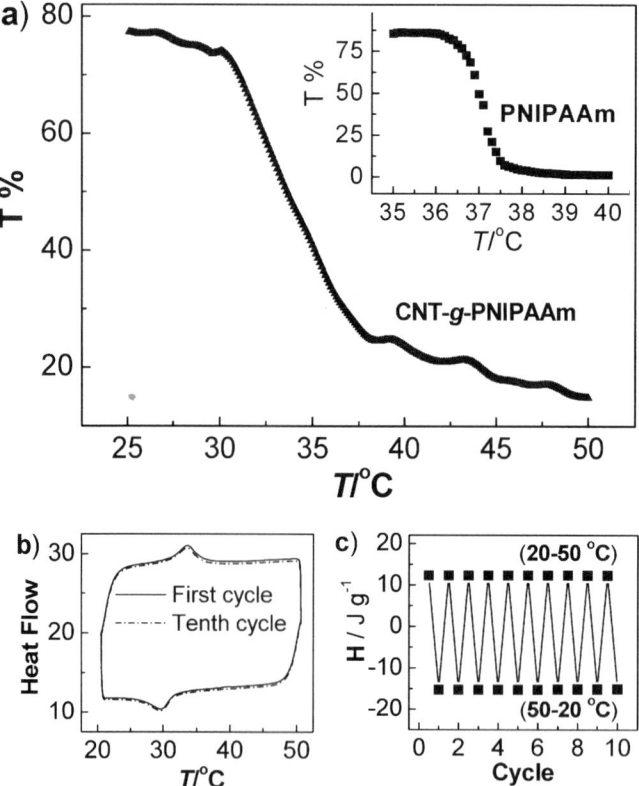

Figure 5.15 (a) Transmittance of PNIPAAm-grafted MWCNTs (84 wt% of polymer) or neat PNIPAAm (M_n = 15 200, PDI = 1.45) (inset) in water as a function of temperature with a concentration of 1 mg mL^{-1}. (b) DSC curves of the first and the tenth cycles for the PNIPAAm-grafted MWCNTs with a sample concentration of ca. 1 mg mL^{-1}. (c) Enthalpy of transition for the PNIPAAm-grafted MWCNTs as a function of cycle in DSC measurements. Reprinted with permission from Kong et al.[51]

behavior at the interface of water and chloroform at 20 °C and moved into chloroform phase at 50 °C.

The CNT surface-initiated ATRP technique can be also extended to other carbon surfaces such as carbon fibers,[54] carbon spheres,[55] carbon onions[56] and carbon black[57–59] to grow various polymers from their surfaces.

5.1.2 Reversible-addition Fragmentation Chain-transfer (RAFT) Polymerization Approach to Polymer-grafted CNTs

RAFT polymerization is another powerful controlled/living radical polymerization technique for surface modification.[60–62] The generally accepted mechanism of the RAFT polymerization is depicted in Scheme 5.4.[63–67] The

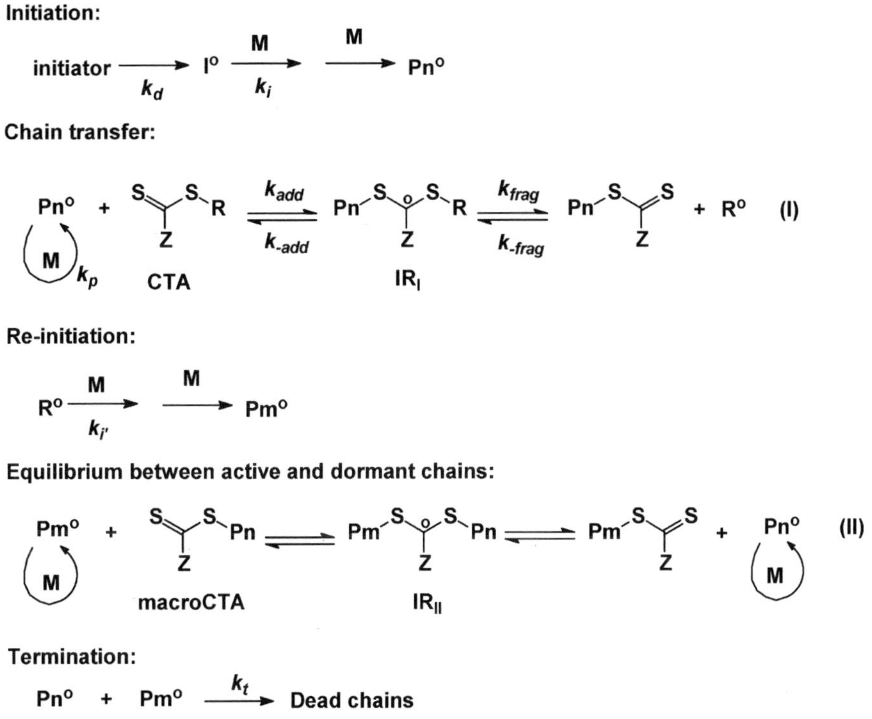

Initiation:

$$\text{initiator} \xrightarrow{k_d} I^\circ \xrightarrow{M}{k_i} \xrightarrow{M} Pn^\circ$$

Chain transfer:

$$Pn^\circ \;(M)\!\!\downarrow\!k_p \; + \; \underset{CTA}{\overset{S}{\underset{Z}{\diagup}}\!\!=\!\!\overset{S}{\diagdown}\!R} \;\underset{k_{-add}}{\overset{k_{add}}{\rightleftarrows}} \; \underset{IR_I}{Pn\!-\!S\!-\!\underset{Z}{C}\!-\!S\!-\!R} \;\underset{k_{-frag}}{\overset{k_{frag}}{\rightleftarrows}} \; Pn\!-\!S\!-\!\underset{Z}{\overset{S}{C}} \; + \; R^\circ \quad (I)$$

Re-initiation:

$$R^\circ \xrightarrow{M}{k_{i'}} \xrightarrow{M} Pm^\circ$$

Equilibrium between active and dormant chains:

$$Pm^\circ \;(M)\!\!\downarrow \; + \; \underset{macroCTA}{\overset{S}{\underset{Z}{\diagup}}\!\!=\!\!\overset{S}{\diagdown}\!Pn} \;\rightleftarrows\; \underset{IR_{II}}{Pm\!-\!S\!-\!\underset{Z}{C}\!-\!S\!-\!Pn} \;\rightleftarrows\; Pm\!-\!S\!-\!\underset{Z}{\overset{S}{C}} \; + \; Pn^\circ \;(M)\!\!\downarrow \quad (II)$$

Termination:

$$Pn^\circ \; + \; Pm^\circ \xrightarrow{k_t} \text{Dead chains}$$

Scheme 5.4 General mechanism of RAFT polymerization.

process is initiated by a free-radical source that used in conventional free radical polymerization (*e.g.*, thermoinitiator and photoinitiator). After initiating, an active propagating radical $Pn\bullet$ is produced. Addition of the $Pn\bullet$ to a chain transfer agent (CTA) possessing a thiocarbonyl-thio group produces an adduct radical which fragments to a dormant species of polymeric thiocarbonyl-thio compound and a new fragment radical $R\bullet$ (equilibrium I, sometimes called pre-equilibrium). Then, a new propagating radical $Pm\bullet$ forms by reinitiation of the radical $R\bullet$. The dormant species containing a thiocarbonyl-thio function (macroCTA) induces another reversible addition–fragmentation equilibrium (equilibrium II). The polymer chains successively pass from a dormant state to an active state during which they can add monomer units. Equilibration of the growing chains leads to a narrow molecular weight distribution and controlled molecular weight. In such a RAFT polymerization, the amount of initiator (the only free-radical source), and thus the concentration of active species are low and the termination reactions are minimized.[66]

In order to graft polymer from CNTs surfaces by RAFT polymerization, the CTAs should be immobilized on the surface firstly (which is quite similar to the cases of ATRP to produce the CNT-based macroinitiator). Various approaches to macroCTAs are shown in Scheme 5.5, and the results are

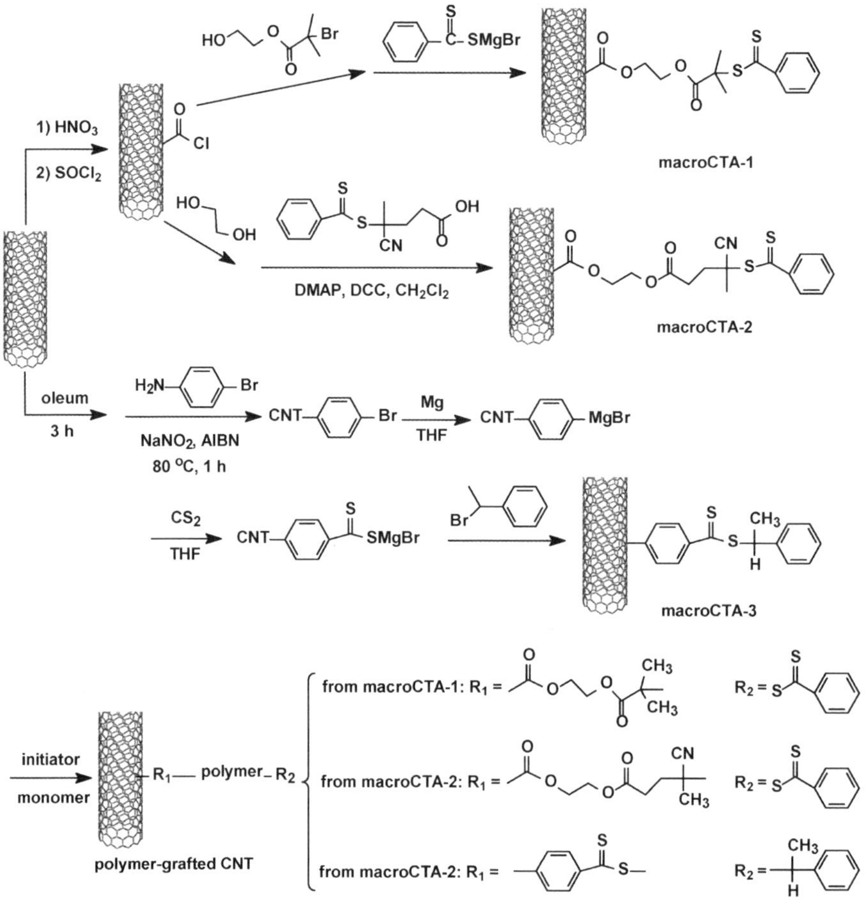

Scheme 5.5 Preparation of thiocarbonylthio group-functionalized CNTs (macroCTA) and polymer-grafted CNTs by RAFT polymerization.[6]

summarized in Table 5.2. By reaction of acyl chloride-functionalized MWCNTs (MWCNT-COCl) with hydroxyethyl-2-bromoisobutyrate and then PhC(S)SMgBr, Wang and co-workers[68] prepared dithiobenzoateimmobilized (macroCTA-1), and grafted PS from CNT surfaces using azobisisobutyronitrile (AIBN) as the initiator in THF at 100 °C for 24 h. Various polymers have been successfully grafted from CNTs in the presence of macroCTA-1 by *in situ* RAFT polymerization (Table 5.2). PS was also grafted from vertically aligned SWCNTs covalently attached to a silicon substrate (macroCTA-5) by Macdonald *et al.*[69] Block copolymers could be grafted from CNTs because of the living nature of RAFT polymerization. For instance, PMMA-*b*-PS[70] and PS-*b*-PNIPAAm[71] were grafted from MWCNTs by subsequent RAFT polymerization of corresponding monomers. The polymer-grafted MWCNTs were characterized by TGA, Fourier transform infrared (FTIR) spectroscopy,

Table 5.2 Functionalization of CNTs with polymers by RAFT polymerization initiated by AIBN.[6]

MacroCTA	Monomer	Solvent/temperature/time	Highest f_{wt}%	Ref.
MacroCTA-1	Styrene	THF/100 °C/24 h	80	68
	NIPAAm	Benzene/100 °C/36 h	87	71,78
	HPMAm	Methanol/60 °C/28 h	80	79
	DMAEMA	60 °C	65	80
	Acrylic acid	60 °C	40	80
	MDMAS	60 °C/20 h	70	80
	Styrene + maleic anhydride	THF/80 °C/–	42	81
	MMA+PS	THF/90 °C/48 h + DMF/90 °C/24 h	75	70
	PS+NIPAAm	THF/90 °C/24 h + benzene/90 °C/36 h	86	72
MacroCTA-2	HEMA	DMF/70 °C/32 h	78.4	72
MacroCTA-3	Acrylamide	Benzene/60 °C/50 h	77.9	74
MacroCTA-4	Styrene	DMF/100 °C /24 h	68.7	73
MacroCTA-5	Styrene	Toluene/70 °C /48 h	–	69
MacroCTA-6	N-Vinylcarbazole	THF/70 °C /24 h	96.4	82

SEM and TEM. The grafted polymer amount also increased with the reaction time, demonstrating efficient control over the RAFT process with the dithiobenzoate-immobilized CNTs as the CTAs.

Another MWCNT-based macroCTA (macroCTA-2) with a CTA group concentration of 0.244 mmol g^{-1} was synthesized by Pei et al.[72] using different chemistry; PHEMA was grafted from MWCNTs via RAFT polymerization. The grafted polymer amount increased from 26.3 to 78.4 wt% when the reaction time increased from 8 to 32 h. Hydrolysis of the PHEMA chains in hydrochloride aqueous solution for 12 h gave rise to poly(methacrylic acid) (PMAA)-grafted MWCNTs. MWCNT–Ag nanohybrids were then prepared using the PMAA-grafted MWCNTs as the templates.[72] They also prepared another macroCTA (macroCTA-4) for the RAFT grafting of PS.[73]

Huang and co-workers[74] prepared dithiobenzoate-immobilized SWCNTs (macroCTA-3) through an alternative reaction route as shown in Scheme 5.5, and then enwrapped the SWCNTs covalently with water-soluble poly(acrylamide) by RAFT polymerization in benzene at 60 °C. The grafted poly(acrylamide) amount increased linearly with the reaction time, and rose from 46.3 to 77.9 wt% when the polymerization time was prolonged from 12 to 50 h.

Therefore, RAFT polymerization is also a powerful technique to functionalize both MWCNTs and SWCNTs.[75–77] Conventional monomers, such as NIPAAm,[71,78] N-(2-hydroxypropyl)methacrylamide (HPMAm),[79] DMAEMA,[80] acrylic acid,[80] 3-[N-(3-methacrylamidopropyl)- N,N-dimethyl] ammoniopropane sulfonate (MDMAS),[80] styrene,[68,69,73] maleic anhydride[81] and N-vinylcarbazole (macroCTA-6)[82] were used in RAFT polymerization in the presence of a macroCTA or a CNT-based RAFT agent. However, it is noteworthy that a monomer always needs a specific CTA to achieve better control over the polymerization. In other words, a CTA cannot work well for all monomers. This should be considered in the molecular design to functionalize CNTs by the RAFT technique.

5.1.3 Nitroxide-mediated Radical Polymerization (NMRP) Approach to Polymer-grafted CNTs

In addition to ATRP and RAFT, NMRP is also a useful technique to synthesize well-defined polymers and block copolymers and to modify solid surfaces.[83–86] Functionalization of CNTs by the NMRP "grafting from" approach has also been developed (see Scheme 5.6 and Table 5.3). Dehonor et al.[87] grafted PS from nitrogen-doped MWCNTs (N-d-MWCNT) by NMRP at 130 °C for 24 h using 2,2,6,6-tetramethylpiperidinyl-1-oxyl (TEMPO)-immobilized N-d-MWCNT as the macroinitiator. Measurements by TEM, electron energy-loss spectroscopy (EELS), FTIR spectroscopy and Raman spectroscopy demonstrated the grafting on to the CNT sidewalls. TGA analyses showed 10 and 35 wt% weight losses for the macroinitiator and PS-grafted nanotubes, respectively.

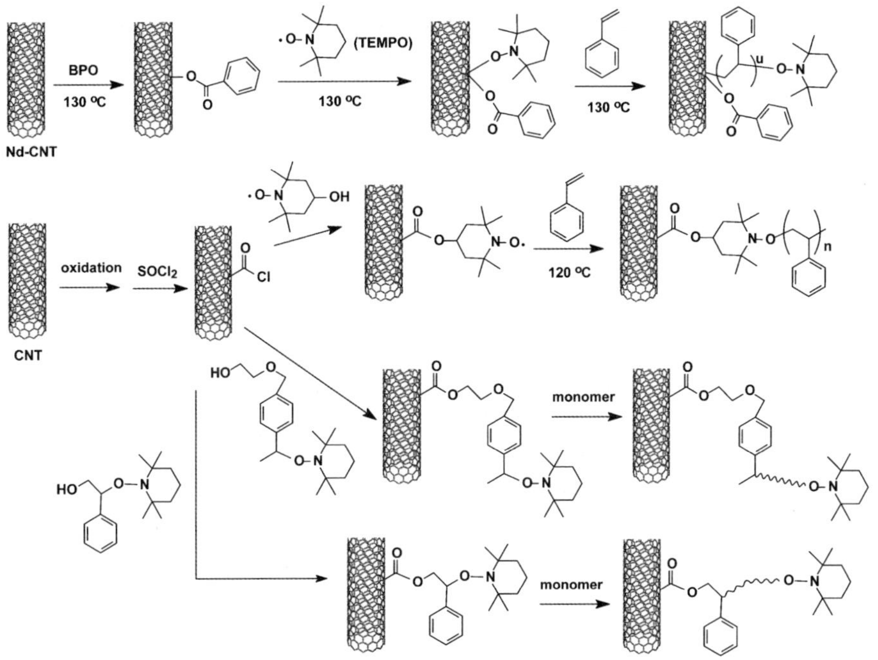

Scheme 5.6 Preparation of TEMPO-functionalized CNTs (macroinitiator of NMRP) and polymer-grafted CNTs by NMRP.[6]

Alternatively, Chang et al.[88] prepared MWCNT-PE-TEMPO by the reaction of HO-PE-TEMPO with MWCNT-COCl, and used it to initiate bulk polymerization of styrene at 130 °C for 4 h, yielding copolymer-grafted MWCNT.

Zhao et al.[89] prepared TEMPO-functionalized MWCNTs with a TEMPO concentration of 0.46 mmol g^{-1} and subsequently grafted PS from the CNT surfaces through NMRP in chlorobenzene at 125 °C for 48 h. Because of the living nature of NMRP, the grafted polymer amount can also be adjusted by the feed ratio of monomer to the macroinitiator; the PS-grafted MWCNTs can be employed as the macroinitiator to initiate the polymerization of 4-vinylpyridine (4VP), giving rise to P4VP-b-PS-grafted MWCNTs.[89] Using the same macroinitiator, water-soluble polymers of P4VP and PSS were grown from the MWCNT surfaces successfully.[90] By reaction of 4-hydroxyl-TEMPO with MWCNT-COCl, He et al. also obtained TEMPO-linked MWCNTs with a TEMPO concentration of 0.36 mmol g^{-1}, and conducted the surface-initiated NMRP of styrene in bulk at 120 °C for 24 h.[91] It was shown that the controllability of the heterogenous reaction is not as good as that of homogeneous systems. Recently, polymer-grafted MWCNTs made by NMRP were used as sensors to detect organic vapors,[92] demonstrating the potential application of the polymer-functionalized CNTs.

Table 5.3 Grafting polymer from CNT surfaces by NMRP technique.[6]

Macroinitiator	Monomer	Solvent/temperature/time	Highest fwt%	Ref.
	Styrene	Xylene/130 °C/24 h	35	87
	Styrene	Chlorobenzene/125 °C/48 h	61.4	89
	Styrene + 4-vinylpyridine	DMF/125 °C/48 h	61	89
	4-vinylpyridine	DMF/125 °C/24 h	40	90
	Sodium 4-styrenesulfonate	DMF/125 °C/24 h		90
	Styrene	Bulk/120 °C/24 h	30.7	91
	Styrene	DMF/130 °C/4 h	20	88

5.1.4 Ring-opening Polymerization (ROP) Approach to Polymer-grafted CNTs

Through ROP, both linear polymers and hyperbranched polymers (HPs) can be covalently grafted from CNT surfaces. The used monomers and grafting results are summarized in Table 5.4. By *in situ* cationic ROP of THF, polytetramethylene ether (PTME) was chemically grafted from MWCNTs starting from the acyl chloride-functionalized MWCNTs with silver perchlorate as a catalyst. The study on interfacial polymer dynamics of PTME-functionalized MWCNTs showed an unusual rate dependence of T_g, that is, T_g occurred at lower temperatures in a faster heating process.[93] By anionic ROP of ε-caprolactam in bulk catalyzed by sodium in the presence ε-caprolactam-functionalized SWCNTs, Sun and co-workers[94] prepared polyamide 6 (PA6)- or nylon 6-grafted SWCNTs with a polymer content of 60 wt%. The fine solubility of the nylon 6-grafted SWCNTs in selected organic solvents promised potential applications in high-performance nanocomposites. Similarly, Li and co-workers[95] synthesized nylon 6-grafted MWCNTs with a polymer content of 65 wt% by anionic ROP in bulk at 170 °C for 6 h in the presence of sodium caprolactamate as a catalyst and caprolactam-functionalized MWCNTs as the initiator. The initiator precursor, isocyanate-functionalized MWCNTs, was prepared by reacting commercial hydroxyl-functionalized MWCNTs with excess toluene 2,4-di-isocyanate. Alternatively, MWCNTs were covalently functionalized with copoly(styrene-maleic anhydride) via free radical polymerization, and nylon 6 was subsequently grafted from MWCNTs by anionic ROP of ε-caprolactam at 170 °C for 6 h.[96] The grafting efficiency was confirmed by TEM.

Choi and co-workers[97] synthesized 6-amino-1-hexanol-immobilized SWCNTs by immersing the acyl chloride-functionalized SWCNTs into a DMF solution of 6-amino-1-hexanol for 12 h, and grafted poly(p-dioxanone) (PDX) from the SWCNT surfaces by in situ ROP in toluene at 100 °C for 3 days with tin(II) 2-ethylhexanoate [$Sn(Oct)_2$] as the catalyst. The 10% weight-loss temperature of grafted polymer is higher than that of free polymer by approximately 20 °C.

Due to its biodegradability and biocompatibility, poly(ε-caprolactone) (PCL) was used to functionalize CNTs. Resasco and co-workers[98] synthesized 4-hydroxymethyl aniline-functionalized SWCNTs via the diazonium salt method, and then grafted PCL from the nanotubes surfaces with a polymer content of 63 wt% by in situ ROP with $Sn(Oct)_2$ as the catalyst in DCB at 130 °C for 24 h. With the same catalyst, Gao and co-workers grafted PCL from MWCNT surfaces in bulk at 120 °C using the glycol-functionalized MWCNTs (MWCNT-OH) as the macroinitiator, and investigated the biodegradability of the PCL grafted on the tubes.[99] The grafted PCL can be completely enzymatically degraded within 4 days in a phosphate buffer solution in the presence of Pseudomonas lipase (PS lipase), and the CNTs still retain their tube-like morphologies, as observed by SEM and TEM (see Figures 5.16 and 5.17).[99] The PCL-grafted MWCNTs (MWCNT-g-PCL) have potential applications in biomedicine and artificial bones. Similarly, PCL could be grafted from eosin Y-functionalized MWCNTs by bulk polymerization at 140 °C for 1 h.[100] MWCNT-g-PCL could be added into a polyurethane (PU) matrix as a nanofiller to fabricate MWCNT–PU composite fibers by electrospinning.[101] Lee and co-workers[102] prepared azide-functionalized PCL grafted CNTs by surface-initiated ROP of poly(α-chloro-ε-caprolactone) (PαClCL) followed by reaction with NaN_3.

Yang et al.[103] developed a green chemical functionalization method using water and room temperature ionic liquids as the reaction media. By in situ ROP of ε-caprolactone from hydroxyl-modified MWCNTs in ionic liquids of 1-butyl-3-methylimidazolium tetrafluoroborate, MWCNT-g-PCL was obtained. When the reaction time was increased from 2 to 8 h, the PCL fraction improved from 30.6 to 62.7 wt%.[103]

Another biodegradable polymer, poly(L-lactide) (PLLA), was also grafted from MWCNT surfaces successfully using $Sn(Oct)_2$ as the catalyst.[104,105] Upon

Table 5.4 Selected conditions and results of polymer-grafted CNTs *via* the ROP approach.[6]

Macroinitiator	Monomer	Catalyst	Solvent/ temperature/time	f_{wt}%	Ref.
CNT-C(O)-O-CH₂CH₂-OH	oxetane-CH₂OH (with ethyl)	BF₃·OEt₂	CH₂Cl₂/ −10 °C- RT/24 h	87	163
CNT-C(O)-Cl	THF	AgClO₄	CH₂Cl₂/ 0 °C/42 h	33	93
CNT-C(O)-N-caprolactam	caprolactam	Na	Bulk/ 140 °C/ 24 h	60	94
CNT-C(O)-NH(CH₂)₆-OH	1,4-dioxan-2-one	Sn(Oct)₂	Toluene/ 90-100 °C/ 3 days	40	97
CNT-C₆H₄-CH₂OH	ε-caprolactone	Sn(Oct)₂	DCB/ 130 °C/ 24 h	63	98
CNT-C(O)-O-CH₂CH₂-OH	ε-caprolactone	Sn(Oct)₂	Bulk/ 120 °C/ 24 h	52.7	99
CNT-C(O)-O-CH₂CH₂-OH	lactide	Sn(Oct)₂			104
CNT-C(O)-O(CH₂)₄-OH	lactide	Sn(Oct)₂	DMF/ 140 °C/ 20 h	35	106
CNT-C(O)-NH-CH₂CH₂-NH₂	BLG-NCA	–	CHCl₃/ 30 °C/ 7 days	85	109

Table 5.4 (*Continued*)

Macroinitiator	Monomer	Catalyst	Solvent/ temperature/time	$f_{wt}\%$	Ref.
CNT-Fe$_3$O$_4$-OH	(lactide)	Sn(Oct)2	Bulk/ 130 °C/ 48 h	33.7	108
CNT-COO(CH$_2$)$_4$OH	Macrocyclic ester	Sb$_2$O$_3$	Bulk/ 200 °C/ 30 min		111
CNT-(naphthyl)-OH	(caprolactone) or (lactide)	Sn(Oct)$_2$	Toluene/ 200 °C/ 24 h	43	105
CNT-(CN)$_2$-OH	(caprolactone)	Sn(Oct)$_2$	BmimBF4/ 200 °C/ 8 h	62.7	103
Caprolactam-functionalized CNTs	(caprolactam)	Sodium caprolactamate	Bulk/ 170 °C/ 6 h	65	95
Eosin Y-functionalized CNTs	(caprolactone)	Stannous (II)-2-ethyl hexanoate	Bulk/ 140 °C/ 4 h	92.3	100
CNT-COOH	(epichlorohydrin)	NaOH	H$_2$O/80 °C/ 3 h	79.1	110
CNT-OH	αClCL	Sn(Oct)$_2$	DMF/ 140 °C/ 24 h	85.6	102
CNT-Sn	Macrocyclic ester	Bu$_2$SnO	Bulk/ 200 °C/ 30 min	59.3	112

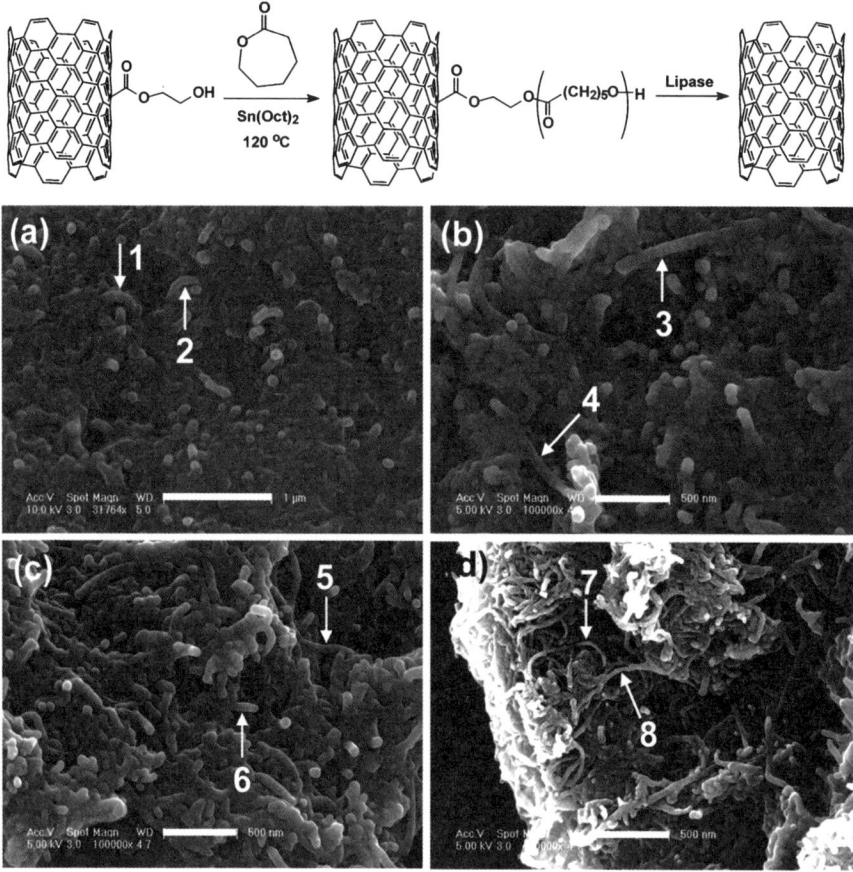

Figure 5.16 Synthesis of PCL-grafted CNTs by the ROP approach and degradation of PCL chains by PS lipase (top), and representative SEM images of the control experiment sample which was collected from the phosphate buffer solution of MWCNT-*g*-PCL (52.7 wt% of PCL) without PS lipase after 96 h (a), and the samples of MWCNT-*g*-PCL degraded with PS lipase for 24 h (b), 48 h (c) and 96 h (d). The scale bars in (a), (b), (c) and (d) represent 1000, 500, 500 and 500 nm, respectively. The marked tubes 1 and 2 have diameters of *ca.* 90.3 and 85.5 nm, respectively. After 24 h of degradation with PS lipase (b), the diameters of tubes 3 and 4 are *ca.* 83.3 and 75.7 nm, respectively. After 48 h of degradation (c), the diameters of tubes 5 and 6 are only *ca.* 54.1 and 50.5 nm, respectively. After 96 h of degradation (d), little residual polymer on the surfaces of MWCNTs can be detected, and many tubes with diameters of *ca.* 20–30 nm are found (tubes 7 and 8 have diameters of 22.7 and 28.2 nm, respectively). In this case, the average diameter of the degraded tubes is almost equal to that of the oxidized tubes. Reprinted with permission from Zeng *et al.*[99]

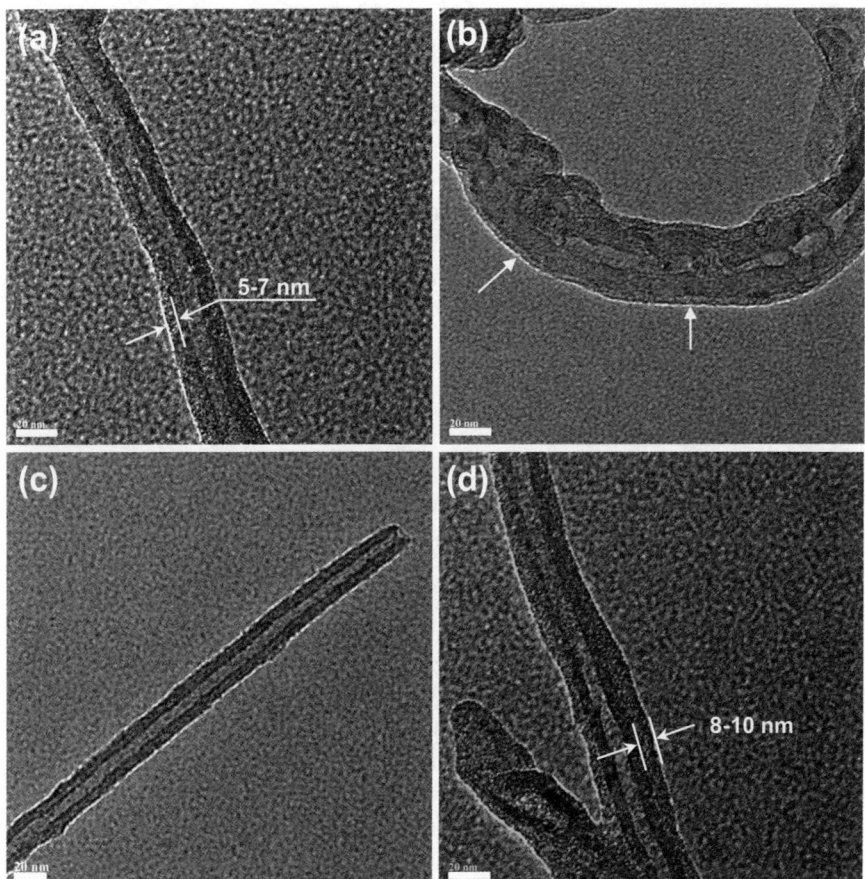

Figure 5.17 Representative HRTEM images of the samples of MWCNT-g-PCL (52.7 wt% of PCL) degraded with PS lipase for 24 h (a), 48 h (b) and 96 h (c), and the control experiment sample (d). The scale bars in (a–d) represent 20 nm. The polymer shell of MWCNT-g-PCL had a thickness of ca. 5–7 nm after 24 h of degradation (a), and decreased to 3–4 nm after 48 h (b). No polymer shell was observed in the sample after 96 h of degradation with the PS lipase (c). In contrast, the thick polymer shell (8–10 nm) still wraps the tubes for the sample from the control experiment without the lipase (d). Reprinted with permission from Zeng et al.[99]

incorporation of 1 wt% of the MWCNT-g-PLLA (35 wt% of PLLA), the tensile modulus of PLLA increased from ~2.46 to 4.71 GPa and the strength was enhanced from 56.4 to 85.6 MPa, which indicates 91% and 52% improvements, respectively.[106] The initial moduli and tensile strengths of PLLA/MWCNT-g-PLLA composites increased with increasing the PLLA chain length due to the better dispersibility.[107] PLLA was grafted from magnetic MWCNTs (m-MWCNTs) by in situ ROP of lactide using Sn(Oct)$_2$ as the catalyst.[108] The grafted polymer weight percentage was ca. 25.6–33.7 wt%.

Yao et al.[109] prepared polypeptide-grafted MWCNTs by ROP of benzyl-L-glutamate N-carboxyanhydride (BLG-NCA) initiated with amino-functionalized MWCNTs. TEM measurements showed that the thickness of the polypeptide shell was about 4.5–22 nm, depending on the grafted polypeptide amount.

Park and co-workers[110] reported surface-initiated epoxide ROP. The grafted polymer amount varied from 14 to 74 wt% when the reaction temperature was increased from 60 to 80 °C. Nanocomposites of poly(hexamethylene terephthalate) (PHT) and MWCNTs by dispersing MWCNT or MWCNT-OH into PHT through in situ ROP of cyclic hexamethylene oligomers and melt blending were prepared by Martinez and co-workers.[111] The thermal stability of the nanocomposites was slightly improved and nanocomposites crystallized at higher rates than neat PHT, depending on the amount of nanotubes loading.

Wu and Yang[112] prepared poly(butylene terephthalate) (PBT)-grafted MWCNTs by ROP of cyclic butylene terephthalate (CBT) oligomers initiated with Sn-functionalized MWCNTs in bulk at 200 °C for 30 min. TEM measurements showed that the thickness of the PBT shell was approximately 6 nm. The resulting MWCNT-g-PBT could be well dispersed in phenol/tetrachloroethane (w/w = 60:40) mixed solvent.

Sakellariou et al.[113] grafted poly(ethylene oxide) (PEO)-grafted MWCNTs by surface-initiated anionic ROP of ethylene oxide with 4-hydroxyethyl benzocyclobutene (BCB-EO)-functionalized MWCNTs as the initiator (Scheme 5.7). The initiation was found to be very slow due to the heterogeneous reaction medium. Nevertheless, the MWCNT material was gradually dispersed into the reaction medium during the polymerization, and a pale green color was observed in the case of ethylene oxide polymerization. The conversion of ethylene oxide and the weight fraction of PEO grafted on MWCNTs increased with increasing reaction time, indicating a controlled polymerization. PEO-grafted nanocomposites, MWCNTs-g-(BCB-PEO)n, containing a very high percentage of hairy polymer with a small fraction of MWCNTs (<1 wt%) were obtained. The same initiator was also used to prepare MWCNT-g-PCL by anionic polymerization of ε-caprolactone.[114] TGA measurements showed that the polymerization proceeded very fast as compared with the case of ethylene oxide, and that both polymerizations could be controlled through the reaction time. DSC results showed that a remarkable nucleation effect was produced by MWCNTs that reduced the supercooling needed for crystallization of both PCL and PEO, and that the isothermal crystallization kinetics of the grafted PCL and PEO was substantially accelerated compared with the neat polymers. The strong impact on the nucleation and crystallization kinetics was attributed to the covalent linkage between polymers and CNTs.

The ROP technique has also been extended to functionalize other carbon materials such as C60,[115] carbon black[116] and carbon fibers.[117,118]

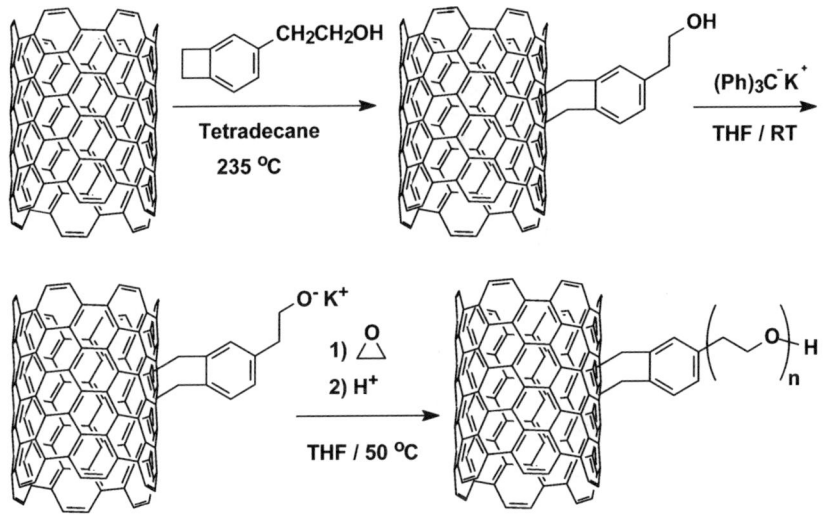

Scheme 5.7 Syntheis of PEG-grafted CNTs by the anionic ROP approach.[113]

5.1.5 Ring-opening Metathesis Polymerization (ROMP) Approach to Polymer-grafted CNTs

ROMP has been demonstrated as a versatile and effective technique to prepare polymers from cyclic olefins.[119–124] Dai and co-workers[125] first ran ROMP on CNT surfaces by physisorbing the ruthenium alkylidene catalyst (or initiator) on to the SWCNTs using pyrene functionality as an anchor. Liu and Adronov[126] functionalized SWCNTs with the ruthenium alkylidene catalyst benzylidenebis(tricyclohexylphosphine)dichloro ruthenium and then grafted polynorbornene from SWCNTs by ROMP in $CHCl_3$ at room temperature, as depicted in Scheme 5.8. NMR, AFM and TEM measurements demonstrated the structure and morphology of the polynorbornene-grafted SWCNTs. The molecular weight M_n of the polynorbornene—cleaved from the nanotubes by KOH/18-crown-6—increased linearly with the reaction time, showing the living nature of the surface-initiated ROMP.

Kessler and co-workers[127] synthesized norbornene-functionalized MWCNT (f-MWCNTs) (Scheme 5.9), and prepared bulk composites of f-MWCNTs/poly(dicyclopentadiene) (DCPD) by *in situ* ROMP of f-MWCNTs in the presence of DCPD and Grubbs' catalyst for 2 h at 70 °C followed by 90 min at 170 °C. A cross-linked polymeric network was instantaneously formed, and the mechanical properties with low loadings (<0.5 wt%) of f-MWCNTs were studied. A remarkable increase in tensile toughness (925% increase over the neat polyDCPD) with only 0.4 wt% f-MWCNTs was found, and the corresponding elongation at break significantly increased from 5.75 to 51.8%.[127] The DSC results demonstrated an increase in T_g with the addition of f-MWCNTs.[127] Using the same method, they subsequently

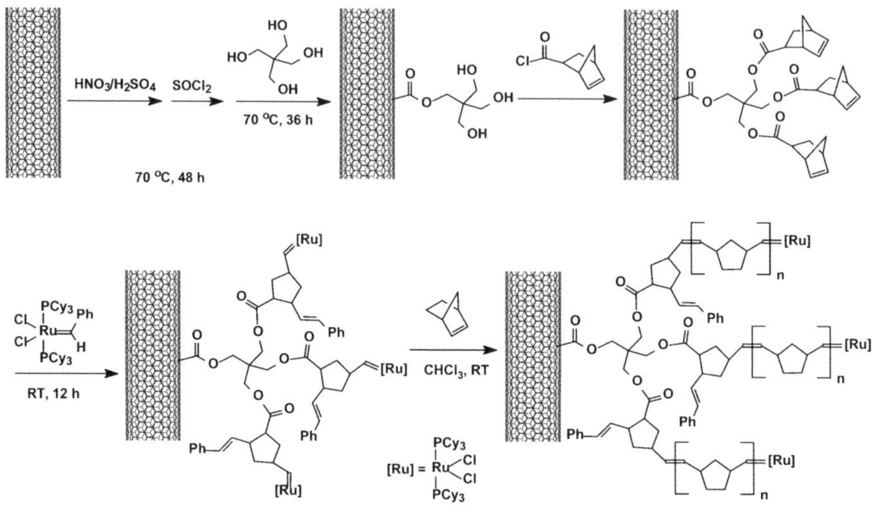

Scheme 5.8 Functionalization of SWNTs with polynorbornene by ROMP.[126]

prepared reinforced poly(5-ethylidene-2-norbornene) (ENB) f-MWCNTs composites.[128]

Izuhara and Swager[129] functionalized MWCNTs with Grubbs' second generation (G2) Ru catalyst and subsequently grafted electroactive block copolymer from the MWCNTs via ROMP in CH_2Cl_2 at room temperature by adding monomer **1** and monomer **2** in sequence (Scheme 5.10). The NMR, FTIR spectroscopy and TEM measurements confirmed the functionalization and morphology of MWCNTs. Gel permeation chromatography (GPC)

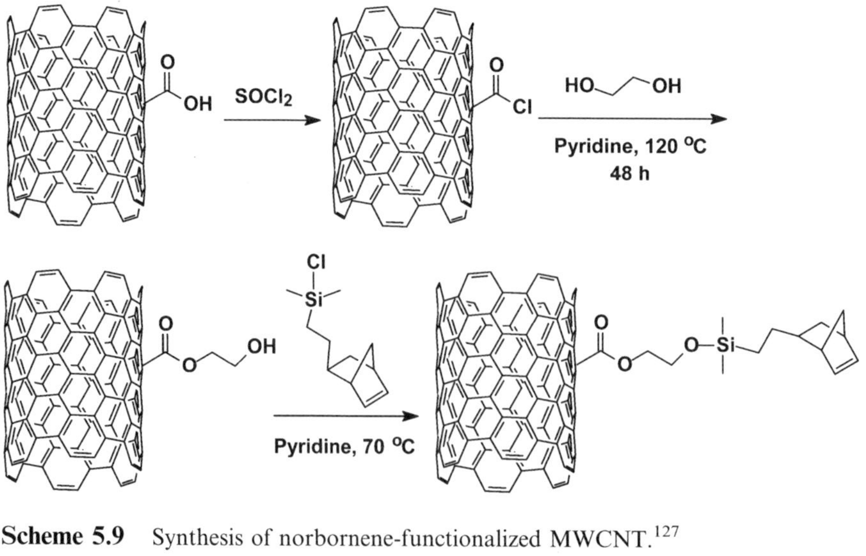

Scheme 5.9 Synthesis of norbornene-functionalized MWCNT.[127]

analysis of the filtrate through a 0.2 μm Teflon filter did not reveal the presence of unbound polymer. The M_n of the polynorbornene—cleaved from the nanotubes by KOH/18-crown-6—displayed similar molecular weights to those obtained from the homogeneously polymerized system. The electrochemical behavior and electrical conductivities were measured.[129]

Recently, Sannino and co-workers[130] studied the activity of Grubbs' catalyst- functionalized MWCNTs by the ROMP synthesis of polynorbornene. The activities of the 1st and 2nd generation Grubbs' catalysts loaded on to CNTs were found to be similar in the ROMP of 2-norbornene. The characterization of polynorbornene–CNT composites confirmed the good dispersibility of the nanotubes in the polymer matrix.[130]

5.1.6 Anionic Polymerization Approach to Polymer-grafted CNTs

Viswanathan et al.[131] prepared SWCNT-based carbanions by treating pristine SWCNTs with sec-butyllithium in purified cyclohexane by sonication, and subsequently grafted PS from the tube surfaces at 48 °C for 2 h under sonication by anionic polymerization. TGA results indicated that the content of PS grafted on to SWNTs is 10 wt%. The T_g value of PS increased up to 15 °C, with a very low nanotube loading of only 0.05 wt% in the PS-grafted composites. Similarly, Chen et al.[132] grafted P*t*BA and P*t*BA-*b*-PMMA from SWCNT surfaces using the carbanions as the initiator at –78 °C for 3 h, and the grafted polymer contents approached 29.1 and 47.4 wt%, respectively. The structure and morphology of the resulting polymer-functionalized SWCNTs were characterized by ^{1}H NMR, FTIR spectroscopy, TGA and AFM.

Billups and co-workers[133] prepared PMMA-grafted SWCNTs by *in situ* anionic polymerization in the presence of debundled nanotube salts serving as

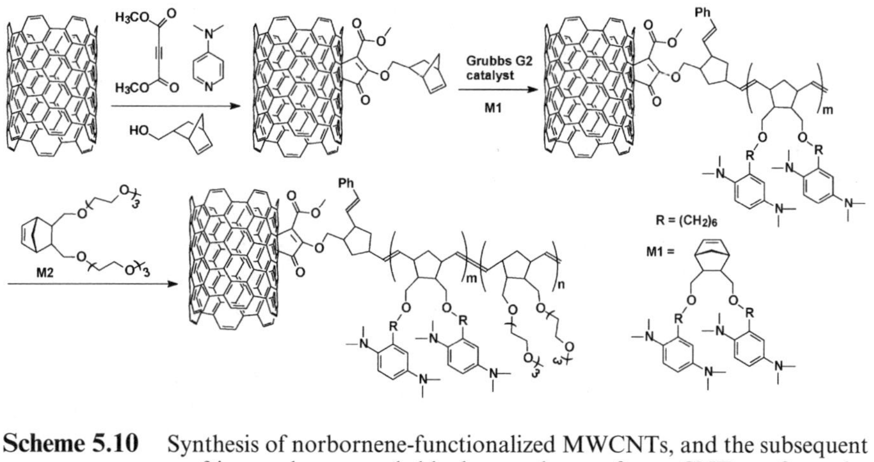

Scheme 5.10 Synthesis of norbornene-functionalized MWCNTs, and the subsequent grafting polymer and block copolymer from CNT surfaces by ROMP.[129]

initiators. The degree of polymerization was proportional to the amount of monomer that was used. The resulting PMMA-grafted SWCNTs were dispersible in chloroform, TH, and acetone. Monomers such as acrylonitrile and *t*BA were also tried to be grafted from SWCNTs by this process. In principle, this *in situ* polymerization is feasible for a wide range of vinyl monomers.[133]

Li *et al.*[134] synthesized poly(*N*-vinyl carbazole) grafted SWCNTs (SWCNT-*g*-PVK) by *in situ* anionic polymerization under the conditions of ultrasonic irradiation for 90 min, sonicating at 70 °C for 4 h and refluxing for 10 h in THF. SWCNT-*g*-PVK possesses excellent solubility in a range of common solvents. This soluble material exhibited an outstanding optical limiting response at 532 nm, making it a suitable candidate for viable optical limiting devices.

Baskaran and co-workers[113] performed anionic surface-initiated polymerization of styrene in the presence of anionic initiator-immobilized MWCNTs prepared by Diels–Alder cycloaddition of 1-benzocyclobutene-1'-phenylethylene (BCB-PE) and CNTs at 235 °C. The PS-grafted CNTs were characterized by FTIR spectroscopy, ^{1}H NMR, TGA and TEM. The thickness of grafted PS layer approached 30 nm (Figure 5.18).

Surface-initiated anionic polymerization provides an alternative approach to polyolefin-functionalized CNTs that would have potential applications in the reinforcement of widely used polyolefins such as PE and PP.

5.1.7 Other "Grafting From" Methods to Polymer-grafted CNTs

Other polymerization methods such as coordination polymerization,[135,136] surface electro-initiated emulsion polymerization,[137] radiation-induced graft polymerization (RIGP),[138–141] surface thiol-lactam-initiated radical polymerization (TLIRP)[142] and LbL click grafting[143] have been employed to graft polymers from CNT surfaces.

Bonduel *et al.*[135] prepared PE-coated MWCNTs by *in situ* coordination polymerization, using bis(pentamethyl-η5-cyclopentadienyl)zirconium(IV) dichloride ($Cp_2^*ZrCl_2$) as a typical polymerization catalyst from methylaluminoxane (MAO)-functionalized nanotubes at 50 °C, 2.7 bar for 1 h. The grafted PE amount was 72 wt%. The PE-grafted CNTs could be well dispersed in high-density polyethylene (HDPE) matrix by melt blending.[135]

Priftis *et al.*[136] used titanium alkoxide catalyst-functionalized MWCNTs for the surface-initiated titanium-mediated coordination polymerizations of L-lactide (LLA), ε-caprolactone (ε-CL) and *n*-hexyl isocyanate (HIC) at 120 °C, room temperature and 130 °C in toluene for 24 h, 20 h and 20 h, respectively. The grafted PLLA content could be controlled by the reaction time. In contrast with the other two monomers, HIC had very low conversions.

Irradiation generates free radicals *via* radiolysis of monomers or energy transfer through initiators and can also produce radicals on polymer surfaces. RIGP has been demonstrated as an effective technique to graft polymers from substrates.[138–141] Liu and co-workers[141] developed a single-step method of *in*

Polymer-grafted Carbon Nanotubes via "Grafting From" Approach

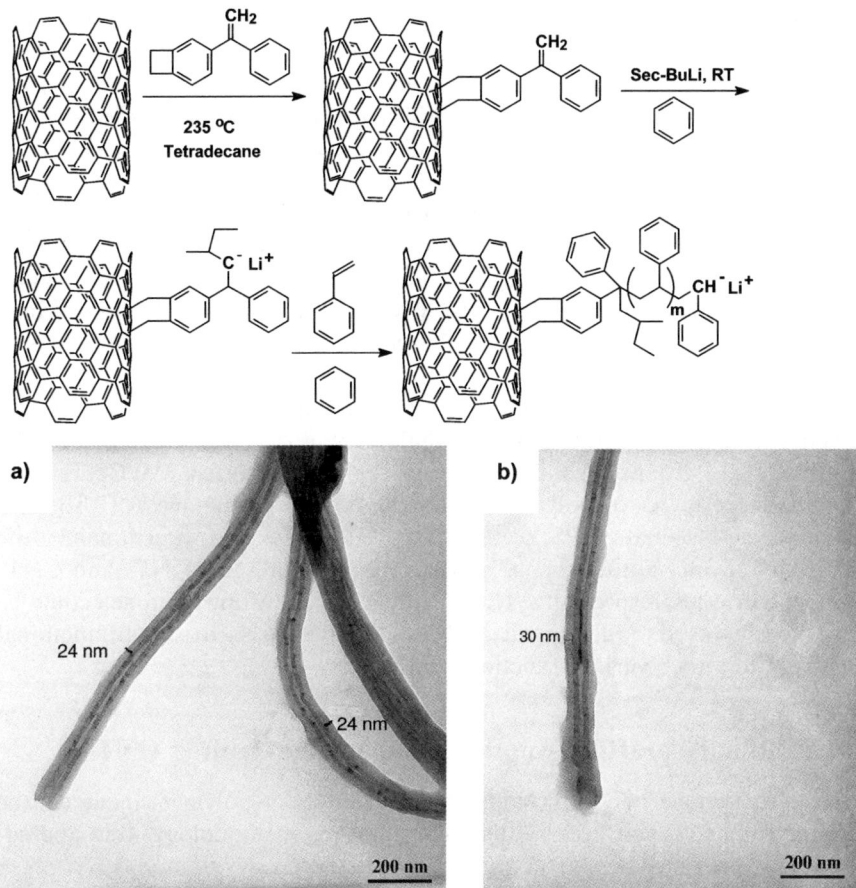

Figure 5.18 CNT surface-initiated anionic polymerization of styrene (top), and representative TEM images of PS-grafted MWCNTs with PS layer thickness up to 30 nm.[113]

situ ^{60}Co γ-ray irradiation, which induced PS grafting from MWCNT surface. Dispersed pristine MWCNTs in pure styrene were subjected to γ-ray irradiation, and PS was then covalently attached to the surface of the MWCNTs. Depending on the duration of irradiation, the grafted PS content could be up to ∼15 wt%. The PS-functionalized MWCNTs could be well dispersed in common organic solvents such as THF, toluene and CHCl$_3$.[141] By the same strategy, Choi and co-workers[139] conducted RIGP of various monomers including acrylic acid (AA), methacrylic acid (MA), GMA maleic anhydride (MAn) and 4-vinylphenyl-boronic acid (VPBA) from MWCNTs, and applied these MWCNTs in enzyme-free biosensing experiments. Petrov *et al.*[138] reported a UV-initiated free-radical polymerization strategy, yielding PAA-, PNIPAAm- poly[poly(ethylene glycol) methacrylate]- and poly(sodium methacrylate)-grafted MWCNTs.

Park and co-workers[142] presented a surface TLIRP to synthesize PS-functionalized MWCNTs in toluene at 100 °C for 24 h. FTIR spectroscopy, XPS, SEM and TEM measurements were employed to investigate the as-prepared core-shell nanostructure hybrid nanocomposites. The MWCNT–PS nanocomposites possess good dispersibility in common organic solvents.

Gao and co-workers[143] synthesized the clickable polymers of poly(2-azidoethyl methacrylate) and poly(propargyl methacrylate). The two polymers were then alternately coated on alkyne-modified MWCNTs using Cu(I)-catalyzed click reaction of Huisgen 1,3-dipolar cycloaddition between azides and alkynes (Scheme 5.11). TGA, SEM and TEM measurements confirmed that the quantity and thickness of the clicked polymer shell on MWCNTs could be well controlled by adjusting the cycles or numbers of click reaction and the polymer shell was uniform and even (Figure 5.19). XPS and FTIR spectroscopy measurements showed that there was still a great amount of residual azido groups on the surfaces of the functionalized MWCNTs after clicking three layers of polymers. Alkyne-modified rhodamine B (RhB) and monoalkyne-terminated PS were subsequently used to functionalize the clickable polymer-grafted MWCNTs, affording fluorescent CNTs and CNT-based PS brushes, respectively. Thus, the LbL click grafting approach could be not only employed to functionalize CNTs, but also was used as multifunctional platform to attach various functional molecules.

5.1.8 Binary-grafting Approach to Polymer-grafted CNTs

Apart from single polymer chains, different kinds of polymer chains can be grafted from CNT surfaces by the binary-grafting methodology. Gao and co-authors[144] presented a versatile "gemini-grafting" strategy to modify SWCNTs and MWCNTs by the combination of conventional "grafting to" and "grafting from" strategies (Figure 5.20). A 'clickable' macroinitiator poly[3-azido-2-(2-bromo-2-methylpropanoyloxy)propyl methacrylate] (polyBrAzPMA) with alkylbromo groups for initiating ATRP and azido groups for click reaction was first synthesized by post-modification of polyGMA with sodium azide followed by 2-bromoisobutyryl bromide. The macroinitiator was clicked on to alkyne-containing CNTs *via* a Cu(I)-catalyzed reaction of Huisgen 1,3-dipolar cycloaddition between azides and alkynes, producing a CNT-based clickable macroinitiator. PnBMA, PS and PEG were subsequently bound on CNTs *via* ATRP "grafting from" and click "grafting to" approaches, producing CNT-supported amphiphilic polymer brushes (Scheme 5.12). The functionalized CNTs were characterized by TGA, FTIR spectroscopy, Raman spectroscopy, XPS, SEM and TEM imaging. This versatile strategy can be readily extended to prepare other Janus/bifunctional polymer brushes, opening an avenue for building complex polymer architectures and tailoring surface properties. Moreover, the resulting amphiphilic brushes could self-assemble into micelle-like particles attached on to CNTs and Janus-like film at the interface of water–chloroform (Figure 5.21).

Polymer-grafted Carbon Nanotubes via "Grafting From" Approach 161

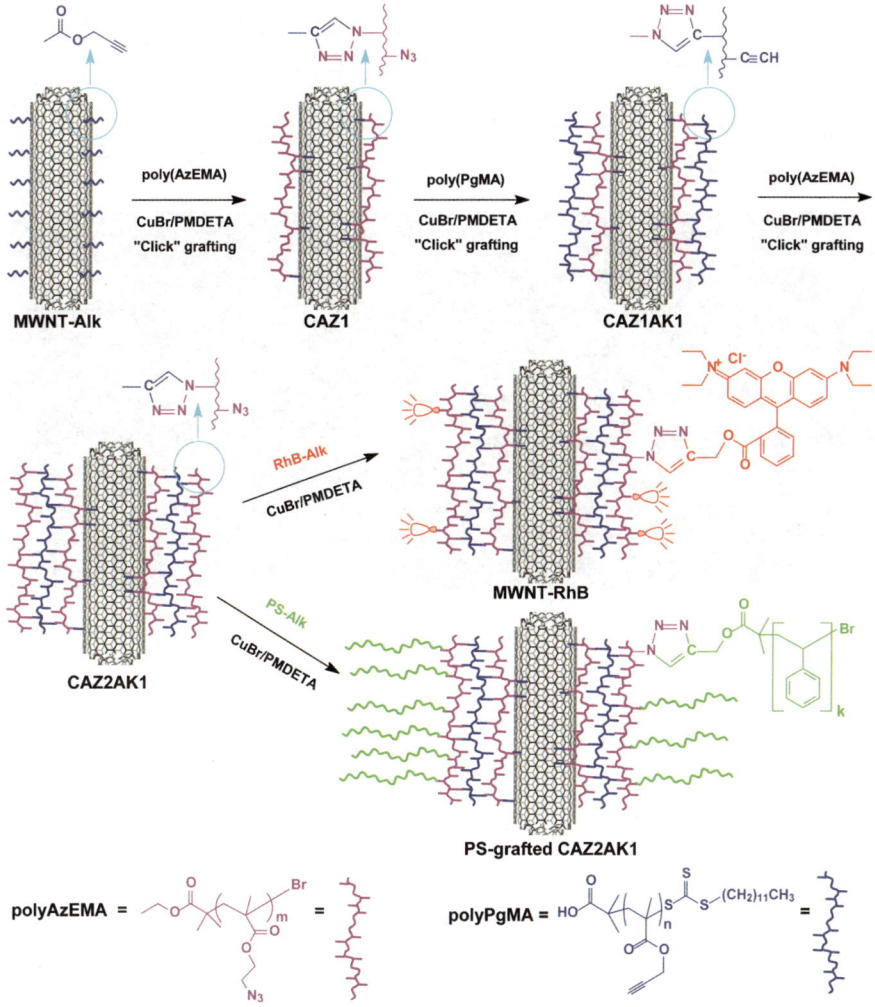

Scheme 5.11 Functionalization of MWCNTs by the layer-by-layer click chemistry (LbL-CC) approach and further modification of the functionalized MWCNTs with fluorescent dye and polystyrene by click chemistry. Reprinted with permission from Zhang *et al.*[143]

Priftis *et al.*[145] skillfully used the [4 + 2] Diels–Alder cycloaddition reaction to functionalize MWCNTs with two different precursor initiators, one for ROP and another one for ATRP (Scheme 5.13). The binary-functionalized MWCNTs were subsequently employed for the simultaneous surface-initiated polymerizations of different monomers, affording binary polymers (*e.g.*, PCL + PMMA, PCL + PS, PLLA + PMMA, PLLA + PS) grafted on to MWCNTs, that could form Janus-type structures under appropriate conditions. Figure 5.22 shows the representative TEM images of PLLA- and PS-grafted MWCNTs, indicating the highly efficient polymer grafting.

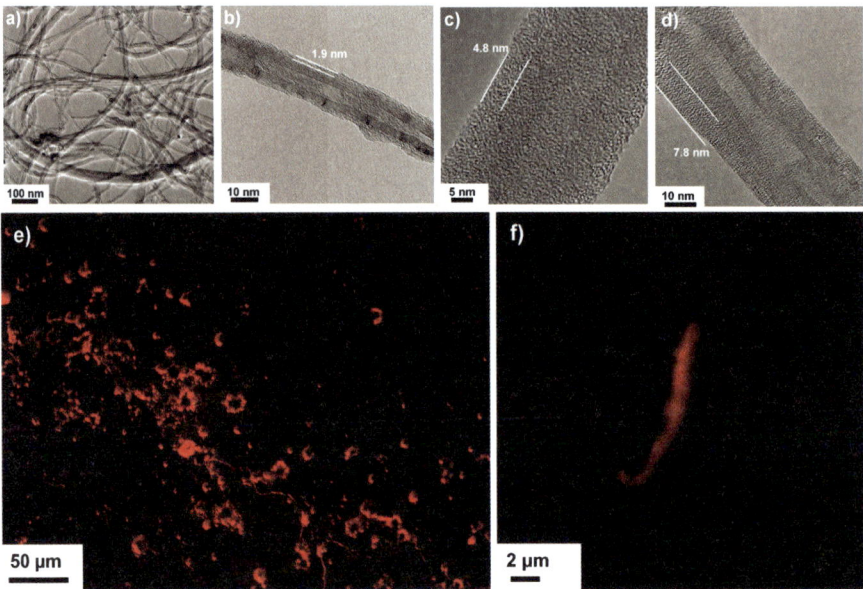

Figure 5.19 TEM images of pristine MWCNTs (a), one polymer layer-clicked WMCNT (b), two polymer layer-clicked MWCNT (c) and three polymer layer-clicked MWCNT (d). (e) Fluorescence microscopy image of RhB-clicked MWCNTs (λex = 520–550 nm), and (f) confocal fluorescence image of singular tube of RhB-functionalized MWCNT (λex=543 nm). Reprinted with permission from Zhang et al.[143]

5.2 Dendritic Polymer-functionalized CNTs

In addition to linear polymers, HPs can also be grafted on CNTs by the "grafting from" approach. HPs are globular, highly branched macromolecules

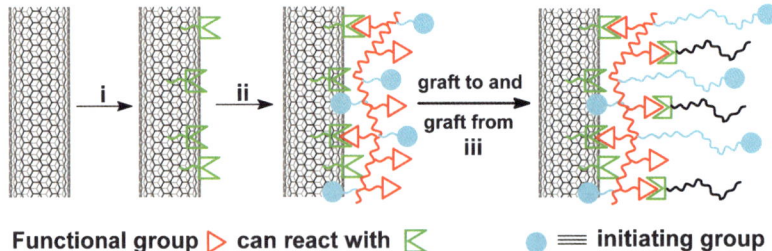

Figure 5.20 Gemini-grafting strategy to functionalize CNTs by combination of "grafting to" and "grafting from" approaches. (i) Premodification of pristine CNTs to introduce functional groups on CNTs; (ii) linkage of polymer with two types of functional groups on CNTs; and (iii) grafting polymer on CNTs via either one-pot orthogonal multi-grafting or sequential multi-grafting. Reprinted with permission from Zhang et al.[144]

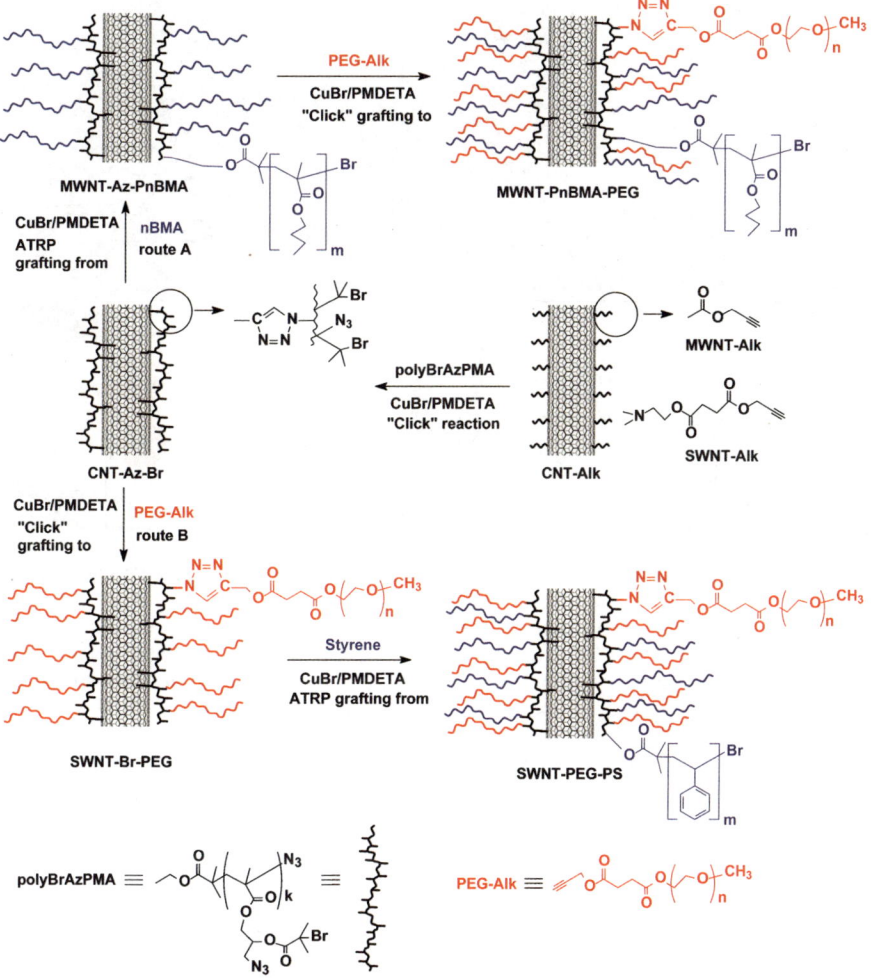

Scheme 5.12 Functionalization of CNTs by the combination of "grafting to" click chemistry and "grafting from" ATRP approach. Reprinted with permission from Zhang et al.[144]

with a three-dimensional dendritic architecture.[146–150] Due to their unique properties, such as high solubility, low viscosity, an abundance of functional groups and large-scale availability with a cost-effective manner, HPs have potential applications in a wide range of fields covered from drug delivery to material coatings.[151–157] Therefore, HP-linked CNTs, tree-like structures, are of special interest, as the novel nano-objects may act as highly effective and highly reactive platforms or templates for the fabrication of complex functional nanodevices and drug carriers.[158–160] Probably, CNT-based bio-nanomachines can be realized using dendritic macromolecules as amplifiers and converters of multiple functions. HPs were covalently grafted from

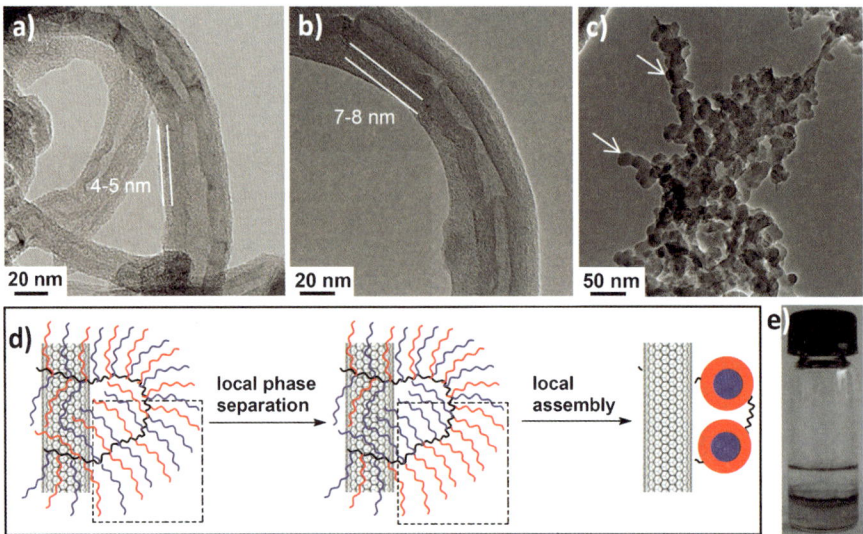

Figure 5.21 TEM images of poly(*n*-butyl methacrylate)-grafted MWCNTs with azido groups (MWCNT-Az-PnBMA) (a), MWCNTs grafted with both poly(*n*-butyl methacrylate) and poly(ethylene glycol) brushes (MWNT-PnBMA-PEG) (b, c). (d) Cartoon for the local phase separation and assembly of amphiphilic polymer brushes into Janus polymer structures on CNTs as shown in (c) (marked by arrows). (e) Photograph of MWCNT-PnBMA-PEG dispersed in a mixed solvent of water (upper layer) and chloroform (bottom layer). Reprint with permission from Zhang *et al.*[144]

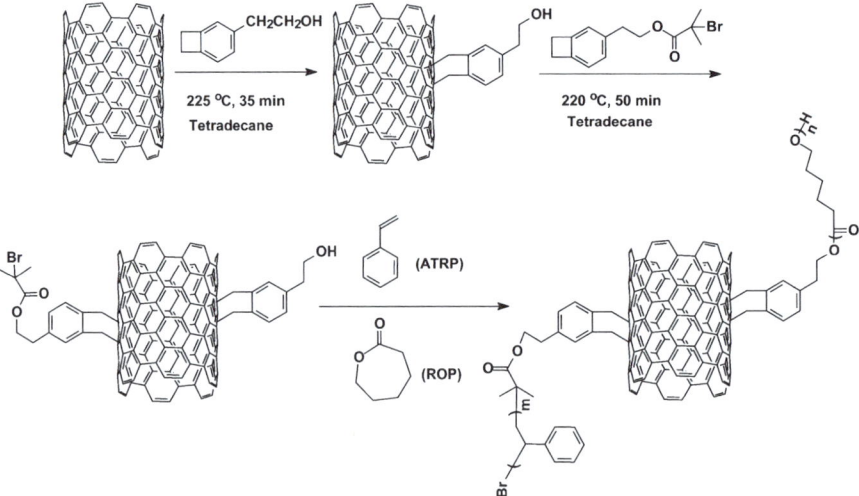

Scheme 5.13 Synthesis of binary polymer-grafted CNTs by the combination of the ROP and ATRP techniques.[145]

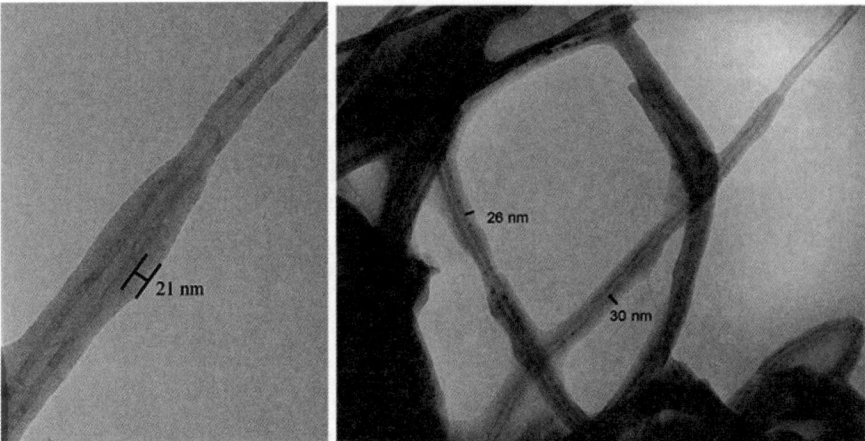

Figure 5.22 Representative TEM images of PCL/PS binary polymer-grafted CNTs. Reprinted with permission from Priftis et al.[145]

MWCNTs by self-condensing vinyl polymerization (SCVP),[40,161] ROP[162–164] and polycondensation approaches.[165–170]

5.2.1 Self-condensing Vinyl (Co)Polymerization (SCVP/SCVCP)

The SCVP approach to HPs was first reported by Fréchet and co-workers[171] through polymerization of an AB*-type monomer. Normally, the AB* contains a vinyl group and a functional group that can initiate the polymerization of the vinyl moiety. Hence, the AB* is also named as an inimer (inimer = initiator + monomer).[172,173] Up to date, SCVP has been extended to cationic, anionic and radical polymerization systems, preparing various libraries of HPs and HP-modified solid surfaces.[174–181] If the AB* inimer possesses a halogen atom that can act as the initiator of ATRP, HPs can be prepared by ATRP of the inimer, accordingly. Likewise, HPs can also be grafted from the nanosurface if there are initiating groups on CNTs, giving rise to nanoforest-like structures.

Gao et al.[40] functionalized MWCNTs with ~38 wt% content of HPs by SCVP of the AB* inimer 2-(2-bromoisobutyryloxy)ethyl methacrylate (BIEM) via the ATRP process using CuBr/bis(triphenylphosphine)nickel(II) bromide [(PPh$_3$)$_2$NiBr$_2$] as the catalyst and CNT-Br-1 as the macroinitiator in ethyl acetate at 100 °C for 4.5 h. In order to functionalize MWCNTs with biocompatible HPs, they extended the SCVP to self-condensing vinyl copolymerization (SCVCP) by adding MAIG as a co-monomer.[40] By variation of the mole feed ratio of MAIG to BIEM from 0.5:1 to 5:1, the degree of branching (DB) of the grafted HPs decreased from 0.49 to 0.21 and the grafted HP content increased from 40 to 53 wt%. After deprotection of MAIG units, hyperbranched glycopolymer-functionalized MWCNTs were obtained, as shown in Scheme 5.14.

Hong et al.[161] grafted HPs from the MWCNT surfaces by SCVP of another AB* inimer, 2-[(bromobutyryl)oxy]ethyl acrylate (BBEA), in toluene at 100 °C for 24 h via the ATRP process using CNT-Br-4 as the macroinitiator and CuBr/PMDETA as the catalyst.[161] The grafted HP content increased with increasing the reaction time, and approached 80 wt% at 24 h. They also carried out successfully SCVCP of BBEA and tBA via the ATRP process.[161]

Kong et al.[182] functionalized CNTs with HPs by SCVP of vinyl benzyl chloride (VBC) in toluene at 110–130 °C for 24–48 h via ATRP using CuBr/PMDETA or CuCl/bpy as the catalyst, and the grafted content of hyperbranched polystyrene reached 50 wt%.

5.2.2 ROP Approach

HPs can also be synthesized by self-condensing ROP of AB*-type cyclic monomers.[159,183–186] Likewise, HPs can be grafted from nanosurfaces by an *in situ* ROP strategy. Gao and co-workers[163] prepared hydroxy-functionalized

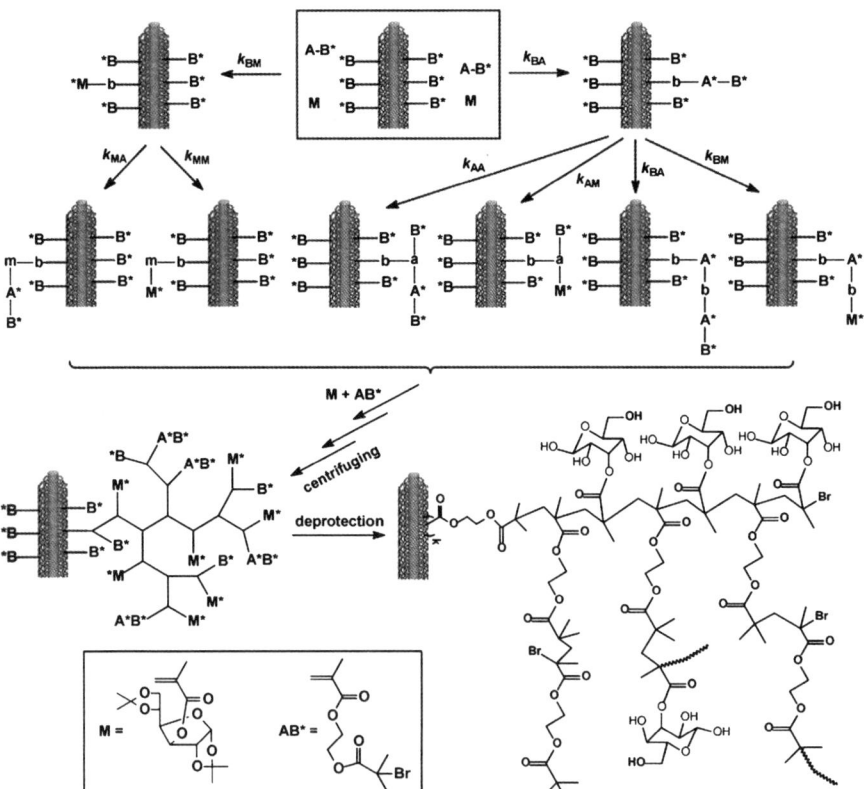

Scheme 5.14 Synthetic strategy for grafting hyperbranched glycopolymer from surfaces of MWCNTs by SCVCP of the inimer (AB*) and monomer (M) via ATRP. Reprinted with permission from Gao et al.[40]

MWCNTs (MWCNT-OH) by reaction of excess of glycol with acyl chloride-functionlized MWCNTs. Multi-hydroxyl HPs were subsequently grown from CNTs by *in situ* cationic ROP of 3-ethyl-3-(hydroxymethyl)oxetane (EHOX) in CH_2Cl_2 at –10 °C for 24 h with BF_3OEt_2 as the catalyst (Scheme 5.15). The grafted HP amount increased from 20 to 87 wt% when the weight feed ratio of monomer (EHOX) to MWCNT-OH varied from 5:1 to 50:1. The resulting products were characterized by TGA, FTIR spectroscopy, NMR, DSC, TEM, AFM and SEM. TEM indicates that the MWCNTs were covered evenly by HPs, including the CNT tips (Figure 5.23). The DB of the grafted hyperbranched polyether peaked at 0.42, which is quite close to the DB of free polymers (0.42–0.44). This implies that molecular trees with a high DB can be grown from CNT surfaces *via* the *in situ* ROP technique.

Through a modified strategy, Gao and co-workers[162] synthesized hyperbranched polyglycerol (HPG) grafted MWCNTs *via in situ* anionic ROP of glycidol with potassium methoxide as catalyst (Scheme 5.16). The macroinitiator of hydroxyl-functionalized MWCNTs (MWCNT-OH) were prepared by one-pot nitrene chemistry in *N*-methyl-2-pyrrolidone (NMP) at 160 °C for 12 h in order to avoid the side reaction between anions and ester bonds for conventional hydroxyl group-functionalized MWCNTs. The polymer content varied from 22.8 to 90.8 wt% when tuning the feed ratio of glycidol to MWCNT-OH macroinitiator (Figure 5.24). The HPG-grafted CNTs showed good dispersibility in polar solvents such as water, DMF, DMSO and

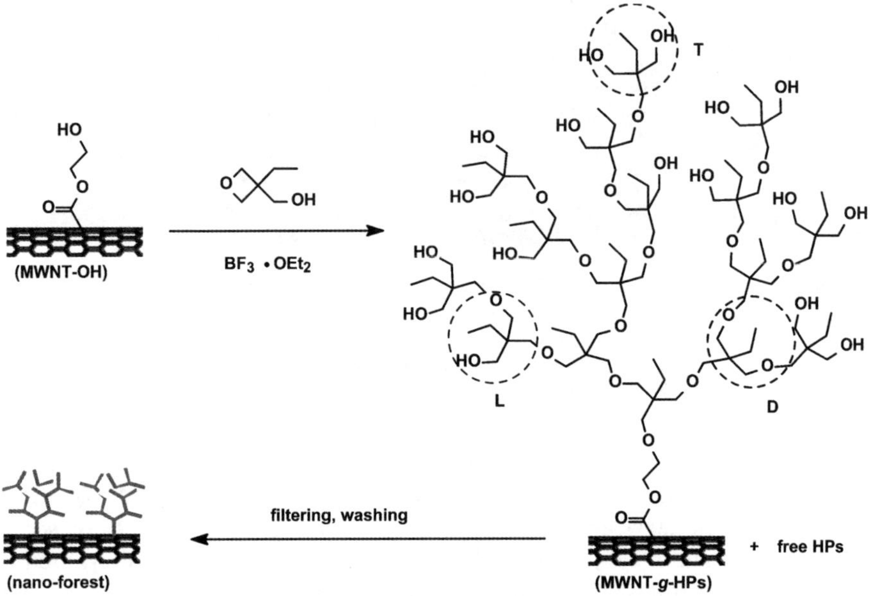

Scheme 5.15 Grafting multihydroxyl hyperbranched polyether from CNT surfaces by ROP approach. D, L and T represent dendritic, linear and terminal units, respectively. Reprinted with permission from Xu *et al.*[163]

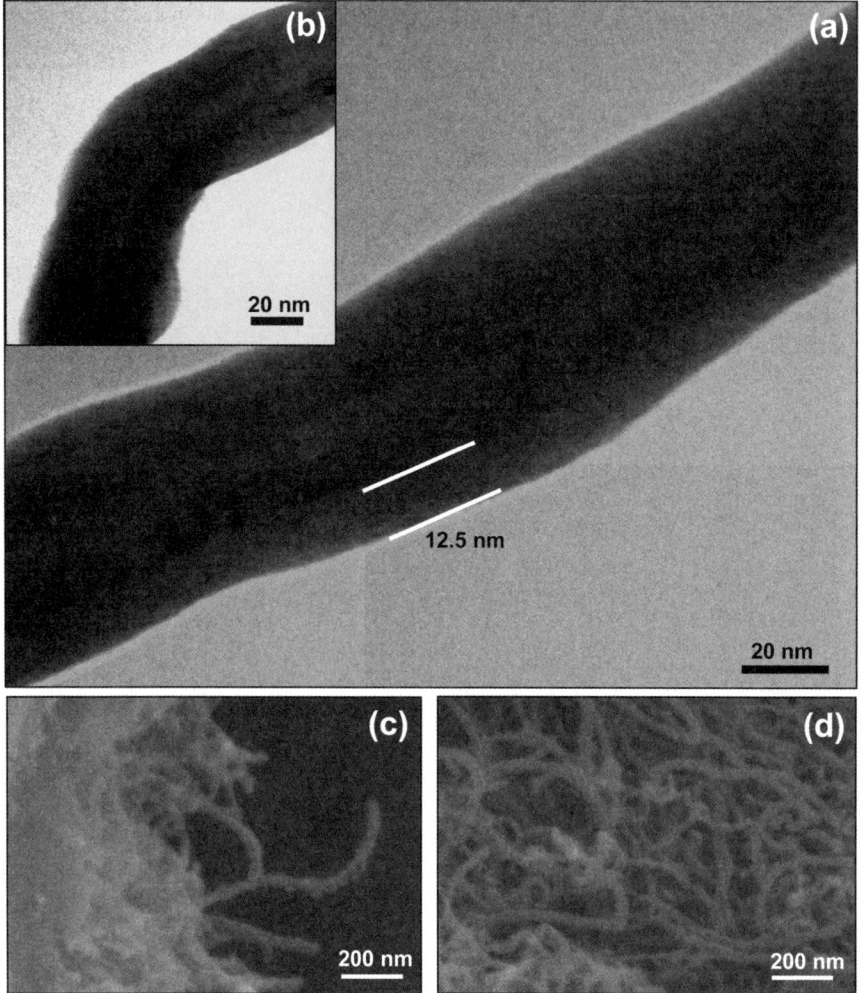

Figure 5.23 Representative TEM image of hyperbranched poly[3-ethyl-3-(hydroxymethyl)oxetane]-grafted MWCNT with 87 wt% of polymer at the straight (a) and bend area of a functionalized nanotube (b). SEM images of HP-grafted MWCNT with 74 wt% of polymer (c) and pristine MWCNTs (d). Reprinted with permission from Xu et al.[163]

methanol. On the basis of multi-hydroxyl groups of HPG grown on MWNTs, fluorescent molecules of rhodamine 6B were conjugated to MWCNT-g-HPG by N,N'-dicyclohexylcarbodiimide (DCC) coupling, affording fluorescent MWCNTs (Figure 5.25). The multifunctional, biocompatible HPG-functionalized CNTs promised potential applications in drug delivery, cell imaging and bioprobing.[162,14,187]

Similarly, Adeli et al.[164] also grafted HPG from opened MWCNTs in methanol at 80 °C for 2 h using potassium methoxide as the catalyst. *In vitro*

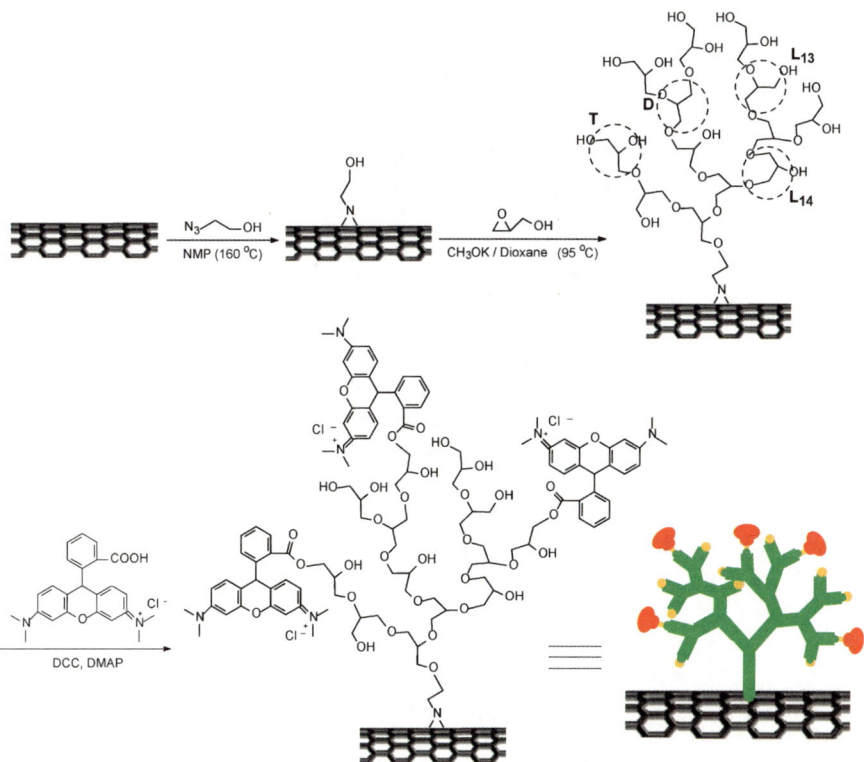

Scheme 5.16 Functionalization of MWCNTs with HPG by anionic ROP and conjugation of rhodamine 6B with hydroxyl groups by N,N'-dicyclohexylcarbodiimide (DCC) coupling chemistry. Reprinted with permission from Zhou et al.[162]

cytotoxicity tests and hemolysis assay results demonstrated high biocompatibility due to the HPG grafting.[164]

5.2.3 Polycondensation Approach

Condensed type HPs can be grown from CNTs by *in situ* one-step polycondensation or step-by-step condensation approach. Gao et al.[165] first presented the *in situ* one-pot polycondensation approach to synthesis of linear and hyperbranched poly(urea urethane) on amino-functionalized MWCNTs (MWCNT-NH$_2$). By adjusting the feed ratio of the isocyanate and amino groups, the grafted polymer content can be easily controlled. FTIR spectroscopy, NMR, Raman spectroscopy, confocal Raman spectroscopy, TEM, EDS and SEM measurements were performed to confirm the structure and morphology of the resulting nanocomposites. Because of intra- and intermolecular hydrogen-bonding interactions, self-assembled arc-, flat- or rose flower-like structures of MWCNT-hyperbranched polyurethane (HPU)

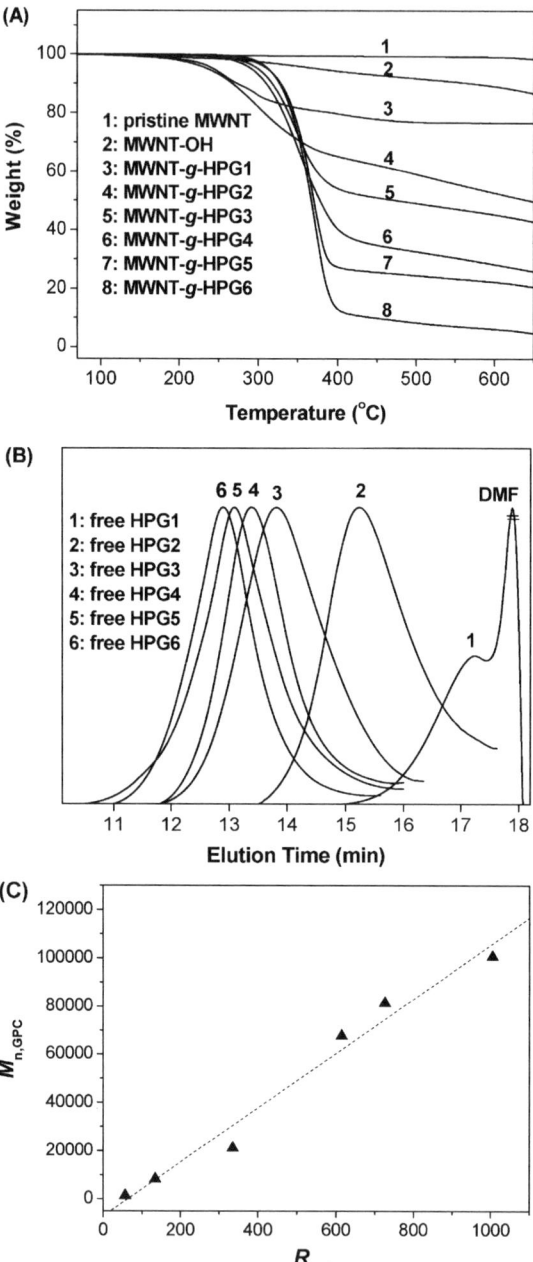

Figure 5.24 (A) TGA thermograms of pristine MWCNTs, MWCNT-OH and HPG-grafted MWCNTs. (B) GPC curves showing the evolution of molecular weight with the elution time for the free HPGs. (C) Average molecular weight of the free HPG measured by GPC as a function of molar feed ratio (R_{mole}). Reprinted with permission from Zhou et al.[162]

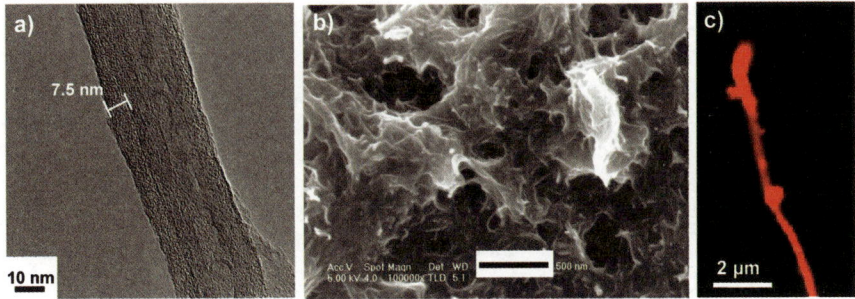

Figure 5.25 Representative TEM image (a) and SEM image (b; scale bar 500 nm) of HPG-functionalized MWCNT and confocal fluorescence image of RhB-modified HPG-functionalized MWCNT (c). Reprinted with permission from Zhou et al.[162]

were observed, providing a new element for supramolecular chemistry.[165] Similarly, Yang et al.[188] also synthesized MWCNT-HPU bearing terminal hydroxy groups, as well as urea and urethane groups from hydroxy-functionalized MWCNTs *via in situ* polycondensation of diethylolamine (DEOA) and toluene di-isocyanate (TDI) (Scheme 5.17). The thickness of the grafted-HPU can be varied by adjusting the feed ratio of monomers. The structure and morphology of MWCNT-HPUs were confirmed by FTIR spectroscopy, Raman spectroscopy, ^{1}H NMR, ^{13}C NMR, TGA, TEM, SEM and AFM measurements. The resultant MWCNT-HPU could be well dispersed in highly polar solvents such as DMF, NMP and DMSO. The solution rheology of the MWCNT-HPU was studied.[188] The results revealed a strong dependence of the rheological behaviors of MWNT-HPU solutions on temperature and shear rate due to intra- and intermolecular hydrogen bonds of MWCNT-HPU.[167] Grafting of hyperbranched poly(urea urethane) improved the compatibility between MWCNT-HPU and PA6, and thus MWCNT-HPU/PA6 nanocomposites were prepared.[189] The dynamic mechanical properties of the nanocomposites were increased progressively with increasing the MWCNT-HPUs content. The study of crystallization of MWCNTs-HPU/PA6 nanocomposites showed that the presence of MWCNTs enhanced the crystallization temperature of PA6 and decreased the crystallization rate and the degree of crystallization of PA6 in the nanocomposites. Furthermore, Mai and co-workers[190] fabricated MWCNT-HPU/PU nanocomposite and investigated its thermal conductive and electrical properties. T_g and thermal stability of PU in PU/HPU-MWCNT composites were higher than those of neat PU. The nanocomposite showed relatively high thermal conductivity and high electrical resistivity due to the strong interface adhesion.

Wang and co-workers[191] synthesized hyperbranched poly(amido amine) (PAMAM)-grafted MWCNTs *via* a step-by-step divergent methodology from amino-functionalized CNTs. The resulting adduct was well dispersible in polar solvents. Further modification of grafted PAMAM with fluorescein

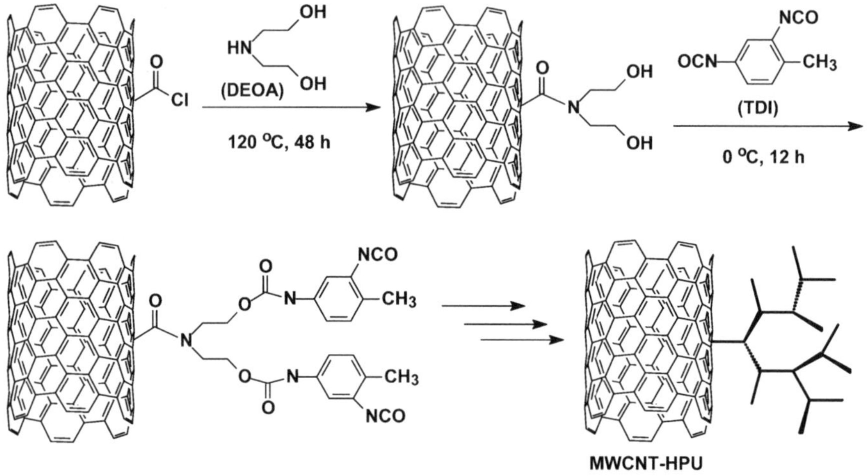

Scheme 5.17 Synthesis of HPU-grafted MWCNTs.[188]

isothiocyanate (FITC) gave rise to functionalized CNTs that could emit strong fluorescence.

Chan-Park and co-workers[166] prepared third-generation dendritic PAMAM-grafted MWCNT nanohybrids encapsulated with silver nanoparticles via an in situ deposition method. As an antimicrobial agent in solution, PAMAM-MWCNT/Ag hybrid showed effective antimicrobial activity for killing both Gram-positive and Gram-negative bacteria.[166] By an ultrasound-induced assembly strategy, Qu and co-workers[192] constructed a switchable gel composed of PAMAM-grafted MWCNTs. The sol–gel switching of this system was very fast and could be easily controlled via heating and ultrasonication. Due to the weak hydrogen-bonding interactions, the formed gels were multi-responsive, which is useful in the design and development of ultrasound-driven devices. Neelgund and Oki[193] constructed two ternary nanocatalysts, f-MWCNT-CdS and f-MWCNT-Ag_2S, by deposition of CdS or Ag_2S QDs in the presence of fourth generation PAMAM-functionalized MWCNTs. Due to the charge transfer processes through the interface of f-MWCNTs and QDs, both f-MWCNT-QDs assemblies presented superior catalytic performance and were promising as reusable photocatalysts. Zhu and co-workers[194] prepared enzyme-immobilized and magnetic PAMAM-functionalized MWCNTs via adsorption, covalent bonding and metallic ion affinity interactions. By the recovery process of magnetic separation, the enzyme-MWCNT biocatalyst offered superior stability and reusability.

Baek and co-workers[168,169] grafted hyperbranched poly(ether ketone)s (HPEKs) on to MWCNTs through an in situ polycondensation of A_3 (trimesic acid) and B_2 (phenyl ether) monomers in the presence of poly(phosphoric acid) and P_2O_5. Due to the globular molecular architecture of HPs, the morphology of the nanocomposites resembled "mushroom-like clusters on MWCNT stalks". The HPEK-g-MWCNT nanocomposites were soluble in polar aprotic solvents because of the numerous carboxylic acids on their surfaces. The nanocomposites could form very stable suspensions when dispersed in 1 M LiOH aqueous solutions. The resulting lithiated nanocomposites were investigated in various applications, such as ion conductivity and energy capacitance.

The HP-functionalized CNTs possess a high density of functional groups on their surfaces and afford a versatile platform for multifunctional integration.

5.3 Concluding Remarks

Various linear and dendritic polymers have been grafted from CNT surfaces *via in situ* controlled/living radical polymerizations of vinyl monomers, ROPs of cyclic monomers and polycondensation methodology. The resulting polymer brushes and nanoforests promised wide applications in the fields of nanoplatforms, nanocomposites, supramolecular self-assembly, sensors, biomineralization, magnetic manipulation, *etc.* However, the control of density and chain length of grafted polymers is still a challenge. Although the dispersibility and wettability of CNTs have been significantly improved by polymer grafting, it is still difficult to readily obtain individually dispersed CNTs. As a parallel action, the aforementioned functionalization strategies could be extended to other nanocarbons such as the emerging graphene, producing 2D polymer brushes. The synergetic effect between 1D and 2D brushes would bring multifunctional, novel materials and structures.

Acknowledgements

This work is supported by the National Natural Science Foundation of China (No. 50473010, No. 50773038, No. 51173162 and No. 20974093), National Basic Research Program of China (No. 2007CB936000), Qianjiang Talent Foundation of Zhejiang Province (No. 2010R10021), the Fundamental Research Funds for the Central Universities (No. 2011QNA4029), the Foundation for the Author of National Excellent Doctoral Dissertation of China (No. 200527), Research Fund for the Doctoral Program of Higher Education of China (No. 20100101110049) and Zhejiang Provincial Natural Science Foundation of China (No. R4110175).

References

1. S. Iijima, *Nature*, 1991, **354**, 56–58.
2. S. Iijima and T. Ichihashi, *Nature*, 1993, **363**, 603–605.
3. D. Tasis, N. Tagmatarchis, A. Bianco and M. Prato, *Chem. Rev.*, 2006, **106**, 1105–1136.
4. N. G. Sahoo, S. Rana, J. W. Cho, L. Li and S. H. Chan, *Prog. Polym. Sci.*, 2010, **35**, 837–867.

5. Z. Spitalsky, D. Tasis, K. Papagelis and C. Galiotis, *Prog. Polym. Sci.*, 2010, **35**, 357–401.
6. C. Gao, in *Encyclopedia of Nanoscience and Nanotechnology*, ed. H. S. Nalwa, American Scientific Publishers, Stevenson Ranch, CA, 2011, pp. 305–355.
7. C. Gao, in *Surface Modification of Nanotube Fillers*, ed. V. Mittal, Wiley-VCH, Weinheim, 2011.
8. N. Roy, R. Sengupta and A. K. Bhowmick, *Prog. Polym. Sci.*, 2012, **37**, 781–819.
9. J. S. Wang and K. Matyjaszewski, *J. Am. Chem. Soc.*, 1995, **117**, 5614–5615.
10. M. Kato, M. Kamigaito, M. Sawamoto and T. Higashimura, *Macromolecules*, 1995, **28**, 1721–1723.
11. T. E. Patten, J. Xia, T. Abernathy and K. Matyjaszewski, *Science*, 1996, **272**, 866–868.
12. T. E. Patten and K. Matyjaszewski, *Adv. Mater.*, 1998, **10**, 901–915.
13. K. Matyjaszewski and J. Xia, *Chem. Rev.*, 2001, **101**, 2921–2990.
14. V. Coessens, T. Pintauer and K. Matyjaszewski, *Prog. Polym. Sci.*, 2001, **26**, 337–377.
15. M. Kamigaito, T. Ando and M. Sawamoto, *Chem. Rev.*, 2001, **101**, 3689–3746.
16. M. K. Mishra and Y. Yagci, *Handbook of Radical Vinyl Polymerization*, CRC, New York, 1998.
17. K. Matyjaszewski, *Recherche*, 2003, **67**, 02.
18. J. Pyun, T. Kowalewski and K. Matyjaszewski, *Macromol. Rapid Commun.*, 2003, **24**, 1043–1059.
19. W. A. Braunecker and K. Matyjaszewski, *Prog. Polym. Sci.*, 2007, **32**, 93–146.
20. N. V. Tsarevsky and K. Matyjaszewski, *Chem. Rev.*, 2007, **107**, 2270–2299.
21. H. Kong, C. Gao and D. Yan, *J. Am. Chem. Soc.*, 2003, **126**, 412–413.
22. H. Kong, C. Gao and D. Yan, *Macromolecules*, 2004, **37**, 4022–4030.
23. C. Gao, C. D. Vo, Y. Z. Jin, W. Li and S. P. Armes, *Macromolecules*, 2005, **38**, 8634–8648.
24. R. Narain, A. Housni and L. Lane, *J. Polym. Sci., Part A: Polym. Chem.*, 2006, **44**, 6558–6568.
25. S. Qin, D. Qin, W. T. Ford, D. E. Resasco and J. E. Herrera, *J. Am. Chem. Soc.*, 2003, **126**, 170–176.
26. D. Baskaran, J. W. Mays and M. S. Bratcher, *Angew. Chem. Int. Ed.*, 2004, **43**, 2138–2142.
27. Z. Yao, N. Braidy, G. A. Botton and A. Adronov, *J. Am. Chem. Soc.*, 2003, **125**, 16015–16024.
28. V. Georgakilas, K. Kordatos, M. Prato, D. M. Guldi, M. Holzinger and A. Hirsch, *J. Am. Chem. Soc.*, 2002, **124**, 760–761.
29. V. Georgakilas, N. Tagmatarchis, D. Pantarotto, A. Bianco, J. P. Briand and M. Prato, *Chem. Commun.*, 2002, 3050–3051.

30. J. H. Choi, S. B. Oh, J. Chang, I. Kim, C. S. Ha, B. G. Kim, J. H. Han, S. W. Joo, G. H. Kim and H. J. Paik, *Polym. Bull.*, 2005, **55**, 173–179.
31. T. Matrab, J. Chancolon, M. M. L'hermite, J. N. Rouzaud, G. Deniau, J. P. Boudou, M. M. Chehimi and M. Delamar, *Colloids Surf., A*, 2006, **287**, 217–221.
32. B. Fragneaud, K. Masenelli Varlot, A. Gonzalez Montiel, M. Terrones and J. Y. Cavaillé, *Chem. Phys. Lett.*, 2006, **419**, 567–573.
33. C. Zhi, Y. Bando, C. Tang, H. Kuwahara and D. Golberg, *J. Phys. Chem. C*, 2006, **111**, 1230–1233.
34. C. Gao, H. He, L. Zhou, X. Zheng and Y. Zhang, *Chem. Mater.*, 2008, **21**, 360–370.
35. H. He and C. Gao, *Chem. Mater.*, 2010, **22**, 5054–5064.
36. W. Li, C. Gao, H. Qian, J. Ren and D. Yan, *J. Mater. Chem.*, 2006, **16**, 1852–1859.
37. W. W. Li, H. Kong, C. Gao and D. Y. Yan, *Chin. Sci. Bull.*, 2005, **50**, 2276–2280.
38. C. Gao, W. Li, H. Morimoto, Y. Nagaoka and T. Maekawa, *J. Phys. Chem. B*, 2006, **110**, 7213–7220.
39. W. W. Li, Master degree thesis, Shanghai Jiao Tong University, 2007.
40. C. Gao, S. Muthukrishnan, W. Li, J. Yuan, Y. Xu and A. H. E. Müller, *Macromolecules*, 2007, **40**, 1803–1815.
41. H. Kong, P. Luo, C. Gao and D. Yan, *Polymer*, 2005, **46**, 2472–2485.
42. C. Gao, *Macromol. Rapid Commun.*, 2006, **27**, 841–847.
43. G. Chao, W. Li, Y. Z. Jin and H. Kong, *Nanotechnology*, 2006, **17**, 2882.
44. H. He and C. Gao, *Molecules*, 2010, **15**, 4679–4694.
45. W. Li and C. Gao, *Langmuir*, 2007, **23**, 4575–4582.
46. T. Hu, H. Xie, L. Chen, G. Zhong and H. Zhang, *Polym. Int.*, 2011, **60**, 93–101.
47. M. Liu, Y. Yang, T. Zhu and Z. Liu, *J. Phys. Chem. C*, 2007, **111**, 2379–2385.
48. H. Kong, C. Gao and D. Yan, *J. Mater. Chem.*, 2004, **14**, 1401–1405.
49. S. Qin, D. Qin, W. T. Ford, D. E. Resasco and J. E. Herrera, *Macromolecules*, 2004, **37**, 752–757.
50. A. M. Shanmugharaj, J. H. Bae, R. R. Nayak and S. H. Ryu, *J. Polym. Sci., Part A: Polym. Chem.*, 2007, **45**, 460–470.
51. H. Kong, W. Li, C. Gao, D. Yan, Y. Jin, D. R. M. Walton and H. W. Kroto, *Macromolecules*, 2004, **37**, 6683–6686.
52. T. Sun, H. Liu, W. Song, X. Wang, L. Jiang, L. Li and D. Zhu, *Angew. Chem. Int. Ed.*, 2004, **43**, 4663–4666.
53. Y. L. Liu and W. H. Chen, *Macromolecules*, 2007, **40**, 8881–8886.
54. L. Li and C. M. Lukehart, *Chem. Mater.*, 2005, **18**, 94–99.
55. Y. Z. Jin, C. Gao, H. W. Kroto and T. Maekawa, *Macromol. Rapid Commun.*, 2005, **26**, 1133–1139.
56. L. Zhou, C. Gao, D. Zhu, W. Xu, F. F. Chen, A. Palkar, L. Echegoyen and E. S.-W. Kong, *Chem. Eur. J.*, 2009, **15**, 1389–1396.

57. T. Liu, S. Jia, T. Kowalewski, K. Matyjaszewski, R. Casado-Portilla and J. Belmont, *Langmuir*, 2003, **19**, 6342–6345.
58. T. Velasco-Santos, A. L. Martinez-Hernandez, F. Fisher, R. Fisher and V. M. Castano, *J. Phys. D. -Appl. Phys.* 2003, **36**, 1423–1428.
59. T. Liu, S. Jia, T. Kowalewski, K. Matyjaszewski, R. Casado-Portilla and J. Belmont, *Macromolecules*, 2005, **39**, 548–556.
60. J. Chiefari, Y. K. Chong, F. Ercole, J. Krstina, J. Jeffery, T. P. T. Le, R. T. A. Mayadunne, G. F. Meijs, C. L. Moad, G. Moad, E. Rizzardo and S. H. Thang, *Macromolecules*, 1998, **31**, 5559–5562.
61. R. Barbey, L. Lavanant, D. Paripovic, N. Schuewer, C. Sugnaux, S. Tugulu, H. A. Klok, *Chem. Rev.*, 2009, **109**, 5437–5527.
62. G. Moad, E. Rizzardo and S. H. Thang, *Polymer*, 2008, **49**, 1079–1131.
63. G. Moad, J. Chiefari, Y. K. Chong, J. Krstina, R. T. A. Mayadunne, A. Postma, E. Rizzardo and S. H. Thang, *Polym. Int.*, 2000, **49**, 993–1001.
64. S. Perrier and P. Takolpuckdee, *J. Polym. Sci., Part A: Polym. Chem.*, 2005, **43**, 5347–5393.
65. G. Moad, *Aust. J. Chem.*, 2006, **59**, 661–662.
66. A. Favier and M. T. Charreyre, *Macromol. Rapid Commun.*, 2006, **27**, 653–692.
67. A. B. Lowe and C. L. McCormick, *Prog. Polym. Sci.*, 2007, **32**, 283–351.
68. J. Cui, W. P. Wang, Y. Z. You, C. H. Liu and P. H. Wang, *Polymer*, 2004, **45**, 8717–8721.
69. T. Macdonald, C. T. Gibson, K. Constantopoulos, J. G. Shapter and A. V. Ellis, *Appl. Surf. Sci.*, 2012, **258**, 2836–2843.
70. G. Y. Xu, W. T. Wu, Y. S. Wang, W. M. Pang, Q. R. Zhu, P. H. Wang and Y. Z. You, *Polymer*, 2006, **47**, 5909–5918.
71. G. Y. Xu, W. T. Wu, Y. S. Wang, W. M. Pang, P. H. Wang, G. R. Zhu and F. Lu, *Nanotechnology*, 2006, **17**, 2458–2465.
72. X. Pei, J. Hao and W. Liu, *J. Phys. Chem. C* 2007, **111**, 2947–2952.
73. X. Pei, W. Liu and J. Hao, *J. Polym. Sci., Part A: Polym. Chem.*, 2008, **46**, 3014–3023.
74. G. J. Wang, S. Z. Huang, Y. Wang, L. Liu, J. Qiu and Y. Li, *Polymer*, 2007, **48**, 728–733.
75. G. Xu, Y. Wang, W. Pang, W. T. Wu, Q. Zhu and P. Wang, *Polym. Int.*, 2007, **56**, 847–852.
76. S. A. Curran, D. H. Zhang, W. T. Wondmagegn, A. V. Ellis, J. Cech, S. Roth and D. L. Carroll, *J. Mater. Res.*, 2006, **21**, 1071–1077.
77. Y. Z. You, C. Y. Hong and C. Y. Pan, *Macromol. Rapid Commun.*, 2006, **27**, 2001–2006.
78. C. Y. Hong, Y. Z. You and C. Y. Pan, *Chem. Mater.*, 2005, **17**, 2247–2254.
79. C. Y. Hong, Y. Z. You and C. Y. Pan, *J. Polym. Sci., Part A: Polym. Chem.*, 2006, **44**, 2419–2427.
80. Y. Z. You, C. Y. Hong and C. Y. Pan, *Nanotechnology*, 2006, **17**, 2350–2354.

81. C. Y. Hong, Y. Z. You and C. Y. Pan, *Polymer*, 2006, **47**, 4300–4309.
82. B. Zhang, J. Wang, Y. Chen, D. Fruechtl, B. Yu, X. Zhuang, N. He and W. J. Blau, *J. Polym. Sci., Part A: Polym. Chem.*, 2010, **48**, 3161–3168.
83. C. J. Hawker, A. W. Bosman and E. Harth, *Chem. Rev.*, 2001, **101**, 3661–3688.
84. K. Matyjaszewski, *Recherche*, 2003, **67**, 438–451.
85. A. Goto and T. Fukuda, *Prog. Polym. Sci.*, 2004, **29**, 329–385.
86. R. B. Grubbs, *Polym. Rev.*, 2011, **51**, 104–137.
87. M. Dehonor, K. Masenelli Varlot, A. Gonzalez Montiel, C. Gauthier, J. Y. Cavaille, H. Terrones and M. Terrones, *Chem. Commun.*, 2005, 5349–5351.
88. J. H. Chang, Y. W. Lee, B. G. Kim, H.-K. Kim, I. S. Choi and H.-J. Paik, *Compos. Interfac.*, 2007, **14**, 493–504.
89. X. D. Zhao, X. H. Fan, X. F. Chen, C. P. Chai and Q. F. Zhou, *J. Polym. Sci., Part A: Polym. Chem.*, 2006, **44**, 4656–4667.
90. X. Zhao, W. Lin, N. Song, X. Chen, X. Fan and Q. Zhou, *J. Mater. Chem.*, 2006, **16**, 4619–4625.
91. D. Q. Fan, J. P. He, W. Tang, J. T. Xu and Y. L. Yang, *Eur. Polym. J.*, 2007, **43**, 26–34.
92. H. C. Wang, Y. Li and M. J. Yang, *Sens. Actuators B*, 2007, **124**, 360–367.
93. X. Wang, H. Liu and L. Qiu, *Mater. Lett.*, 2007, **61**, 2350–2353.
94. L. W. Qu, L. M. Veca, Y. Lin, A. Kitaygorodskiy, B. L. Chen, A. M. McCall, J. W. Connell and Y. P. Sun, *Macromolecules*, 2005, **38**, 10328–10331.
95. M. Yang, Y. Gao, H. M. Li and A. Adronov, *Carbon*, 2007, **45**, 2327–2333.
96. D. Yan and G. Yang, *Mater. Lett.*, 2009, **63**, 298–300.
97. K. R. Yoon, W. J. Kim and I. S. Choi, *Macromol. Chem. Phys.*, 2004, **205**, 1218–1221.
98. F. Buffa, H. Hu and D. E. Resasco, *Macromolecules*, 2005, **38**, 8258–8263.
99. H. L. Zeng, C. Gao and D. Y. Yan, *Adv. Funct. Mater.*, 2006, **16**, 812–818.
100. H. H. Chen, R. Anbarasan, L. S. Kuo and P. H. Chen, *Mater. Chem. Phys.*, 2011, **126**, 584–590.
101. G. Jiang, J. Ji and H. Lu, *Des. Monomers Polym.*, 2011, **14**, 121–131.
102. R. S. Lee, W. H. Chen and J. H. Lin, *Polymer*, 2011, **52**, 2180–2188.
103. Y. Yang, S. Qiu, C. He, W. He, L. Yu and X. Xie, *Appl. Surf. Sci.*, 2010, **257**, 1010–1014.
104. H. L. Zeng, Doctoral dissertation, Shanghai Jiao Tong University, 2006.
105. J. Ma, X. Cheng, X. Ma, S. Deng and A. Hu, *J. Polym. Sci., Part A: Polym. Chem.*, 2010, **48**, 5541–5548.
106. G. X. Chen, H. S. Kim, B. H. Park and J. S. Yoon, *Macromol. Chem. Phys.*, 2007, **208**, 389–398.

107. J. T. Yoon, S. C. Lee and Y. G. Jeong, *Compos. Sci. Technol.*, 2010, **70**, 776–782.
108. J. Feng, W. Cai, J. Sui, Z. Li, J. Wan and A. N. Chakoli, *Polymer*, 2008, **49**, 4989–4994.
109. Y. Yao, W. Li, S. Wang, D. Yan and X. Chen, *Macromol. Rapid Commun.*, 2006, **27**, 2019–2025.
110. F. L. Jin, K. Y. Rhee and S. J. Park, *J. Solid State Chem.*, 2011, **184**, 3253–3256.
111. N. Gonzalez-Vidal, A. Martinez de Ilarduya, S. Munoz-Guerra, P. Castell and M. T. Martinez, *Compos. Sci. Technol.*, 2010, **70**, 789–796.
112. F. Wu and G. Yang, *Polym. Adv. Technol.*, 2011, **22**, 1466–1470.
113. G. Sakellariou, H. N. Ji, J. W. Mays and D. Baskaran, *Chem. Mater.*, 2008, **20**, 6217–6230.
114. D. Priftis, G. Sakellariou, N. Hadjichristidis, E. K. Penott, A. T. Lorenzo and A. J. Muller, *J. Polym. Sci., Part A: Polym. Chem.*, 2009, **47**, 4379–4390.
115. W. Kai, L. Hua, L. Zhao and Y. Inoue, *Macromol. Rapid Commun.*, 2006, **27**, 1702–1706.
116. Q. Yang, L. Wang, W. Xiang, J. Zhou and G. Jiang, *J. Appl. Polym. Sci.*, 2007, **103**, 2086–2092.
117. K. Wang, W. Li and C. Gao, *J. Appl. Polym. Sci.*, 2007, **105**, 629–640.
118. S. M. Rhodes, B. Higgins, Y. Xu and W. J. Brittain, *Polymer*, 2007, **48**, 1500–1509.
119. S. D. Bergin, V. Nicolosi, P. V. Streich, S. Giordani, Z. Y. Sun, A. H. Windle, P. Ryan, N. P. P. Niraj, Z. T. T. Wang, L. Carpenter, W. J. Blau, J. J. Boland, J. P. Hamilton and J. N. Coleman, *Adv. Mater.*, 2008, **20**, 1876–1881.
120. B. Moon and M. Kang, *Macromol. Symp.*, 2007, **249**, 336–343.
121. M. R. Buchmeiser, *Surface-Initiated Polymerization*, Springer Verlag, Heidelberg, 2006.
122. R. H. Grubbs, *Angew. Chem. Int. Ed.*, 2006, **45**, 3760–3765.
123. C. Slugovc, *Macromol. Rapid Commun.*, 2004, **25**, 1283–1297.
124. T. Sun, V. Nicolosi and R. H. Grubbs, *Acc. Chem. Res.*, 2000, **34**, 18–29.
125. F. J. Gomez, R. J. Chen, D. Wang, R. M. Waymouth and H. Dai, *Chem. Commun.*, 2003, 190–191.
126. Y. Liu and A. Adronov, *Macromolecules*, 2004, **37**, 4755–4760.
127. W. Jeong and M. R. Kessler, *Chem. Mater.*, 2008, **20**, 7060–7068.
128. W. Jeong and M. R. Kessler, *Carbon*, 2009, **47**, 2406–2412.
129. D. Izuhara and T. M. Swager, *Macromolecules*, 2009, **42**, 5416–5418.
130. C. Costabile, F. Grisi, G. Siniscalchi, P. Longo, M. Sarno, D. Sannino, C. Leone and P. Ciambelli, *J. Nanosci. Nanotechnol.*, 2011, **11**, 10053–10062.
131. G. Viswanathan, N. Chakrapani, H. C. Yang, B. Q. Wei, H. S. Chung, K. W. Cho, C. Y. Ryu and P. M. Ajayan, *J. Am. Chem. Soc.*, 2003, **125**, 9258–9259.

132. S. Chen, D. Chen and G. Wu, *Macromol. Rapid Commun.*, 2006, **27**, 882–887.
133. F. Liang, J. M. Beach, K. Kobashi, A. K. Sadana, Y. I. Vega-Cantu, J. M. Tour and W. E. Billups, *Chem. Mater.*, 2006, **18**, 4764–4767.
134. P. P. Li, L. J. Niu, Y. Chen, J. Wang, Y. Liu, J. J. Zhang and W. J. Blau, *Nanotechnology*, 2011, **22**, 015204.
135. D. Bonduel, M. L. Mainil, M. Alexandre, F. Monteverde and P. Dubois, *Chem. Commun.*, 2005, 781–783.
136. D. Priftis, N. Petzetakis, G. Sakellariou, M. Pitsikalis, D. Baskaran, J. W. Mays and N. Hadjichristidis, *Macromolecules*, 2009, **42**, 3340–3346.
137. L. Tessier, J. Chancolon, P. J. Alet, A. Trenggono, M. Mayne-L'Hermite, G. Deniau, P. Jégou and S. Palacin, *Phys. Stat. Sol. A*, 2008, **205**, 1412–1418.
138. P. Petrov, G. Georgiev, D. Momekova, G. Momekov and C. B. Tsvetanov, *Polymer*, 2010, **51**, 2465–2471.
139. D. S. Yang, D. J. Jung and S. H. Choi, *Radiat. Phys. Chem.*, 2010, **79**, 434–440.
140. S. Huang, T. X. Liu, W. D. Zhang, W. C. Tjiu, C. B. He and X. H. Lu, *Polym. Int.*, 2010, **59**, 1346–1349.
141. H. Xu, X. Wang, Y. Zhang and S. Liu, *Chem. Mater.*, 2006, **18**, 2929–2934.
142. M. H.-O. Rashid, J. H. Bae, C. Park and K. T. Lim, *Mol. Cryst. Liq. Cryst.*, 2010, **532**, 98/[514]–105/[521].
143. Y. Zhang, H. He, C. Gao and J. Y. Wu, *Langmuir*, 2009, **25**, 5814–5824.
144. Y. Zhang, H. K. He and C. Gao, *Macromolecules*, 2008, **41**, 9581–9594.
145. D. Priftis, G. Sakellariou, D. Baskaran, J. W. Mays and N. Hadjichristidis, *Soft Matter*, 2009, **5**, 4272–4278.
146. P. J. Flory, *J. Am. Chem. Soc.*, 1952, **74**, 2718–2723.
147. Y. H. Kim and O. W. Webster, *J. Am. Chem. Soc.*, 1990, **112**, 4592–4593.
148. Y. H. Kim and O. W. Webster, *Macromolecules*, 1992, **25**, 5561–5572.
149. Y. Xu, C. Gao, H. Kong, D. Yan, P. Luo, W. Li and Y. Mai, *Macromolecules*, 2004, **37**, 6264–6267.
150. D. Yan, C. Gao and H. Frey, *Hyperbranched Polymers: Synthesis, Properties, and Applications*, John, Wiley & Sons, Hoboken, NJ, 2011.
151. Y. H. Kim, *J. Polym. Sci., Part A: Polym. Chem.*, 1998, **36**, 1685–1698.
152. K. Inoue, *Prog. Polym. Sci.*, 2000, **25**, 453–571.
153. B. Voit, *J. Polym. Sci., Part A: Polym. Chem.*, 2000, **38**, 2505–2525.
154. M. Jikei and M.-a. Kakimoto, *Prog. Polym. Sci.*, 2001, **26**, 1233–1285.
155. C. Gao and D. Yan, *Prog. Polym. Sci.*, 2004, **29**, 183–275.
156. J. Han and C. Gao, *Curr. Org. Chem.*, 2011, **15**, 2–26.
157. X. Hu, L. Zhou and C. Gao, *Colloid Polym. Sci.*, 2011, **289**, 1299–1320.
158. L. Zhou, C. Gao and W. Xu, *Langmuir*, 2010, **26**, 11217–11225.
159. L. Zhou, C. Gao and W. Xu, *ACS Appl. Mater. Interfaces*, 2010, **2**, 1483–1491.
160. L. Zhou, C. Gao, X. Hu and W. Xu, *ACS Appl. Mater. Interfaces*, 2010, **2**, 1211–1219.

161. C. Y. Hong, Y. Z. You, D. Wu, Y. Liu and C. Y. Pan, *Macromolecules*, 2005, **38**, 2606–2611.
162. L. Zhou, C. Gao and W. Xu, *Macromol. Chem. Phys.*, 2009, **210**, 1011–1018.
163. Y. Xu, C. Gao, H. Kong, D. Yan, Y. Z. Jin and P. C. P. Watts, *Macromolecules*, 2004, **37**, 8846–8853.
164. M. Adeli, N. Mirab, M. S. Alavidjeh, Z. Sobhani and F. Atyabi, *Polymer*, 2009, **50**, 3528–3536.
165. C. Gao, Y. Z. Jin, H. Kong, R. L. D. Whitby, S. F. A. Acquah, G. Y. Chen, H. Qian, A. Hartschuh, S. R. P. Silva, S. Henley, P. Fearon, H. W. Kroto and D. R. M. Walton, *J. Phys. Chem. B*, 2005, **109**, 11925–11932.
166. W. Yuan, G. Jiang, J. Che, X. Qi, R. Xu, M. W. Chang, Y. Chen, S. Y. Lim, J. Dai and M. B. Chan-Park, *J. Phys. Chem. C*, 2008, **112**, 18754–18759.
167. Y. Yang, X. Xie, Z. Yang, X. Wang, W. Cui, J. Yang and Y.-W. Mai, *Macromolecules*, 2007, **40**, 5858–5867.
168. J. Y. Choi, S. W. Han, W. S. Huh, L. S. Tan and J. B. Baek, *Polymer*, 2007, **48**, 4034–4040.
169. J.-Y. Choi, S.-J. Oh, H.-J. Lee, D. H. Wang, L.-S. Tan and J.-B. Baek, *Macromolecules*, 2007, **40**, 4474–4480.
170. Y. P. Zheng, J. X. Zhang, P. Y. Yu, L. L. Liu and Y. Gao, *J. Compos. Mater.*, 2009, **43**, 2771–2783.
171. J. M. J. Fréchet, M. Henmi, I. Gitsov, S. Aoshima, M. R. Leduc and R. B. Grubbs, *Science*, 1995, **269**, 1080–1083.
172. A. H. E. Müller, D. Yan and M. Wulkow, *Macromolecules*, 1997, **30**, 7015–7023.
173. D. Yan, A. H. E. Müller and K. Matyjaszewski, *Macromolecules*, 1997, **30**, 7024–7033.
174. C. J. Hawker, J. M. J. Frechet, R. B. Grubbs and J. Dao, *J. Am. Chem. Soc.*, 1995, **117**, 10763–10764.
175. K. Matyjaszewski, S. G. Gaynor and A. H. E. Müller, *Macromolecules*, 1997, **30**, 7034–7041.
176. C. Y. Hong, C. Y. Pan, Y. Huang and Z. D. Xu, *Polymer*, 2001, **42**, 6733–6740.
177. H. Mori, D. C. Seng, M. Zhang and A. H. E. Müller, *Langmuir*, 2002, **18**, 3682–3693.
178. H. Mori, D. C. Seng, H. Lechner, M. Zhang and A. H. E. Müller, *Macromolecules*, 2002, **35**, 9270–9281.
179. H. Mori, A. Walther, X. André, M. G. Lanzendörfer and A. H. E. Müller, *Macromolecules*, 2004, **37**, 2054–2066.
180. Z. Jia and D. Yan, *J. Polym. Sci., Part A: Polym. Chem.*, 2005, **43**, 3502–3509.
181. C. Cheng, K. L. Wooley and E. Khoshdel, *J. Polym. Sci., Part A: Polym. Chem.*, 2005, **43**, 4754–4770.
182. H. Kong, C. Gao, Z. Jia, C. Liu and D. Yan, *Chin. Pat.*, ZL 200310121620.2., 2003.

183. L. Zhou, C. Gao, W. J. Xu, X. Wang and Y. H. Xu, *Biomacromolecules*, 2009, **10**, 1865–1874.
184. L. Zhou, C. Gao and W. Xu, *Langmuir*, 2010, **26**, 11217–11225.
185. L. Zhou, C. Gao, W. Xu, X. Wang and Y. Xu, *Biomacromolecules*, 2009, **10**, 1865–1874.
186. D. Wilms, S.-E. Stiriba and H. Frey, *Acc. Chem. Res.*, 2009, **43**, 129–141.
187. X. Wang, L. Zhou, C. Gao and Y. H. Xu, *Acta Polym. Sin.*, 2009, 717–722.
188. Y. Yang, X. Xie, J. Wu, Z. Yang, X. Wang and Y. W. Mai, *Macromol. Rapid Commun.*, 2006, **27**, 1695–1701.
189. R. H. Zhang, Y. K. Yang, X. L. Xie and R. K. Y. Li, *Compos. Part A – Appl. S.*, 2010, **41**, 670–677.
190. J. C. Zhao, F. P. Du, X. P. Zhou, W. Cui, X. M. Wang, H. Zhu, X. L. Xie and Y.-W. Mai, *Compos. Part B – Eng.*, 2011, **42**, 2111–2116.
191. L. Cao, W. Yang, J. Yang, C. Wang and S. Fu, *Chem. Lett.*, 2004, **33**, 490–491.
192. Y. Z. You, J. J. Yan, Z. Q. Yu, M. M. Cui, C. Y. Hong and B. J. Qu, *J. Mater. Chem.*, 2009, **19**, 7656–7660.
193. G. M. Neelgund and A. Oki, *Appl. Catal. B – Environ.*, 2011, **110**, 99–107.
194. G. H. Zhao, Y. F. Li, J. Z. Wang and H. Zhu, *Appl. Microbiol. Biotechnol.*, 2011, **91**, 591–601.

CHAPTER 6
Metallic Single-walled Carbon Nanotubes for Electrically Conductive Materials and Devices

ANKOMA ANDERSON[a], FUSHEN LU*[b], MOHAMMED J. MEZIANI*[c] AND YA-PING SUN*[a]

[a] Department of Chemistry and Laboratory for Emerging Materials and Technology, Clemson University, Clemson, SC 29634-0973, USA;
[b] Department of Chemistry, Shantou University, Guangdong 515063, China;
[c] Department of Chemistry and Physics, Northwest Missouri State University, Maryville, MO 64468-6001, USA
*E-mail: fslu@stu.edu.cn, meziani@nwmissouri.edu or syaping@clemson.edu

6.1 Introduction

Carbon nanotubes are often described in the literature as being extremely electrically conductive.[1-5] For single-walled carbon nanotubes (SWNTs), however, the widely prescribed high electrical conductivity is associated only with the metallic ones. The currently available production methods for SWNTs generally yield metallic and semiconducting mixtures. Conceptually, the formation of an SWNT may be understood in terms of a graphene sheet being rolled into a cylindrical structure,[6,7] for which the operation requires a chiral vector C_h, consisting of two primitive vectors ($C_h = na_1 + ma_2$), to match the graphene carbon atoms from edge to edge (Figure 6.1). The chiral vector, also commonly referred to as the chiral index (*n,m*) (or chirality, helicity), uniquely defines the diameter (*d*) and chiral angle (*θ*) of an SWNT:

$$d = \frac{\sqrt{3}a_{c-c}}{\pi}\sqrt{n^2+nm+m^2} \tag{1}$$

$$\theta = \tan^{-1}\left[\sqrt{3}m\big/(2n+m)\right] \tag{2}$$

where a_{c-c} (~0.142 nm) is the nearest neighbor C–C distance. Depending on the corresponding chiral vector, an SWNT is either metallic (including semi- or quasi-metallic) or semiconducting, which is often referred to as "metallicity". When $n - m \neq 3q$ (q is an integer), the electronic density of states (DOS) in the SWNT exhibits a significant band-gap near the Fermi level, and the nanotube is thus semiconducting; when $n - m = 3q$, the conductance and valence bands in the SWNT overlap, and the nanotube is thus metallic (or semi-metallic when $n \neq m$). Statistically, there are twice as many ways for rolling a graphene sheet into a semiconducting SWNT as ways for rolling the same sheet into a metallic SWNT. Therefore, the metallic-to-semiconducting ratio of 1:2 should generally be expected in an as-produced mixture of SWNTs.

There have been many predictions and reports in the literature on the superior properties of SWNTs for potentially a wide variety of technological applications.[8–12] Among those based on the high electrical conductivity are the use of SWNTs in high mobility or even ballistic transistors,[13–15] and in the development of electrodes for signal transmission and detectors for sensing chemical and biological materials.[16–21] In the ongoing pursuit of new and/or renewable energy sources, SWNTs are considered as potentially excellent building blocks for a variety of energy conversion and storage technologies,

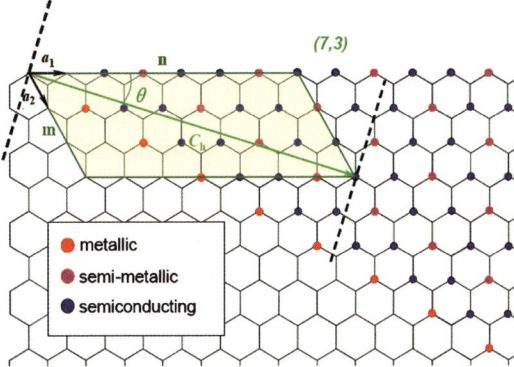

Figure 6.1 The conceptual SWNT formation by rolling up a graphene sheet. As an example, the dashed lines represent the two edges that will merge in the rolling up of a (7,3) semiconducting SWNT.[22]

such as in optoelectronic devices,[1,9,23–35] batteries and supercapacitors,[9,36–43] and various fuel cells.[9,44–48] While these are all potentially exciting applications, what is often missing in many studies and/or development efforts has been an acknowledgment of the fact that as-produced nanotube samples are mixtures of metallic and semiconducting SWNTs, which are distinctively different in electrical conductivity and many other aspects. Indeed, a lot of the widely prescribed and/or predicted potential SWNT applications exploit the properties of metallic SWNTs, which generally speaking represent the minor fraction in the as-produced mixtures. Therefore, the harvesting of sufficient quantities of metallic SWNTs is critical to the relevant technological applications.

Beyond electrical conductivity, metallic and semiconducting SWNTs also differ in many other physical and chemical properties (static polarizibility, doping effect, chemical reactivity and those related to electronic structures). For example, it is known that semiconducting SWNTs are extremely sensitive to electrical gating, and capable of conductance changes by orders of magnitude under various electrostatic gate voltages.[49–52] Conversely, metallic SWNTs are less sensitive to molecular adsorption and chemical gating, as charge transfer does not significantly affect the charge density at the Fermi level.[49] In the widely pursued use of SWNTs in transparent conductive coatings to compete with the currently predominant indium tin oxide (ITO) technology,[4,53] metallic SWNTs are obviously required. As semiconducting SWNTs are more absorptive than their metallic counterparts,[54] the presence of the former in the coatings is negative to performance in terms of both lower electrical conductivity and reduced optical transmittance. In fact, these technological needs have been driving the development of methodologies for the post-production separation of SWNTs, especially for the isolation of the metallic ones for their lower population (thus relatively more valuable) in the as-produced mixtures.[5]

6.2 Harvesting Metallic SWNTs

There are presently no production methods that can selectively yield metallic SWNTs. Therefore, post-production separation has and will continue to represent a viable option for the harvesting of highly enriched metallic SWNTs in sufficient quantities.[22,55–58] The separation methods reported in the literature generally exploit the differences (beyond those in electrical conductivity) between metallic and semiconducting SWNTs, including static polarizability and surface characteristics, chemical reactivity, *etc.* SWNTs defined by different chiral index values are often of different diameters, and the difference is particularly meaningful in nanotube samples of broad size distributions, such as those from high-pressure carbon monoxide (HiPCO) and similar production methods. For post-production separation in various quantity scales, agents from DNA and surfactants (coupled with ultracentrifugation or electrophoresis) to those with more specifically targeted selective interactions and thus for separation in relatively larger quantities have been developed and used with significant successes.[59,60,61]

The wrapping of SWNTs by DNA was found to be selective between the nanotubes of different chiral index values,[55,57,64–67] which has served as the basis for the now well-developed density gradient ultracentrifugation (DGU) method for post-production separation.[54,57,64–67] According to Zheng et al.,[68,69] single-stranded DNAs (ssDNAs) readily adsorb on to the nanotube surface and efficiently disperse SWNTs upon ultrasonication. As ssDNAs are negatively charged species, they express linear negative charges on the wrapped SWNTs. On the other hand, metallic and semiconducting SWNTs have different polarizabilities and also different size profiles, which result in different linear charge densities in the ssDNA-wrapped SWNTs. Zheng and co-workers[69,70] exploited such differences for the separation of metallic and semiconducting SWNTs in an anionic exchange column.

Hersam[57] first used centrifugation to separate ssDNA-wrapped SWNTs in solutions with a density gradient medium. Upon centrifugation, the nanotubes formed discrete colored bands in the centrifuge tubes, from which SWNTs of different diameter distributions were harvested and, in the resulting optical absorption spectra, semiconducting SWNTs with different chiralities were identified. Subsequently, the same group substituted DNAs with surfactants in the DGU method.[65–67] For example, SWNTs were dispersed with a mixture of sodium cholate and sodium dodecyl sulfate (SDS), followed by ultracentrifugation in density gradient media. Kataura and co-workers[59] recently found that SDS plays an integral role in density gradient centrifugation and agarose gel separations (more details below) due to its straight alkyl tail and charged head group, allowing metallic and semiconducting SWNTs to be well-dispersed first and then their discriminations and separation.

Other than the ultracentrifugation, the ssDNA-dispersed HiPCO SWNTs have been separated by ion exchange chromatography.[70] However, the method was reported to be ineffective in the separation of metallic SWNTs. In another method that exploited the dispersion of SWNTs by DNA or conceptually similar surfactant species, agarose gels (originally developed for DNA separation) were used to separate metallic and semiconducting SWNTs.[71–74] For the separation, a nanotube sample from the HiPCO or laser-ablation production was dispersed in a surfactant (SDS) solution, followed by centrifugation to remove impurities and those bundled SWNTs.[71] The resulting dispersion was mixed with liquid agarose gel for gelation. The gel containing SDS-dispersed SWNTs was frozen, thawed and squeezed to yield a solution of enriched (70%) metallic SWNTs, while the semiconducting SWNTs (95%) were left in the gel.[71] The same separation was later demonstrated by column-based gel chromatography.[73,74] It seems that this method is more amenable to scaling-up than the DGU or the ion exchange chromatography discussed above, although the separation efficiency may still be limited by how effectively SWNTs are dispersed to the individual nanotube level by SDS or a similar surfactant. Recently, it was reported that the SWNT–surfactant interaction is a dominating factor in separation.[59] Among the parameters that affect the interactions are pH, temperature and agarose concentration. Specifically, a higher agarose concentration could result in an increase in the amount of un-adsorbed SWNTs.[59] In the subsequent separation with allyl dextran-based size exclusion gel, 13 highly enriched semiconducting SWNT species (corresponding to different chiral indices) were collected, and metallic SWNTs due to their weaker interactions with the gel remained unbound (thus collected after elution).

Metallic and semiconducting SWNTs are apparently wrapped differently by DNA or more generally in the same concept by surfactant molecules, as discussed above for various approaches to exploit such differences for post-production separation purposes. The nanotubes also have different interactions with selected functionalization or solubilization agents. Such selective interactions, sometimes considered as non-covalent functionalizations, have been found to allow relatively facile post-production separation at significant quantities.[58,75–81]

Since Haddon and co-workers[82] initiated the functionalization and solubilization of carbon nanotubes with octadecylamine (ODA), alkyl amines have been used extensively for similar purposes. In the thermal reaction of alkyl amines with purified SWNTs (by oxidative acid treatment, thus the creation of carboxylic acid moieties on the nanotube surface),[82–85] the functionalization was thought to be primarily the formation of ammonium carboxylate zwitterionic bonds.[85,86] In addition to the chemical bonds, the chemical adsorption-like strong interactions of a great quantity of amino molecules on the nanotube surface were also considered as playing a significant role in the functionalization and solubilization of the nanotubes.[87] By recognizing this combination of chemical bonds and strong interactions in the functionalization of purified SWNTs with ODA (Figure 6.2), Papadimitrakopoulos and co-workers[79,80] first reported that the solubilization was preferential towards the semiconducting SWNTs; thus their separation from the metallic counterparts. It was proposed that the observed selectivity toward the solubilization of semiconducting SWNTs could be attributed to the charge redistribution on the surface of semiconducting-enriched small bundles together with the formation of an ordered two-dimensional (2D) arrangement of NO_3/surfactant amine/water layer.[80] The method was apparently more effective for smaller diameter SWNTs from the HiPCO process (0.8–1.3 nm) than larger diameter SWNTs derived from laser ablation production (1.15–1.55 nm).[79]

Sun and co-workers[75–78] exploited the selectivity in the non-covalent functionalization of SWNTs with planar aromatic molecules, such as derivatized porphyrin or pyrene (Figure 6.3), for the post-production separation. The separation method simply "splits" the starting nanotube mixture by selectively solubilizing semiconducting SWNTs and leaving their metallic counterparts behind, and consequently is capable of handling significant sample quantities. Experimentally, as-produced samples of

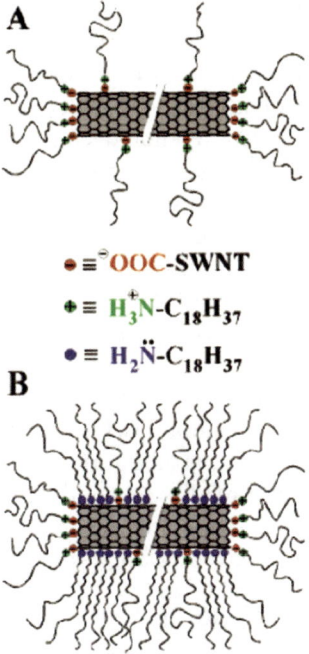

Figure 6.2 Pictorial representation of possible ODA interactions with oxidative acid purified SWNTs through (A) zwitterions formation and (B) physisorption-assisted organization of ODA on SWNT sidewalls.[79]

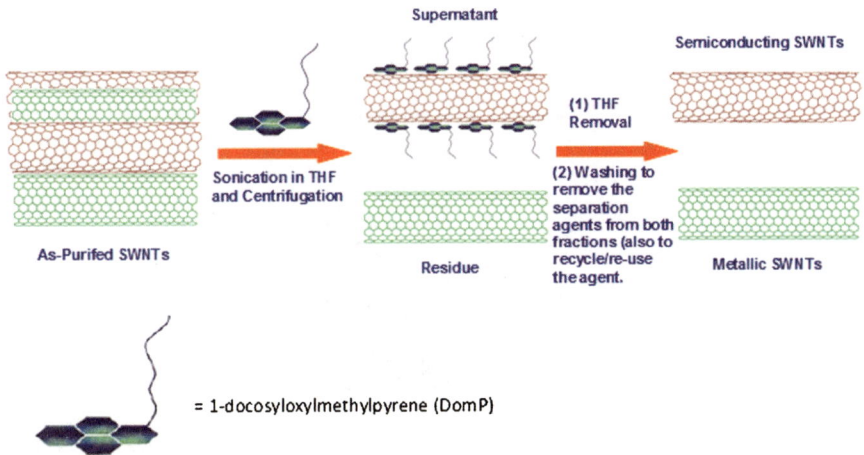

Figure 6.3 The post-production separation scheme with the use of derivatized pyrene and porphyrin compounds.[75,76]

SWNTs from the arc discharge or laser ablation production were purified in the widely used procedure involving nitric acid treatment.[76] The purified sample was dispersed in an organic solvent such as tetrahydrofuran (THF) with the selected planar aromatic molecule, derivatized porphyrin or pyrene with long alkyl chain(s), aided by homogenization and sonication. These conditions were similar to those commonly used in the non-covalent functionalization of carbon nanotubes.[88] The resulting dispersion was separated *via* simple centrifugation to yield the soluble fraction containing non-covalently functionalized semiconducting SWNTs and the insoluble residue enriched with metallic SWNTs.[75,77] The same separation could be improved by repeating the same procedure for a second time, although a third repeat was not necessary. The selective interaction between separation agents and semiconducting SWNTs was reflected by their diminishing of both the S_{11} and S_{22} absorption bands of the nanotubes (doping effects due to "adsorption" of aromatic moieties on to the nanotube surface).[81,89] Separated metallic and semiconducting SWNTs could readily be recovered from the soluble fraction and residue, respectively, by removing the separation agents in repeated solvent washing and/or dialysis.[76,78] The experimental conditions for the recovery were sufficiently mild so as not to change the dispersion characteristics of the separated SWNTs.[78,90] The separation agents could also be recovered nearly quantitatively, as verified by NMR results, and reused for the same separation purpose.[90]

More recently a molecule with a pair of planar aromatic moieties (Figure 6.4) was synthesized and used to exaggerate the difference between metallic and semiconducting SWNTs in the non-covalent functionalization and solubilization.[78] The molecule representing essentially the "molecular tweezers" approach (Figure 6.4) exhibited significant selectivity toward

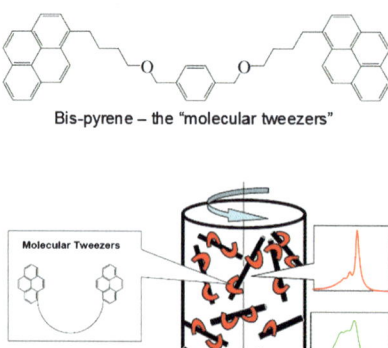

Figure 6.4 Upper panel: the chemical structure of 1,1'-bis-pyrene butyoxyl-*p*-xylene (bis-pyrene); lower panel: schematic illustration of the molecular tweezers approach for the post-production separation of metallic and semiconducting SWNTs with the use of bis-pyrene.[58]

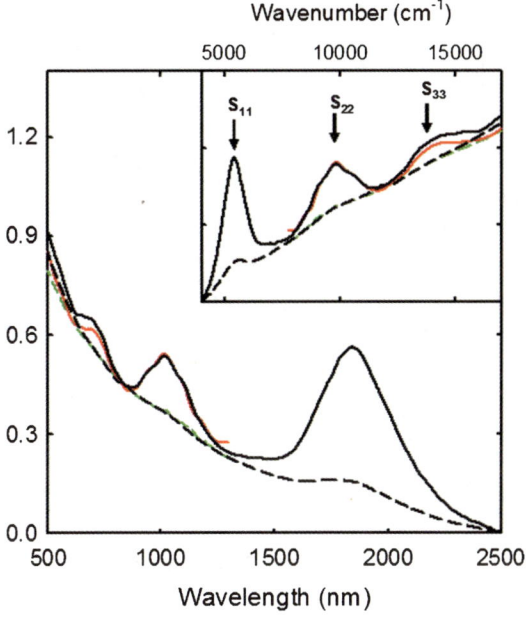

Figure 6.5 Optical absorption spectra of the separated metallic (dashed lines: green, in DMF dispersion; black, in solid-state thin film) and semiconducting (solid lines: red, in DMF dispersion; black, in solid-state thin film) fractions. The same spectra on the wave number scale are shown in the insert, with the S_{11}, S_{22} and S_{33} peaks marked.[58]

semiconducting SWNTs in the solubilization to allow the convenient harvesting of bulk metallic and semiconducting fractions of high purities, up to 92% and 97%, respectively, according to optical absorption spectral results (Figure 6.5).

Electron microscopy analyses of the separated fractions revealed no significant differences in their images from those of the pre-separated mixtures.[58,76,78] Major differences were observed in their near-infrared (IR) absorption (Figure 6.5) and resonance Raman spectra, consistent with the metallic and semiconducting fractions.[58,76,78] In Raman spectra (Figure 6.6), the G-band of the metallic fraction was much broader and more asymmetric, known as the Breit-Wigner-Fano (BWF) feature,[91] as compared with those for the pre-separation nanotube sample and more so the semiconducting fraction.

Lu et al.[92] also applied the same non-covalent functionalization approach to the purification of SWNTs. A water-soluble pyrene derivative 1-pyreneacetic acid was used to solubilize the purified (typical nitric acid treatment) SWNTs in aqueous solution, allowing a nearly complete removal of residual metal catalysts and carbonaceous impurities. According to thermogravimetric analysis (TGA) results, the purified sample of SWNTs contained little other carbonaceous impurities and no more than 3% of residual catalysts by weight

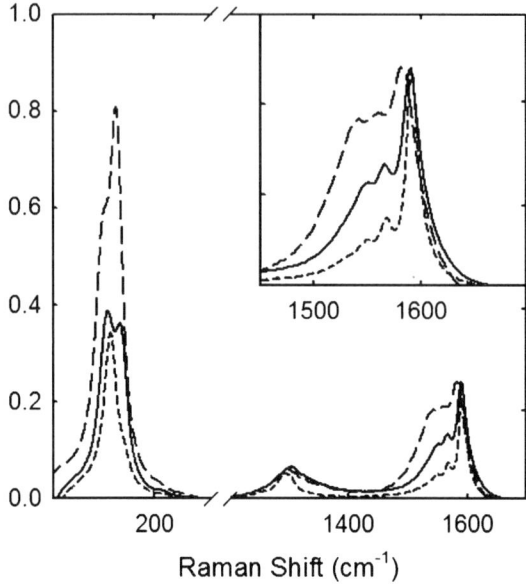

Figure 6.6 Raman spectra (785 nm excitation, with the G-band region featured in the insert) of the pre-separated SWNTs (solid line), and the separated metallic (long dashed line) and semiconducting (short dashed line) fractions.[58]

(obviously much less by volume due to the much higher density of the metals than that of carbon).[92] The non-covalent functionalization of SWNTs with 1-pyreneacetic acid in an aqueous solution was non-selective toward either metallic or semiconducting SWNTs, with hardly any changes in their ratio in the purification according to quantitative Raman results. Another important feature for the purification was such that the highly pure SWNTs remained solvent-dispersible,[92] which is obviously valuable to the targeted use of the purified sample in the separation into metallic and semiconducting fractions. Therefore, a combination of the purification method and the molecular tweezers, both based on planar aromatic molecules or moieties and their non-covalent interactions with the nanotubes, should enable the harvesting of metallic SWNTs with high sample purity as well as high metallicity purity.

There have been computational studies on the selectivity in non-covalent functionalization of SWNTs with planar aromatic molecules.[93] The results, suggesting that the adsorption strength in the most stable configuration of porphyrin and pyrene on a semiconducting (10,0) vs. metallic (6,6) SWNT was different (with the former being larger),[93] were thus consistent with the experimental observation that these aromatic species were selective toward functionalizing semiconducting SWNTs over their metallic counterparts.

The selectivity in the non-covalent functionalization of metallic vs. semiconducting SWNTs is apparently dependent on the specific planar aromatic molecule, positive for one pyrene derivative (1-docosyloxymethyl-pyrene or

DomP, Figure 6.3),[76,89,90] and negative for another (1-pyreneacetic acid).[92] Such dependence actually strengthens the argument for the selectivity. Anthracene derivatives were also found to be generally poor or even incapable of selectively solubilizing semiconducting SWNTs for the intended separation,[22,94] so were molecules with a pair of anthracene moieties. Nevertheless, a number of other aromatic compounds, including coronene tetracarboxylic acid,[95] derivatized pentacene[94,96] and crown ether-terminated pyrene,[97] have been used for the similar separation of metallic and semiconducting SWNTs. An interesting twist was that the crown ether-terminated pyrene was reported to be selective toward solubilizing metallic HiPCO SWNTs in chloroform.[97] With the recent success,[58] the further design and use of molecular tweezers probably represent the most promising for improvements in both the separation operation itself and the separation results (especially the purity of the separated metallic SWNTs).

6.3 Electrically Conductive Nanocomposites

Owing to their remarkable electrical properties and structural uniqueness, SWNTs have been extensively incorporated into polymers for the fabrication of electrically conductive nanocomposites.[76,98] These conductive composites showed promising potentials in diverse applications such as electrostatic dissipation, electrostatic painting, electromagnetic interference (EMI) shielding, printable circuit wiring and transparent conductive coating.[99] The conductivity of SWNT-polymer composites follow the established percolation theory. Namely, electrical conductivity jumps sharply when SWNTs form a three-dimensional conductive network within the matrix.[100] Percolation occurred from an SWNT loading of as low as 0.005 vol% to a few vol%, depending on the processing technique, nanotube structure and morphology, etc.[101] As SWNTs are heavily bundled and tend to aggregate in the polymer matrix, homogeneously dispersing nanotubes in solvents is a prerequisite for solution-based casting techniques. Fortunately, there has been significant progress in the solubization and functionalization of carbon nanotubes to enable relatively simple fabrication of the relevant conductive nanocomposites.[102] Various polymers, including conductive polymers, have been used for the incorporation of SWNTs for nanocomposites from different casting processes.

6.3.1 Composites with Non-enriched SWNTs

In a recent report by Carey and co-workers[103] SWNTs were covalently functionalized with poly(aminobenzene sulfonic acid) and then mixed with poly(vinyl alcohol) (PVA) for spin casting. The conductivity increased with nanotube loading and reached 10^{-3} S m^{-1} at the highest loading of 15 vol%. The relatively low conductivity was likely due to the damage of electrical structure by covalent linkage and the presence of a thin polymer coating surrounding the nanotubes. Non-covalent functionalization might be more advantageous in preserving the electrical properties of SWNTs with minimal damage to the nanotube surface. For examples, SWNTs non-covalently functionalized with poly(vinyl butyral) were injected on to the surface of an ethanol/water solvent mixture to form thin composite films.[104] The films were then transferred on to glass or plastic substrates. The observed sheet resistances decreased with the increase in nanotube loadings, reaching 3.8×10^6 and 4.6×10^4 Ω sq^{-1} (corresponding to 717 S m^{-1}) in films with 50 wt% and 80 wt% SWNTs (68% transmittance), respectively. By using a similar functionalization scheme, Guo et al.[105] fabricated SWNT/polyacrylonitrile composites with various nanotube loadings via drop-casting. The conductivity values were 88, 243 and 5500 S m^{-1} for the composites with 5, 10 and 20 wt% nanotube loadings, respectively.

Sung et al.[106] constructed transparent HiPCO SWNT/poly(styrene-*block*-4 vinyl pyridine) composite films via spin coating. The sheet resistance in the composite film with ~90% transmittance was approximately 700 kΩ sq^{-1} at 7% nanotube loading.[106] Interestingly, upon doping with HAuCl$_4$·3H$_2$O, the sheet resistance decreased to as low as 6 kΩ sq^{-1}. However, the high cost of the gold salt makes this technique less practical for production of conductive composites at large scales. For HiPCO SWNT/polystyrene composite films fabricated by a filtration method, the conductivity was ~100 S m^{-1} at ~22 wt% of nanotube loading (compared with 10^4 S m^{-1} in neat nanotube film).[107] Bryning et al.[108] incorporated laser-oven SWNTs and HiPCO SWNTs into epoxy matrices by dispersing the nanotubes with epoxy resin in N,N-dimethylformamide. Upon curing the composites, SWNTs were allowed to re-aggregate slightly to acquire low percolation thresholds by increasing the local interactions between nanotubes. The results seemed to be contradictory to the common observation that well-dispersed nanotubes generally have a lower percolation threshold. Grunlan et al. reported an emulsion-based process to produce porous SWNT/poly(vinyl acetate) composites.[109] In their work, gum arabic-stabilized SWNTs were combined with a poly(vinyl acetate) emulsion to create an aqueous pre-composite mixture. During drying, microscopic solid polymer particles in the emulsion pushed the nanotubes into an interstitial (or segregated) network, achieving a low percolation threshold (estimated to be approximatley 0.04 wt%, with maximum conductivity of $10^{1.25}$ S m^{-1} at 4 wt% nanotube loading).[109] Very recently, spin-spray layer-by-layer (SSLbL) assembly, a combination of multiprocessing techniques, was used to generate composite films from poly(styrene sulfonate) (PSS)- or Nafion-stabilized SWNTs and PVA.[110] SSLbL-assembled films with integrated carbon nanotubes yielded a better correlation between LbL film growth and conductance than previously seen for these systems.

SWNT/polymer composites could also be made by a two-step fabrication process. For example, SWNT films were first fabricated on glass slides via Mayer rod coating, and then polymers such as PVA, Nafion and polyvinylidenefluoride (PVDF) were infiltrated into SWNT networks.[111] As the polymer occupied the empty space between the nanotubes, a freestanding

SWNT/polymer composite film could be obtained by peeling the film off the glass substrate. In comparison with the performance in pristine SWNT films, the sheet resistance values in the PVA-SWNT and PVDF-SWNT composite films were higher by factors of 8 and 3, respectively, whereas the sheet resistance in Nafion-SWNT was slightly lower. The resistance change was attributed to the difference in the electron-donating and electron-withdrawing capabilities of the polymers.[112] In another report, SWNT-based nanofoams (cross-linked SWNT networks with carbon nanoparticles as binder) were prepared and subsequently used as scaffolds to create poly(dimethylsiloxane) (PDMS) composites *via* the infiltration method.[113] The observed conductivity value was 1 S cm^{-1} in the composite with a low nanotube loading level (1.2 wt%).

Transparent SWNT/insulating polymer nanocomposites have not been made with high conductivity values, mainly because of the intrinsic charge localization arising from insulating polymer dispersant. In a direct comparison, SWNTs were stabilized with insulating surfactant (SDS) or conductive poly(3,4-ethylenedioxythiophene) (PEDOT):PSS.[114] Both SWNT dispersions were introduced into a polystyrene matrix *via* a latex-based route. Composites with PEDOT:PSS-stabilized SWNTs showed a percolation threshold of 0.18 wt% and a conductivity value of 500 S m^{-1}, compared with 3.8 wt% and 20 S m^{-1} for the SDS-based composites, respectively. The authors attributed the conductivity enhancement to "conduction bridge" formed by the conductive polymer between adjacent SWNTs.[114,115]

Conjugated polymers are conductive upon doping and capable of effectively dispersing SWNTs in terms of π–π stacking between their multi-aromatic moieties and the nanotube surface. Therefore, various conjugated polymers including polypyrrole, polyaniline (PANI), polythiophene and their derivatives have been successfully incorporated into conductive SWNT composites to minimize the electrical resistance. Xian *et al.*[116] used electrochemical approach to prepare layered SWNT/polypyrrole composites. A SWNT array was loaded on to a heavily doped silicon wafer, followed by electro-polymerization of the pyrrole layer on top of SWNTs. A free-standing, transparent and conductive composite film was obtained after peeling off from silicon substrate, achieving sheet resistance of ∼20 kΩ sq^{-1} at a transmittance of 68%. Similarly, Ge *et al.*[117] prepared SWNT films by vacuum filtration method and an electrochemically deposited PANI layer on top of the SWNT film. The SWNT/PANI composite film exhibited a sheet resistance of 1.5 kΩ sq^{-1} at 65% transmittance (500 nm). Ma *et al.*[118] found that only an *in situ* polymerized thin layer of conductive polymer (PANI boronic acid) could effectively interlink the SWNTs and thus increase the conductivity. Composites by simply mixing a preformed conducting polymer with dispersed SWNTs did not exhibit conductivity enhancement. Following this strategy, Yu *et al.*[119] dropped a monomer solution on to SWNT films and photopolymerized the monomer to obtain a SWNT/polymer composite. The sheet resistances of the composite films varied from 500 Ω sq^{-1} (87% transmittance) to 20 Ω sq^{-1} (40%

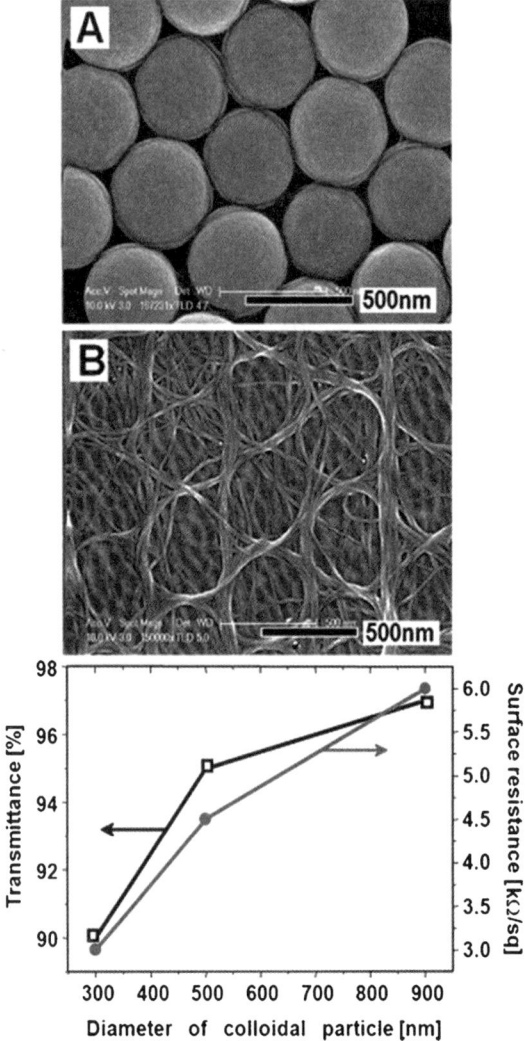

Figure 6.7 Upper: SEM surface images of (A) 2D silica colloid–carbon nanotube complex crystals, and (B) the carbon nanotube network morphology after removing the colloidal particles; lower: transmittance and resistivity tuning using the controlled colloidal templates.[120]

transmittance). The composites could be reversibly stretched by up to 50% strain with little change in sheet resistance after annealing.

Kim et al.[120] fabricated SWNT composite films with ordered voids by using sacrificial 2D silica colloidal template (Figure 6.7). Experimentally, SWNTs stabilized with water-soluble polythiophene were mixed with a suspension of silica colloidal particles, and then assembled into 2D colloid–SWNT array on flexible substrate. Selective removal of silica colloids led to a composite film with regular voids, which showed a surface resistivity of 4.5 kΩ sq^{-1} at 95% transmittance. These films gave superior optical transmittance compared with those films with random network structures because the ordered voids enabled efficient transmittance of light (Figure 6.7).

PEDOT with and without the dopant PSS has been the most commonly used conductive polymer for SWNT composites. De et al.[121] prepared composite films by vacuum filtration from aqueous dispersions with PEDOT:PSS as the matrix and both arc and HiPCO SWNTs as the filler. The optimal performance was observed for a 80 nm thick film containing 60 wt% arc SWNTs, with a sheet resistance of 80 Ω sq^{-1} at 75% transmittance. Electromechanical testing showed these films to be stable under flexing and cycling. Later, a modified PEDOT copolymer with enhanced solubility, perchlorate-doped poly(3,4-ethylenedioxythiophene)-*block*-poly(ethyleneoxide) (P-PEDOT-b-PEO) was used to disperse SWNTs and then to fabricate conductive nanocomposites *via* vacuum filtration method.[122] The sheet resistance of the composite film was approximately 600 Ω sq^{-1} with 80% transmittance.

6.3.2 Composites with Separated Metallic SWNTs

Metallic SWNTs are of extremely high electrical conductivity, with the theoretical value estimated as high as 10^6 S cm^{-1}.[123] It is also well-established that the propagation of electrons in metallic nanotubes is ballistic, largely free from scattering over a distance in thousands of atoms.[5] With their resistance approaching the theoretical lower limits,[124,125] metallic SWNTs may, in principle, carry an electrical current density of 4×10^9 A cm^{-2}, more than 1000 times greater than that in electrically conductive metals such as copper.[126] Indeed, since the first fabrication and investigation of electrical devices based on metallic SWNTs in 1997,[127,128] a number of other potential applications have been pursued,[129] including especially the development of electrically conductive polymeric nanocomposites,[76] and the use of SWNTs to mitigate static charges in thin film materials and related devices.[130]

However, the reported conductive SWNT/polymer composites have mostly been based on the mixture of metallic and semiconducting nanotubes. In addition to the resistance from polymer matrix, the large inter-tube junction resistance, especially between metallic and semiconducting tubes, dramatically compromised electrical performance of the SWNT-based composites. Sun and co-workers[76] unambiguously demonstrated that the separated metallic SWNTs offered a much enhanced performance in nanocomposites with conductive polymers poly(3-hexylthiophene) (P3HT) and other polythiophenes. The separated metallic SWNTs were dispersed in P3HT solution for drop-casting the composite films. Electrical conductivity measurements revealed that the composite films were obviously more conductive than those containing a non-separated mixture of SWNTs. The conductivity improvement was more than one order of magnitude with 20 wt% metallic SWNT loading (Figure 6.8).

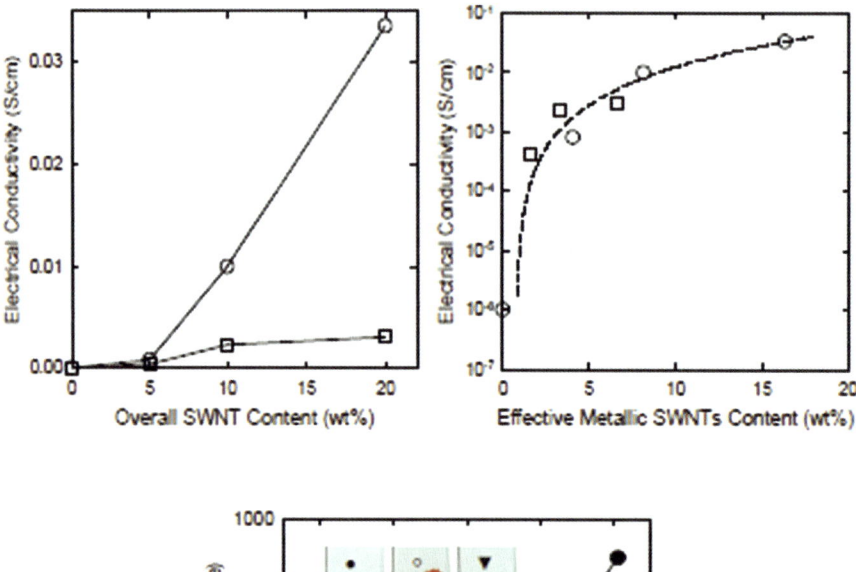

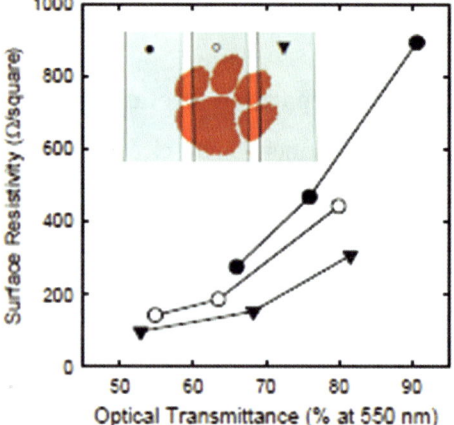

Figure 6.8 Electrical conductivity and surface resistivity comparison. Upper panel: electrical conductivity results of P3HT/SWNT composite films depending on (left) different amounts of pre-separated (□) and separated metallic (○) nanotube samples, and (right) their corresponding effective metallic SWNT contents in the films (dashed line: the best fit in terms of the percolation theory equation). Lower panel: Surface resistivity results of PEDOT:PSS/SWNT films on glass substrate with the same 10 wt% nanotube content (○: pre-separated purified sample and ▼: separated metallic SWNTs; and for comparison, ●: blank PEDOT:PSS without nanotubes) but different film thickness and optical transmittance at 550 nm. Shown in the inset are representative films photographed with tiger paw print as background.[76]

The superior conductivity of separated metallic SWNTs was also verified in the transparent conductive PEDOT:PSS composites. A suspension of the separated metallic SWNTs or the pre-separation nanotube mixture in DMSO was mixed with aqueous PEDOT:PSS for spray coating on to glass substrate.

The thickness of the composite films reflected by transmittance was controlled by spraying different volumes of nanotube suspension. Again, the composite films with enriched metallic SWNTs were consistently and substantially more conductive than those with non-separated SWNTs (and both better than the films with neat PEDOT:PSS, Figure 6.8).[76]

6.4 Transparent Conductive Coatings and Films

In nearly all devices encompassing some type of conversion between electrons and photons, a component that is electrically conductive yet optically transparent is necessary.[131] Although ITO coating technology is the predominant method used to create transparent electrodes,[53,132,133] significant deficiencies have been identified, including their incompatibility with flexible substrates and demanding processing conditions. Thus, the use of carbon nanotubes as potential replacements for ITO in transparent conductive coatings, especially those requiring high flexibility, has been a research endeavor of great interest ever since the discovery of SWNTs.[4,30,134] Again, it is the metallic fraction in nanotube mixtures that is responsible for the extremely high electrical conductivity, so that the use of metallic SWNTs are particularly promising to the desired low resistivity and high optical transparency in the electrodes.[5]

Because of the generally scarcity of separated metallic SWNTs, most of the reported nanotube films as transparent electrodes have been based on nanotube mixtures (containing both metallic and semiconducting SWNTs, with the former being the minor fraction).[5] These nanotube films are quasi-2D interconnected networks, in which the electrical conductivity is controlled by both the intrinsic conductivity of individual SWNTs and tube–tube junctions.[126,135] Many approaches have been applied to the fabrication of transparent conducting coatings (ultra-thin films) from SWNTs, including spraying, filtration, rod/wire coating, LbL deposition, spin coating and dip coating.

Spray coating, a simple method for directly fabricating nanotube films in any size on various substrates, has been employed in many investigations. For example, Lee and co-workers[136] used air spraying to fabricate films of arc discharge-produced SWNTs on flexible substrate. In their experiment, a suspension of surfactant (SDS)-dispersed SWNTs was sprayed on to a preheated (100 °C) poly(ethylene terephthalate) (PET) substrate, followed by repeated rinsing in water to remove SDS. Upon doping in concentrated HNO_3, the sheet resistance of the nanotube coating could reach $\sim 40\ \Omega\ sq^{-1}$ or $70\ \Omega\ sq^{-1}$ for optical transmittance of 70% or 80% at 550 nm, respectively.[136] Blackburn and co-workers[137] modified the spraying process by replacing the airbrush pistol with an ultrasonic spray head, allowing more controlled, uniform and reproducible coating of a specific amount of SWNTs on a large substrate. Generally speaking, however, while spray coating is simple and flexible, the resulting films have an inherent sparse density that may decrease the electrical conductivity, negative to performance.[136] Even with the use of separated metallic SWNTs, according to Lu et al.,[78] the spray-coated films also exhibited relatively poorer performance. The observed surface resistivity values were actually not so different between films of as-purified SWNTs and separated metallic SWNTs, mostly of the order of 20 000 Ω sq^{-1} at 80% optical transmittance (550 nm). The results were attributed to the aggregation of the nanotubes before and during the air spray fabrication.[30,78] The presence of a surfactant in the dispersion of SWNTs would generally mitigate some of the negative aggregation effect.[78] For example, the nanotubes could be better dispersed in a dilute aqueous solution of SDS (0.1 wt%) for the simple air spray coating. After the film fabrication on glass substrate, the removal of surfactant from the film was achieved by dipping the specimen into deionized water several times. The electrical conductive performance in these films was significantly improved. For the films of as-purified SWNTs at 91% and 80% optical transmittance (550 nm), the surface resistivity values were 11 000 Ω sq^{-1} and 3000 Ω sq^{-1}, respectively.

Vacuum filtration is another popular method in the fabrication of transparent conductive films, which are morphologically like "buckypapers".[30,43,138,139] For example, Rinzler and co-workers used filtration to fabricate coatings from laser ablation-produced SWNTs.[30] Experimentally, the nanotube sample was first dispersed in solution with surfactants and then filtered through a porous membrane to form a thin film on the membrane. Upon the removal of surfactants *via* careful washing, the film on the filter membrane could reach electrical sheet resistance down to 30 Ω sq^{-1}, with an estimated optical transmittance of approximately 70% in the visible spectral region.[30] A number of advantages have been discussed regarding the filtration method, including a more homogeneous distribution and improved packing of SWNTs in the coated films.[30,78] The homogeneity is attributed to the compensation effect naturally associated with the filtration process, such that the already deposited nanotubes would reduce the flow of the nanotube suspension, thus depositing additional nanotubes into other less dense areas of the film, while the packing for improved contacts between nanotubes is due to the vacuum-pressing in the filtration.[30,78] However, films thus fabricated are limited by the filter size, a drawback for applications requiring larger size films. As most membrane filters are largely opaque or incompatible with electronic devices, an additional process to transfer films from the filter membrane to the desired substrate is required for characterization and/or device construction. There are primarily two methods for such transfer: one involves dissolving the filter wet-chemically to release the film;[30,138] and the other involves using an adhesive stamp to peel the film from the filter membrane.[33]

Lu *et al.*[78] applied the vacuum filtration fabrication to the comparison of films of as-purified SWNTs and separated metallic SWNTs (approximately 85% purity in metallicity). In the fabrication, the two nanotube samples were each dispersed into an aqueous solution of SDS. A porous alumina membrane was used as filter in the vacuum filtration of the suspended SWNTs. After filtration, the film on the filter was washed repeatedly with deionized water to remove the surfactant SDS, for which the progress in surfactant removal *via*

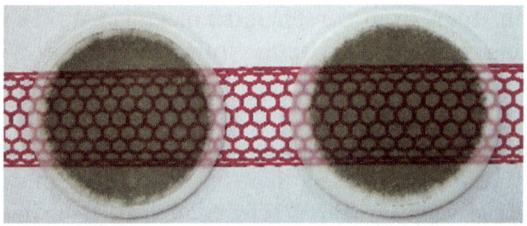

Figure 6.9 Films of SWNTs on alumina filters from the as-purified sample (left) and the separated metallic fraction (right).[78]

repeated washing was monitored in terms of TGA measurements, until no weight loss was observed up to 230 °C under nitrogen atmosphere. The resulting films were characterized by using Raman spectroscopy, from which the results confirmed the higher content of metallic SWNTs in the film corresponding to the separated metallic fraction. The porous alumina membrane as filter and then as substrate for the nanotube film is not optically fully transparent (Figure 6.9), so that a more direct comparison between the films from as-purified and separated metallic SWNTs was performed by keeping fabrication procedures for the films identical and by quantitatively determining the amount of nanotubes used in each film. Variations in the observed surface resistivity against nanotube contents in the films on alumina membrane filter are shown in Figure 6.10. Clearly, the electrical conductive

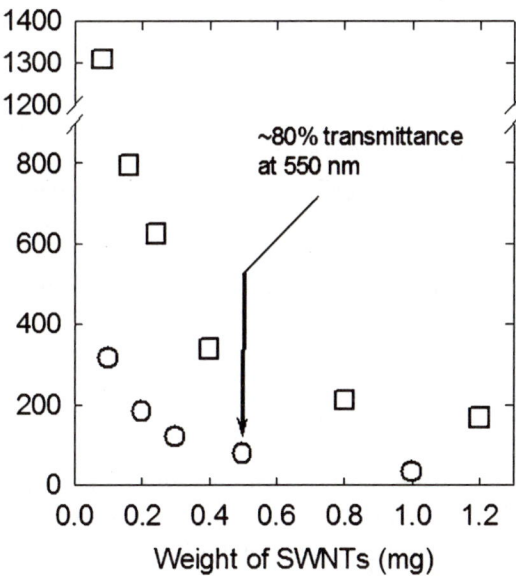

Figure 6.10 A direct comparison of surface resistivity values in films of SWNTs (fabricated *via* vacuum filtration) from the as-purified sample (□) and the separated metallic fraction (○).[78]

performance in the films from the separated metallic SWNTs was consistently much better than that in the films from as-purified SWNTs. Optical transmittances of selected films on the alumina filter were estimated by transferring the films to a transparent substrate. The results thus obtained suggested that at ~80% transmittance (550 nm) the surface resistivity of the film from metallic SWNTs was less than 100 Ω sq^{-1}, significantly lower than that of the film from as-purified SWNTs. Such a performance level achieved with the nanotube films is already competitive to that of ITO coatings for some applications.[140,141]

Rod/wire coating is another widely used fabrication method for transparent conductive films of SWNTs, which is compatible with various substrates and often used in the coating industry for the continuous and scalable production of liquid thin films.[142,143] A critical aspect in the rod coating process is the preparation of coating fluids with specific rheological behavior and wetting properties. For example, a mixture of the sodium dodecylbenzenesulfonate (SDBS) and Triton X-100 surfactants was used to disperse SWNTs for enhanced viscosity of the coating fluid.[142] The nanotube films from the coating, upon acid treatment, exhibited sheet resistances of 80 Ω sq^{-1} and 140 Ω sq^{-1} for 70% and 80% optical transmittances at 550 nm, respectively.[142] LbL deposition is another frequently used method for fabricating SWNT films,[144–146] in which thin films are created by alternately exposing a substrate to a polymer solution and an aqueous suspension of SWNTs (with a negatively charged surfactant such as PSS or SDS as a dispersion agent).[144] For example, a film thus fabricated was treated in the doping process with fuming sulfuric acid to reach a sheet resistance of 86 Ω sq^{-1} for 80% optical transmittance at 550 nm, although additional measures were necessary to preserve the doping effect for the performance stability over time.[144] Kotov and co-workers[112] took advantage of the LbL engineering method to prepare composite coatings by immersing slides in 0.1 wt% hydroethyl cellulose (HOCS) for 1 min, followed by rinsing in water and then drying with compressed air. Subsequently, these slides were dipped into the sulfonated polyetheretherketone (SPEEK)-stabilized SWNT solutions for 2 min, followed by similar rinsing and drying. The films were found to have a conductivity of 1.1×10^5 S m^{-1} and a sheet resistance of 920 Ω sq^{-1} at 86.7% transmittance. The nanotube films were doped on the basis of electron transfer from several valence bands of SWNTs to low lying unoccupied levels.[112] In order to partially align SWNTs in the transparent conductive films for improved performance, spin coating and dip coating have been found to be more effective.[140,147]

All these wet-processing methods for nanotube films seem to share some common features, as determined by the properties of SWNTs and their networks in the coated films. It is known in the literature that the resistance at an inter-tube junction is higher when the junction is between nanotube bundles.[148] Therefore, the homogeneous dispersion and individualization of SWNTs are a prerequisite to the formation of more conductive nanotube films in terms of wet-processing methods. Indeed, dispersion strategies in these fabrication methods range from the use of surfactants or polymers as dispersion agents to aggressive sonication and their various combinations. A potentially negative outcome for overly aggressive sonication is the shortening of SWNTs, which according to percolation theory[100] and related experiments[149,150] is generally less favorable for producing the desired conductive films. A significant downside involving the use of surfactants or other dispersion agents is that their complete removal post-fabrication is most challenging, to say the least, especially regarding the necessity for removing dispersion agents without affecting the morphology and other desired performance characteristics of the coated films.[136,140]

Experimental evidence suggests that the conductivity in nanotube films is dominated by resistance at the tube–tube junctions.[126,135,149–152] Indeed, the contact resistance at metallic–semiconducting junctions is three orders of magnitude higher than that at the metallic–metallic junctions,[153] which justifies the use of metallic SWNTs only. It has been shown that nanotube films post-fabrication may be doped *via* treatment with a strong acid (HNO_3 or $SOCl_2$) to "metallize" semiconducting nanotubes and to decrease the inter-tube resistance at the junctions (the mitigation of a Schottky barrier), thus significantly enhancing the electrical conductivity in the films.[135,154] However, these doped films are generally less stable thermally and chemically, often degrading in performance over time.[144,154] Nevertheless, the demonstrated effect of "metallization" does again point to the great potential of transparent conductive films from separated metallic SWNTs (no need for metallization and thus no associated problems either[155]). According to a direct comparison by Miyata *et al.*,[155] the film (90 nm in thickness) from enriched metallic SWNTs was ~ 1 kΩ sq^{-1} in sheet resistance, whereas the thicker reference film (130 nm in thickness) by the same fabrication from non-separated SWNTs exhibited a much higher sheet resistance of ~ 20 kΩ sq^{-1}.

Yang and co-workers[156] used the separated metallic SWNTs in a systematic evaluation on transparent conductive films, which also included nanotubes from different sources coupled with various fabrication conditions. It is worth noting that the films were on a flexible substrate (PET), for which dip coating was used. Again, those films from the separated metallic SWNTs exhibited sheet resistance down to ~ 130 Ω sq^{-1} for 80% optical transmittance at 550 nm (Table 6.1). The comparison of nanotube films with ITO coatings on the

Table 6.1 Effect of carbon nanotube chirality on the conductivity.[156]

	One coating		Double coating		Triple coating	
Sample	T%	Rs (Ω sq^{-1})	T%	Rs (Ω sq^{-1})	T%	Rs (Ω sq^{-1})
Mixture	78	167 × 10³	72	61 × 10³	69	14 × 10³
Semiconducting	82	175 × 10³	79	95 × 10³	75	1493
Metallic	85	403	82	262	80	130 ± 5

Substrate used: PET (control); T% = 85.

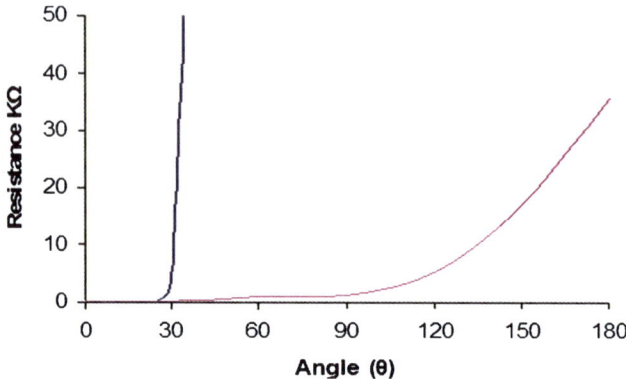

Figure 6.11 Flexibility study of SWNTs on PET vs. ITO on PET with two probe resistances.[156]

flexible substrate was particularly striking, with the latter apparently broke-down at the bending angle beyond approximately 30° (Figure 6.11).[156]

Enriched metallic SWNTs from other sources and separation methods have also been used for transparent conductive films.[66,135,157–159] For example, Hersam and co-workers[65,66] employed the DGU method to harvest metallic SWNTs of different diameter ranges, with the starting nanotube samples from different productions (HiPCO, laser ablation and arc discharge), and used the enriched metallic SWNTs for transparent conductive films. The films were fabricated through vacuum filtration, followed by transfer to transparent hard (glass and quartz) and flexible (PET) substrates.[66] For enriched metallic HiPCO SWNTs, the resulting film exhibited a sheet resistance of $\sim$231 Ω sq^{-1} for 75% optical transmittance at 550 nm, in comparison with $\sim$1340 Ω sq^{-1} in the reference film of the same optical transmittance from non-separated HiPCO SWNTs. The films of enriched metallic SWNTs from laser ablation- and arc discharge-produced nanotube samples generally exhibited better performance, with less than 140 Ω sq^{-1} sheet resistances at optical transmittances over 70% in the visible and near-IR spectral regions.[66] An interesting twist was that the films of enriched metallic SWNTs from different sources exhibited distinctive colors due to the different diameter ranges for the nanotubes, which, as suggested by the authors, might be exploited for applications in conductive optical filters.[66]

In another study, Maeda et al.[158] fabricated transparent conductive films by air spraying enriched metallic SWNTs (from the amine-assisted post-production separation[79,158]) on to both quartz and PET substrates. The sheet resistances in the films on PET were 690 Ω sq^{-1} at an optical transmittance of 81% (550 nm) and 9000 Ω sq^{-1} at 97% (550 nm), which represented reductions by a factor of 20 in comparison with the performance in films of non-separated SWNTs.[158] The same group also found that the effectiveness in the separation for metallic SWNTs is determined by the use of different amines at various concentrations. The proportion of metallic SWNTs against semiconducting SWNTs in the separated sample increased with an increase in the concentration of amines.[160] The results also suggested that metallic SWNTs are more effective for transparent conductive thin films of higher performance.[161]

The improved interfacial properties may be another benefit for the use of nanotube films to replace ITO coatings in certain energy conversion devices. In dye-sensitized solar cells (DSSCs), for example, the photoanode is typically comprised of semiconductor nanoparticles deposited on a transparent electrode (generally ITO-coated glass) and sensitized with a self-assembled monolayer of dye molecules.[162] Several studies on the incorporation of SWNTs into the photoanode have found significant performance improvements in the resulting DSSCs.[39–41,163] The use of a transparent electrode made from metallic SWNTs in the photoanode may further enhance the performance of existing solar cells.

In addition to conductivity, durability and mechanical flexibility are also important in interactive electronics.[164] Bao and co-workers[164] recently spray-deposited SWNT films, which were directly applied on to a substrate of PDMS (activated by exposure to ultraviolet light/O_3) from a solution in N-methylpyrrolidone. The films were stretched by applying strain along each axis, and then released. Correspondingly, at high levels of stress (up to 150%) the membrane conductivity was increased to as high as 2200 S cm^{-1} in the stretched state. In another recent report by Hobbie and co-workers,[165] thin membranes of length-purified SWNTs were uniaxially and isotropically compressed by depositing them on to pre-strained polymer substrates. At higher strains, the membrane conductivity was found to increase due to a compression-induced restoration of conductive pathways.[165] The same group also found that metallic films are generally better flexible transparent conductive coatings, with higher conductivity/transmission ratios and durability.[166]

There is now sufficient evidence to validate, in principle, the long-held expectation that metallic SWNTs may ultimately be used in transparent electrodes, or at least as alternatives to the ITO technology. In practice, many technical issues from materials (separated metallic SWNTs) to fabrication have yet to be fully addressed. Beyond transparent electrodes, metallic SWNTs may find other applications in which extremely high electrical conductivity and excellent optical properties are both required, or even some in which optical transparency is not necessary. Again for DSSCs, as an example, great benefits for using metallic SWNTs to replace the presently used platinum metal in the cathode may be expected on the basis of available experimental results.[156,164]

6.5 Perspective

Metallic and semiconducting SWNTs, which are distinctively different in electrical conductivity and many other aspects, are mixed in as-produced nanotube samples. While much effort has been made to produce either metallic or semiconducting SWNTs directly, with seemingly somewhat more success for the latter, post-production separation methods (including those for destroying one of the two) have seen significant advances and major achievements. The available separation methods are now capable of harvesting separated metallic SWNTs from different production sources, with sufficiently high enrichment and up to gram quantities for satisfying at least the needs in research and technological explorations. Further advances in the separation methods are anticipated, including especially the goals for higher purities (both in terms of nanotube sample and metallic purities) and maturation for scaling up, although the latter might have to be driven by the implementation of one or more technological applications. Among the widely pursued applications, the separated metallic SWNTs are most promising for transparent electrodes on both hard and flexible substrates, with the latter already competitive to the ITO coating technology. Finally, the recent emergence of graphene nanosheets and related materials may offer great opportunities for the development of carbon tube–sheet hybrid nanotechnologies.

Acknowledgements

Financial support from the Air Force Office of Scientific Research (AFOSR) through the program of Dr Charles Lee is gratefully acknowledged. The preparation of this Chapter was also made possible by support from South Carolina Space Grant Consortium (Y.-P.S., including a Graduate Research Fellowship to A.A.), NSF (Y.-P.S.) and Shantou University (F.L., through NTF10022, YR10002 and ITC11002).

References

1. G. Grüner, *J. Mater. Chem.*, 2006, **16**, 3533.
2. Q. Cao and J. A. Rogers, *Adv. Mater.*, 2009, **21**, 29.
3. V. Sgobba and D. M. Guldi, *Chem. Soc. Rev.*, 2009, **38**, 165.
4. A. Kumar and C. Zhou, *ACS Nano*, 2010, **4**, 11.
5. F. Lu, M. J. Meziani, L. Cao and Y.-P. Sun, *Langmuir*, 2011, **27**, 4339.
6. R. Saito, M. Fujita, G. Dresselhaus and M. S. Dresselhaus, *Appl. Phys. Lett.*, 1992, **60**, 2204.
7. M. S. Dresselhaus, G. Dresselhaus and P. C. Eklund, *Science of Fullerenes and Carbon Nanotubes*, Academic Press, San Diego, CA, 1996.
8. R. H. Baughman, A. A. Zakhidov and W. A. de Heer, *Science*, 2002, **297**, 787.
9. A. Jorio, M. S. Dresselhaus and G. Dresselhaus, *Carbon Nanotubes: Advanced Topics in the Synthesis, Structure, Properties and Applications*, Springer, Berlin/Heidelberg, 2008.
10. F. Lu, L. Gu, M. J. Meziani, X. Wang, P. G. Luo, L. M. Veca, L. Cao and Y.-P. Sun, *Adv. Mater.*, 2009, **21**, 139.
11. M. J. Connell, *Carbon Nanotubes Properties and Applications*, CRC Taylor & Francis, Boca Raton, FL, 2006.

12. Z. Liu, L. Jiao, Y. Yao, X. Xian and J. Zhang, *Adv. Mater.*, 2010, **22**, 2285.
13. A. Javey, J. Guo, Q. Wang, M. Lundstrom and H. Dai, *Nature*, 2003, **424**, 654.
14. T. Dulrkop, S. A. Getty, E. Cobas and M. S. Fuhrer, *Nano Lett.*, 2004, **4**, 35.
15. X. Zhou, J. Y. Park, S. Huang, J. Liu and P. L. McEuen, *Phys. Rev. Lett.*, 2005, **95**, 146805.
16. G. Gruener, *Anal. Bioanal. Chem.*, 2006, **384**, 322.
17. X. Liu, C. Lee, J. Han and C. Zhou, *Appl. Phys. Lett.*, 2001, **79**, 3329.
18. A. Javey, Q. Wang, A. Ural, Y. Li and H. Dai, *Nano Lett.*, 2002, **2**, 929.
19. V. Derycke, R. Martel, J. Appenzeller and P. Avouris, *Nano Lett.*, 2001, **1**, 453.
20. Z. Chen, J. Appenzeller, Y. Lin, J. S. Sippel-Oakley, A. G. Rinzler, J. Tang, S. J. Wind, P. M. Solomon and P. Avouris, *Science*, 2006, **311**, 1735.
21. H. Dai, A. Javey, E. Pop, D. Mann, W. Kim and Y. Lu, *Nano*, 2006, **1**, 1.
22. Y. Lin, K. A. S. Fernando, W. Wang and Y.-P. Sun, in *Carbon Nanotechnology: Recent Developments in Chemistry, Physics, Materials Science and Device Applications*, Dai, L. ed., Elsevier, Amsterdam, 2006, p. 255.
23. E. Katz and I. Willner, *ChemPhysChem*, 2004, **5**, 1084.
24. K. Balasubramanian and M. Burghard, *Anal. Bioanal. Chem.*, 2006, **385**, 452.
25. B. L. Allen, P. D. Kichambare and A. Star, *Adv. Mater.*, 2007, **19**, 1439.
26. S. N. Kim, J. F. Rusling and F. Papadimitrakopoulos, *Adv. Mater.*, 2007, **19**, 3214.
27. R. Martel, *ACS Nano*, 2008, **2**, 2195.
28. P. Avouris and R. Martel, *MRS Bull.*, 2010, **35**, 306.
29. E. Kymakis and G. A. J. Amaratunga, *Appl. Phys. Lett.*, 2002, **80**, 112.
30. Z. Wu, Z. Chen, X. Du, J. M. Logan, J. Sippel, M. Nikolou, K. Kamaras, J. R. Reynolds, D. B. Tanner, A. F. Hebard and A. G. Rinzler, *Science*, 2004, **305**, 1273.
31. A. D. Pasquier, H. E. Unalan, A. Kanwal, S. Miller and M. Chhowalla, *Appl. Phys. Lett.*, 2005, **87**, 203511.
32. C. M. Aguirre, S. Auvray, S. Pigeon, R. Izquierdo, P. Desjardins and R. Martel, *Appl. Phys. Lett.*, 2006, **88**, 183104.
33. D. Zhang, K. Ryu, X. Liu, E. Polikarpov, J. Ly, M. E. Tompson and C. Zhou, *Nano Lett.*, 2006, **6**, 1880.
34. E. Artukovic, M. Kaempgen, D. S. Hecht, S. Roth and G. Grüner, *Nano Lett.*, 2005, **5**, 757.
35. A. K. Feldman, M. L. Steigerwald, X. Guo and C. Nuckolls, *Acc. Chem. Res.*, 2008, **41**, 1731.
36. P. Avouris, M. Freitag and V. Perebeinos, *Nat. Photonics*, 2008, **2**, 341.
37. T. Umeyama and H. Imahori, *Energy Environ. Sci.*, 2008, **1**, 120.

38. M. Bansal, R. Srivastava, C. Lal, M. N. Kamalasanan and L. S. Tanwar, *Nanoscale*, 2009, **1**, 317.
39. K. Suzuki, M. Yamaguchi, M. Kumagai and S. Yanagida, *Chem. Lett.*, 2003, **32**, 28.
40. S.-R. Jang, R. Vittal and K.-J. Kim, *Langmuir*, 2004, **20**, 9807.
41. A. Kongkanand, R. M. Dominguez and P. V. Kamat, *Nano Lett.*, 2007, **7**, 676.
42. H. Shimoda, B. Gao, X. P. Tang, A. Kleinhammes, L. Fleming, Y. Wu and O. Zhou, *Phys. Rev. Lett.*, 2002, **88**, 015502.
43. A. S. Claye, J. E. Fischer, C. B. Huffman, A. G. Rinzler and R. E. Smalley, *J. Electrochem. Soc.*, 2000, **147**, 2845.
44. C. Niu, E. K. Sickel, R. Hoch, D. Moy and H. Tennent, *Appl. Phys. Lett.*, 1997, **70**, 1480.
45. K. H. An, W. S. Kim, Y. S. Park, J.-M. Moon, D. J. Bae, S. C. Lim, Y. S. Lee and Y. H. Lee, *Adv. Funct. Mater.*, 2001, **11**, 387.
46. D. N. Futaba, K. Hata, T. Yamada, T. Hiraoka, Y. Hayamizu, Y. Kakudate, O. Tanaike, H. Hatori, M. Yumura and S. Iijima, *Nat. Mater.*, 2006, **5**, 987.
47. V. L. Pushparaj, M. M. Shaijumon, A. Kumar, S. Murugesan, L. Ci, R. Vajtai, R. J. Linhardt, O. Nalamasu and P. M. Ajayan, *Proc. Natl. Acad. Sci. U.S.A.*, 2007, **104**, 13574.
48. C. Yu, C. Masarapu, J. Rong, B. Wei and H. Jiang, *Adv. Mater.*, 2009, **21**, 4793.
49. J. Kong, N. Franklin, C. Zhou, M. Chapline, S. Peng, K. Cho and H. Dai, *Science*, 2000, **287**, 622.
50. S. J. Tans, A. R. M. Verschueren and C. Dekker, *Nature*, 1998, **393**, 49.
51. R. Martel, T. Schmidt, H. R. Shea, T. Hertel and P. Avouris, *Appl. Phys. Lett.*, 1998, **73**, 2447.
52. C. Zhou, J. Kong and H. Dai, *Appl. Phys. Lett.*, 2000, **76**, 1597.
53. R. G. Gordon, *MRS Bull.*, 2000, **25**(8), 52.
54. M. S. Strano, C. A. Dyke, M. L. Usrey, P. W. Barone, M. J. Allen, H. Shan, C. Kittrell, R. H. Hauge, J. M. Tour and R. E. Smalley, *Science*, 2003, **301**, 1519.
55. J. Liu and M. C. Hersam, *MRS Bull.*, 2010, **35**, 315.
56. C. N. R. Rao, R. Voggu and A. Govindaraj, *Nanoscale*, 2009, **1**, 96.
57. M. C. Hersam, *Nat. Nanotech.*, 2008, **3**, 387.
58. F. Yang, P. Anilkumar, A. Anderson, L. Cao, M. J. Meziani, A. Parenzan and Y.-P. Sun, *J. Phys. Chem. C*, 2012, **116**, 6800.
59. T. Tanaka, Y. Urabe, D. Nishide and H. Kataura, *J. Am. Chem. Soc.*, 2011, **133**, 17610.
60. H. Liu, Y. Feng, T. Tanaka, Y. Urabe and H. Kataura, *J. Phys. Chem. C*, 2010, **114**, 9270.
61. M. S. Arnold, A. A. Green, J. F. Hulvat, S. I. Stupp and M. C. Hersam, *Nat. Nanotechnol.*, 2006, **1**, 60.
62. C. Fantini, A. Jorio, A. P. Santos, V. S. T. Peressinotto and M. A. Pimenta, *Chem. Phys. Lett.*, 2007, **439**, 138.

63. X. Y. Huang, R. S. McLean and M. Zheng, *Anal. Chem.*, 2005, **77**, 6225.
64. M. S. Arnold, S. I. Stupp and M. C. Hersam, *Nano Lett.*, 2005, **5**, 713.
65. M. S. Arnold, A. A. Green, J. F. Hulvat, S. I. Stupp and M. C. Hersam, *Nat. Nanotechnol.*, 2006, **1**, 60.
66. A. A. Green and M. C. Hersam, *Nano Lett.*, 2008, **8**, 1417.
67. A. A. Green, M. C. Duch and M. C. Hersam, *Nano Res.*, 2009, **2**, 69.
68. M. Zheng, A. Jagota, M. S. Strano, A. P. Santos, P. Barone, S. G. Chou, B. A. Diner, M. S. Dresselhaus, R. S. McLean, G. B. Onoa, G. G. Samsonidze, E. D. Semke, M. Usrey and D. J. Walls, *Science*, 2003, **302**, 1545.
69. M. Zheng, A. Jagota, E. D. Semke, B. A. Diner, R. S. McLean, S. R. Lustig, R. E. Richardson and N. G. Tassi, *Nat. Mater.*, 2003, **2**, 338.
70. X. Tu, S. Manohar, A. Jagota and M. Zheng, *Nature*, 2009, **460**, 250.
71. T. Tanaka, H. Jin, Y. Miyata, S. Fujii, H. Suga, Y. Naitoh, T. Minari, T. Miyadera, K. Tsukagoshi and H. Kataura. H. *Nano Lett.*, 2009, **9**, 1497.
72. T. Tanaka, H. Jin, Y. Miyata and H. Kataura, *Appl. Phys. Express*, 2008, **1**, 114001.
73. T. Tanaka, Y. Urabe, D. Nishide and H. Kataura, *Appl. Phys. Express*, 2009, **2**, 125002.
74. H. Liu, Y. Feng, T. Tanaka, Y. Urabe and H. Kataura, *J. Phys. Chem. C*, 2010, **114**, 9270.
75. H. Li, B. Zhou, L. Gu, W. Wang, K. A. S. Fernando, S. Kumer, L. F. Allard and Y.-P. Sun, *J. Am. Chem. Soc.*, 2004, **126**, 1014.
76. W. Wang, K. A. S. Fernando, Y. Lin, M. J. Meziani, L. M. Veca, L. Cao, P. Zhang, M. M. Kimani and Y.-P. Sun, *J. Am. Chem. Soc.*, 2008, **130**, 1415.
77. Y.-P. Sun, *U.S. Pat.*, 7374685, 2008.
78. F. Lu, W. Wang, K. A. S. Fernando, M. J. Meziani, E. Myers and Y.-P. Sun, *Chem. Phys. Lett.*, 2010, **497**, 57.
79. D. Chattopadhyay, I. Galeska and F. Papadimitrakopoulos, *J. Am. Chem. Soc.*, 2003, **125**, 3370.
80. S.-Y. Ju, M. Utz and F. Papadimitrakopoulos, *J. Am. Chem. Soc.*, 2009, **131**, 6775.
81. P. Anilkumar, K. A. S. Fernando, L. Cao, F. Lu, F. Yang, W. Song, S. Sahu, H. Qian, T. J. Thorne, A. Anderson and Y.-P. Sun, *J. Phys. Chem. C*, 2011, **115**, 11010.
82. J. Chen, M. A. Hamon, H. Hu, Y. Chen, A. M. Rao, P. C. Eklund and R. C. Haddon, *Science*, 1998, **282**, 95.
83. Y.-P. Sun, K. Fu, Y. Lin and W. Huang, *Acc. Chem. Res.*, 2002, **35**, 1096.
84. J. Liu, A. G. Rinzler, H. Dai, J. H. Hafner, R. K. Bradley, P. J. Boul, A. Lu, T. Iverson, K. Shelimov, C. B. Huffman, F. Rodriguez-Macias, Y.-S. Shon, T. R. Lee, D. T. Colbert and R. E. Smalley, *Science*, 1998, **280**, 1253.
85. S. Niyogi, M. A. Hamon, H. Hu, B. Zhao, P. Bhowmik, R. Sen, M. E. Itkis and R. C. Haddon, *Acc. Chem. Res.*, 2002, **35**, 1105.

86. J. Chen, A. M. Rao, S. Lyuksyutov, M. E. Itkis, M. A. Hamon, H. Hu, R. W. Cohn, P. C. Eklund, D. T. Colbert, R. E. Smalley and R. C. Haddon, *J. Phys. Chem. B*, 2001, **105**, 2525.
87. E. V. Basiuk, V. A. Basiuk, J.-G. Banuelos, J.-M. Saniger-Blesa, V. A. Pokrovskiy, T. Y. Gromovoy, A. V. Mischanchuk and B. G. Mischanchuk, *J. Phys. Chem. B*, 2002, **106**, 1588.
88. Y.-L. Zhao and J. F. Stoddart, *Acc. Chem. Res.*, 2009, **42**, 1161.
89. K. A. S. Fernando, Y. Lin, W. Wang, S. Kumar, B. Zhou, S.-Y. Xie, L. T. Cureton and Y.-P. Sun, *J. Am. Chem. Soc.*, 2004, **126**, 10234.
90. K. A. S. Fernando, PhD thesis, Clemson University, 2007.
91. S. D. M. Brown, A. Jorio, P. Corio, M. S. Dresselhaus, G. Dresselhaus, R. Saito and K. Kneipp, *Phys. Rev. B*, 2001, **63**, 155414.
92. F. Lu, X. Wang, M. J. Meziani, L. Cao, L. Tian, M. A. Bloodgood, J. Robinson and Y.-P. Sun, *Langmuir*, 2010, **26**, 7561.
93. J. Lu, L. Lai, G. Luo, J. Zhou, R. Qin, D. Wang, L. Wang, W. N. Mei, G. Li, Z. Gao, S. Nagase, Y. Maeda, T. Akasaka and D. Yu, *Small*, 2007, **3**, 1566.
94. C.-H. Liu, Y.-Y. Liu, Y.-H. Zhang, R.-R. Wei and H.-L. Zhang, *Phys. Chem. Chem. Phys.*, 2009, **11**, 7257.
95. R. Voggu, K. V. Rao, S. J. George and C. N. R. Rao, *J. Am. Chem. Soc.*, 2010, **132**, 5560.
96. C.-H. Liu, Y.-Y. Liu, Y.-H. Zhang, R.-R. Wei, B.-R. Li, H.-L. Zhang and Y. Chen, *Chem. Phys. Lett.*, 2009, **471**, 97.
97. X. Pan, Q. J. Cai, C. M. Li, Q. Zhang and M. B. Chan-Park, *Nanotechnology*, 2009, **20**, 305601.
98. F. Du, R. C. Scogna, W. Zhou, S. Brand, J. E. Fischer and K. I. Winey, *Macromolecules*, 2004, **37**, 9048.
99. R. H. Baughman, A. A. Zakkidov and W. A. de Heer, *Science*, 2002, **297**, 787.
100. L. Hu, D. S. Hecht and G. Grüner, *Nano Lett.*, 2004, **4**, 2513.
101. M. Moniruzzaman and K. I. Winey, *Macromolecules*, 2006, **39**, 5194.
102. M. T. Byrne and Y. K. Gun'ko, *Adv. Mater.*, 2010, **22**, 1672.
103. T. Connolly, R. C. Smith, Y. Hernandez, Y. Gun'ko, J. N. Coleman and J. D. Carey, *Small*, 2009, **5**, 826.
104. Y. Li, T. Yu, T. Pui, P. Chen, L. Zhengand and K. Liao, *Nanoscale*, 2011, **3**, 2469.
105. H. Guo, M. L. Minus, S. Jagannathanand and S. Kuma, *ACS Appl. Mater. Inter.*, 2010, **2**, 1331.
106. J. Sung, P. S. Jo, H. Shin, J. Huh, B. G. Min, D. H. Kim and C. Park, *Adv. Mater.*, 2008, **20**, 1505.
107. F. M. Blighe, Y. R. Hernandez, W. J. Blau and J. N. Coleman, *Adv. Mater.*, 2007, **19**, 4443.
108. M. B. Bryning, M. F. Islam, J. M. Kikkawa and A. G. Yodh, *Adv. Mater.*, 2005, **17**, 1186.
109. J. C. Grunlan, A. R. Mehrabi, M. V. Bannon and J. L. Bahr, *Adv. Mater.*, 2004, **16**, 150.

110. F. S. Gittleson, D. J. Kohn, X. Li and A. D. Taylor, *ACS Nano*, 2012, **6**, 3703.
111. X. Li, F. S. Gittleson, M. Carmo, R. C. Sekol and A. D. Taylor, *ACS Nano*, 2012, **6**, 1347.
112. J. Zhu, B. S. Shim, M. D. Prima and N. A. Kotov, *J. Am. Chem. Soc.*, 2011, **133**, 7450.
113. M. A. Worsley, S. O. Kucheyev, J. D. Kuntz, A. V. Hamza, J. H. Satcher and T. F. Baumann, *J. Mater. Chem.*, 2009, **19**, 3370.
114. M. C. Hermant, B. Klumperman, A. V. Kyrylyuk, P. V. Schoot and C. E. Koning, *Soft Matter*, 2009, **5**, 878.
115. M. C. Hermant, P. V. Schoot, B. Klumperman and C. E. Koning, *ACS Nano*, 2010, **4**, 2242.
116. X. Xian, L. Jiao, T. Xue, Z. Wu and Z. Liu, *ACS Nano*, 2011, **5**, 4000.
117. J. Ge, G. Cheng and G. Chen, *Nanoscale*, 2011, **3**, 3084.
118. Y. Ma, W. Cheung, D. Wei, A. Bogozi, P. L. Chiu, L. Wang, F. Pontoriero, R. Mendelsohn and H. He, *ACS Nano*, 2008, **2**, 1197.
119. Z. Yu, X. Niu, Z. Liu and Q. Pei, *Adv. Mater.*, 2011, **23**, 3989.
120. M. H. Kim, J. Y. Choi, H. K. Choi, S. M. Yoon, O. O. Park, D. K. Yi, S. J. Choi and H. J. Shin, *Adv. Mater.*, 2008, **20**, 457.
121. S. De, P. E. Lyons, P. Sorel, E. M. Doherty, P. J. King, W. J. Blau, P. N. Nirmalraj, J. J. Boland, V. Scardaci, J. Joimel and J. N. Coleman, *ACS Nano*, 2009, **3**, 714.
122. H. S. Park, B. G. Choi, W. H. Hong and S. Y. Jang, *J. Phys. Chem. C*, 2012, **116**, 7962.
123. P. L. McEuen and J. Y. Park, *MRS Bull.*, 2004, **29**, 272.
124. W. Liang, M. Bockrath, D. Bozovic, J. H. Hafner, M. Tinkham and H. Park, *Nature*, 2001, **411**, 665.
125. C. T. White and T. N. Todorov, *Nature*, 1998, **393**, 240.
126. S. Hong and S. Myung, *Nat. Nanotech.*, 2007, **2**, 207.
127. S. J. Tans, M. H. Devoret, H. Dai, A. Thess, R. E. Smalley, L. J. Georliga and C. Dekker, *Nature*, 1997, **386**, 474.
128. M. Bockrath, D. H. Cobden, P. L. McEuen, N. G. Chopra, A. Zettl, A. Thess and R. E. Smalley, *Science*, 1997, **275**, 1922.
129. A. Maffucci, G. Miano and F. Villone, *Int. J. Circ. Theor. Appl.*, 2008, **36**, 31.
130. D. M. Delozier, K. A. Watson, J. G. Smith and J. W. Connell, *Compos. Sci. Technol.*, 2005, **65**, 749.
131. S. Roth and H. J. Park, *Chem. Soc. Rev.*, 2010, **39**, 2477.
132. E. Fortunato, D. Ginley, H. Hosono and D. C. Paine, *MRS Bull.*, 2007, **32**, 242.
133. C. G. Granqvist and A. Hultaker, *Thin Solid Films*, 2002, **411**, 1.
134. J. Opatkiewicz, M. C. LeMieux and Z. Bao, *ACS Nano*, 2010, **4**, 2975.
135. R. K. Jackson, A. Munro, K. Nebesny, N. Armstrong and S. Graham, *ACS Nano*, 2010, **4**, 1377.
136. H. Z. Geng, K. K. Kim, K. P. So, Y. S. Lee, Y. Chang and Y. H. Lee, *J. Am. Chem. Soc.*, 2007, **129**, 7758.

137. R. C. Tenent, T. M. Barnes, J. D. Bergeson, A. J. Ferguson, B. To, L. M. Gedvilas, M. J. Heben and J. L. Blackburn, *Adv. Mater.*, 2009, **21**, 3210.
138. Y. Wang, C.-A. Di, Y. Liu, H. Kajiura, S. Ye, L. Cao, D. Wei, H. Zhang, Y. Li and Noda, K. *Adv. Mater.*, 2008, **20**, 4442.
139. J. Li, L. Hu, L. Wang, Y. Zhou, G. Grüner and T. J. Marks, *Nano Lett.*, 2006, **6**, 2472.
140. M. C. LeMieux, M. E. Roberts, S. Barman, Y. W. Jin, J. M. Kim and Z. Bao, *Science*, 2008, **321**, 101.
141. A. Bachtold, P. Hadley, T. Nakanishi and C. Dekker, *Science*, 2001, **294**, 1317.
142. B. Dan, G. C. Irvin and M. Pasquali, *ACS Nano*, 2009, **3**, 835.
143. T. Kitano, Y. Maeda and T. Akasaka, *Carbon*, 2009, **47**, 3559.
144. B. S. Shim, J. Zhu, E. Jan, K. Critchley and N. A. Kotov, *ACS Nano*, 2010, **4**, 3725.
145. Y. T. Park, A. Y. Ham and J. C. Grunlan, *J. Phys. Chem. C*, 2010, **114**, 6325.
146. N. I. Kovtyukhova and T. E. Mallouk, *J. Phys. Chem. B*, 2005, **109**, 2540.
147. E. Y. Jang, T. J. Kang, H. W. Im, D. W. Kim and Y. H. Kim, *Small*, 2008, **4**, 2255.
148. P. N. Nirmalraj, P. E. Lyons, S. De, J. N. Coleman and J. J. Boland, *Nano Lett.*, 2009, **9**, 3890.
149. D. Simien, J. A. Fagan, W. Luo, J. F. Douglas, K. B. Migler and J. Obrzut, *ACS Nano*, 2008, **2**, 1879.
150. D. Hecht, L. Hu and G. Grüner, *Appl. Phys. Lett.*, 2006, **89**, 133112.
151. H. Xu, S. M. Anlage, L. Hu and G. Grüner, *Appl. Phys. Lett.*, 2007, **90**, 183119.
152. A. Behnam and A. Ural, *Phys. Rev. B*, 2007, **75**, 125432.
153. M. S. Fuhrer, J. Nygrd, L. Shih, M. Forero, Y.-G. Yoon, M. S. C. Mazzoni, H. J. Choi, J. Ihm, S. G. Louie, A. Zettl and P. L. McEuen, *Science*, 2000, **288**, 494.
154. R. Jackson, B. Domercq, R. Jain, B. Kippelen and S. Graham, *Adv. Funct. Mater.*, 2008, **18**, 2548.
155. Y. Miyata, K. Yanagi, Y. Maniwa and H. Kataura, *J. Phys. Chem. C*, 2008, **112**, 3591.
156. A. Rahy, P. Bajaj, I. H. Musselman, S. H. Hong, Y.-P. Sun and D. J. Yang, *Appl. Surf. Sci.*, 2009, **255**, 7084.
157. D. H. Shin, J.-E. Kim, H. C. Shim, J.-W. Song, J.-H. Yoon, J. Kim, S. Jeong, J. Koang, S. Baik and C. S. Han, *Nano Lett.*, 2008, **8**, 4380.
158. Y. Maeda, M. Hashimoto, S. Kaneko, M. Kanda, T. Hasegawa, T. Tsuchiya, T. Akasaka, Y. Naitoh, T. Shimizu, H. Tokumoto, J. Lu and S. Nagase, *J. Mater. Chem.*, 2008, **18**, 4189.
159. T. M. Barnes, J. L. Blackburn, J. van de Lagemaat, T. J. Coutts and M. J. Heben, *ACS Nano*, 2008, **2**, 1968.
160. Y. Maeda, K. Komoriya, K. Sode, M. Kanda, M. Yamada, T. Hasegawa, T. Akasaka, J. Lu and S. Nagase, *Phys. Status Solidi B*, 2010, **247**, 2641.

161. Y. Maeda, K. Komoriya, K. Sode, J. Higo, T. Nakamura, M. Yamada, T. Hasegawa, T. Akasaka, T. Saito, J. Lu and S. Nagase, *Nanoscale*, 2011, **3**, 1904.
162. B. O'Regan and M. Gratzel, *Nature*, 1991, **353**, 737.
163. F. Bonaccorso, *Int. J. Photoenergy*, 2010, 727134.
164. D. J. Lipomi, M. Vosgueritchian, B. C. K. Tee, S. L. Hellstrom, J. A. Lee, C. H. Fox and Z. Bao, *Nat. Nanotechnol.*, 2011, **6**, 788.
165. J. M. Harris, G. R. S. Iyer, D. O. Simien, J. A. Fagan, J. Y. Huh, J. Y. Chung, S. D. Hudson, J. Obrzut, J. F. Douglas, C. M. Stafford and E. K. Hobbie, *J. Phys. Chem. C*, 2011, **115**, 3973.
166. J. M. Harris, G. R. S. Iyer, A. K. Bernhardt, J. Y. Huh, S. D. Hudson, J. A. Fagan and E. K. Hobbie, *ACS Nano*, 2012, **6**, 881.

CHAPTER 7

Characterization of Dispersability of Industrial Nanotube Materials and their Length Distribution Before and After Melt Processing

B. KRAUSE[1], M. MENDE[2], G. PETZOLD[2], R. BOLDT[3] AND P. PÖTSCHKE*[1]

Leibniz Institute of Polymer Research Dresden; [1] Department of Polymer Reactions and Blends; [2] Department of Polyelectrolytes and Dispersions; [3] Department of Processing, Hohe Str. 6, 01069 Dresden, Germany
*E-mail: poe@ipfdd.de

7.1 Introduction

Due to the exceptional properties of carbon nanotubes (CNTs), such as high electrical and thermal conductivity and excellent mechanical properties, they are expected to have great potential as fillers for polymeric matrices. CNTs are incorporated in electrically insulating polymer materials to achieve electrostatic dissipative behavior or electrical conductivity and improved mechanical properties. Current applications for such nanocomposites include electrostatically dissipative plastic housing or fuel lines, as well as lightweight and electrostatically paintable plastic components replacing metals in car panel applications. Incorporation of CNTs in polymers on an industrial scale is usually performed by melt processing. Industrial CNT materials are, in most cases, synthesized in an agglomerated and entangled state and have to be dispersed into single tubes during melt compounding in order to obtain the greatest benefit from intrinsic CNT properties. Electrically percolated networks are most effectively formed by secondary agglomeration of suitably dispersed nanotubes. Processes of wetting and diffusion of polymer chains into agglomerates, as well as rupture and erosion of the CNT agglomerates, occur before isolated CNTs can be distributed in the matrix. Both the properties of the matrix polymer (viscosity, polarity, surface energy and functional groups) and of the CNTs (wall number, purity, diameter, length, bulk density, waviness, dispersability and functionalization) play a role in dispersion. To achieve good separation of the CNTs and a low electrical percolation threshold in polymers, the surface properties of the polymer and nanotubes should be in the same range, and the melt processing parameters should be optimized according the desired nanocomposite property profile. For this purpose, the specific mechanical energy (SME) input during melt processing is of importance.

Due to the high heterogeneity within different types of CNTs, generalized conclusions about their properties (*e.g.* functionalization, dispersability, agglomerate strength, aspect ratio, degree of defects) can be only drawn to a limited extent. In the present study, two properties of CNTs are considered. First, a method for characterizing the dispersability of CNT materials in aqueous surfactant solutions is presented, which also allows conclusions to be drawn regarding the dispersability in other media, such as polymers. Second, a method was developed to quantify the length of CNTs using the same experimental approach before and after melt mixing with thermoplastics. As the aspect ratio (length divided by diameter) of a nanotube significantly affects the achievable electrical percolation threshold in the composites, knowledge of CNT lengths and their reduction during processing is of interest. Furthermore, the relations between the SME, CNT dispersion, CNT length, and electrical properties of melt mixed composites are discussed.

7.2 Experimental

7.2.1 Materials

The nanotubes used represent industrial multi-walled CNTs (MWCNTs). They were Nanocyl^{TM} NC7000 (Nanocyl SA, Sambreville, Belgium; >90% purity C, bulk density 66 kg m^{-3}), C150P Baytubes® (Bayer MaterialScience AG, Leverkusen, Germany; >95% purity C, bulk density 120–170 kg m^{-3}), FutureCarbon CNT-MW (as grown, FutureCarbon GmbH, Bayreuth, Germany; >90% purity C, bulk density 28 kg m^{-3}), and Graphistrength® C100 (Arkema, Colombes, France; >90% purity C, bulk density of 50–150 kg m^{-3}).[1]

7.2.2 Centrifugal Separation Analysis (CSA)

The experimental conditions for the characterization of the dispersability and particle size distribution of CNTs in aqueous surfactant solutions using CSA have been described in detail previously.[1] Based on previous investigations,[2] the anionic surfactant sodium dodecylbenzene sulfonate (SDDBS) was selected. An amount of 0.0025 g of nanotubes was dispersed in 35 ml of SDDBS solution (0.7 g SDDBS L^{-1}) in a beaker glass at room temperature using an ultrasonic processor UP 200S (Hielscher Ultrasonics GmbH, Teltow, Germany) working at a frequency of 24 kHz and 200 W, which was equipped with a Sonotrode S14 made of titanium. The amplitude was adjusted to 20% and the dispersion time was varied between 1 and 30 min to vary the total mechanical energy input.

The sedimentation behavior of the CNT dispersions was investigated under centrifugation forces using a LUMiSizer® LS611 (L.U.M. GmbH, Berlin, Germany), which is a microprocessor-controlled analytical centrifuge allowing the determination of space- and time-resolved transmission profiles during the centrifugation.[3] Centrifugation at high speed results in an accelerated migration of the dispersed particles. Quadratic synthetic cells made of glass with an optical path of 10 mm were used. The evaluation of the time dependence of the transmission profiles recorded at a wavelength of 880 nm and measured between the bottom and the fluid level in time intervals of 10 s allows the quantification of the dispersion stability. The LUMiSizer® experiments were carried out at 3000 rev min^{-1} (rpm) for 45 min at room temperature. To evaluate the dispersion stability, the integration of the transmission profiles was performed in the middle region of the cell between the positions of 106 and 124 mm. The LUMiSizer® experiments were shown to be reproducible and the results shown are based on two measurements.

Particle size distributions of the dispersed CNTs were determined by two different methods. Results obtained using the LUMiSizer® LS611 were used for quantification of the particle size according to ISO 13318-2. The obtained transmission values were transformed into extinction values and subsequently the velocity distribution was obtained by using the software SEPView. The mode "constant position" was used and three notes were set at 120, 122, and 124 mm. Based on the velocity distribution, the software allows the calculation of the intensity-weighted particle size distribution and the average particle sizes (x_{10}, x_{50}, and x_{90}). In addition, dynamic light scattering (DLS) measurements were used to determine the z-average hydrodynamic particle diameter $d_{h,z\ ave}$ as a function of ultrasonic treatment time. This method provides very quick and efficient statements about the treatment time at which a marginal change in the sample state can be observed. The instrument (Zetasizer Nano S, Malvern Instr., UK) was equipped with a monochromatic coherent 4 mW Helium Neon laser ($\lambda = 633$ nm) as a light source and the so-called NIBS® technology (non-invasive back-scattering; patent from ALV GmbH, Germany) was applied in accordance to ISO 22412. However, due to the anisotropy of a single nanotube, it is not possible to give a clear indication from the determined particle diameters if they characterize single CNTs and/or agglomerates.

7.2.3 Melt Processing

In order to establish relationships between the dispersability of CNTs in aqueous dispersions and polymer melts, different polymer–CNT composites were prepared using the small-scale conical, co-rotating twin-screw Microcompounder Xplore DSM15 (volume 15 ml). The polycarbonate (PC) composites based on Makrolon® 2600 (Bayer MaterialScience AG) were prepared at 280 °C, 250 rpm and 5 min. Low-density polyethylene (LDPE) composites based on BPD2000 (INEOS Polyolefins) were mixed at 200 °C, 200 rpm and 5 min. Composites based on polyamide 12 (PA12, VESTAMID® L grade, Evonik Industries) were prepared using a DACA Microcompounder (DACA Instruments, USA; volume, 4.5 cm^3) at 210 °C, 250 rpm and 5 min.[4]

For the illustration of the method for quantification of CNT lengths after processing, two composites based on PC (Makrolon® 2600, Bayer MaterialScience AG, Germany) with either 3 wt% Baytubes® C150P or Nanocyl™ NC7000 were melt extruded using a twin-screw extruder ZE25 (Berstorff) at 260 °C and 500 rpm, 5 kg h^{-1}, using a distributive mixing screw with a length-to-diameter ratio of 48. In addition, the effect of dry ball milling treatment of Nanocyl™ NC7000 on CNT length was studied and composites with PC (Lexan™ 141R, Sabic Innovative Plastics, The Netherlands) were prepared by masterbatch dilution as described in detail by Krause *et al.*[5]

The experiments with regards to the effect of SME on dispersion, CNT length, and electrical properties have been, in part, described in previous studies. The preparation of polyamide 6 composites containing 5 wt% Nanocyl™ NC7000 by melt mixing in a DACA Microcompounder at different mixing speeds (50–300 rpm) and mixing times (5–15 min) is described by Krause *et al.*[6] Melt mixing of 1 wt% Baytubes® C150HP in PC (Makrolon® 2205, Bayer MaterialScience AG, Germany) using a DACA Microcompounder under various mixing speeds (50–300rpm) and mixing times (0.5–40 min) is discussed by Kasaliwal *et al.*[7] The composites based on polycaprolactone (PCL; CAPA 6800, Perstorp) containing 0.5 wt% Nanocyl™ NC7000 were melt compounded using a Microcompounder Xplore DSM 5 (volume 5 ml) with different rotation speeds, varying from 25 to 400 rpm, at a mixing temperature of 90°C for 2 min.[8] Composites with different matrix viscosities were prepared using the incorporation of Baytubes® C150P into low-viscous PC Makrolon® 2205 (PC1; Bayer MaterialScience AG) and high viscous PC Makrolon® 3108 (PC2; Bayer MaterialScience AG) using a DACA Microcompounder at 280°C, 250 rpm for 5 min.[9]

Finally, PC-based composites (Makrolon® 2600, Bayer MaterialScience AG) containing 3 wt% Baytubes® C150P were prepared using a twin-screw extruder ZE25 (Berstorff) at 260 °C. The screw rotation speed was varied at 100, 300, 500, and 1000 rpm and the throughput at 5, 10, and 15 kg h^{-1} (at 300 rpm) using the screw described earlier.

7.2.4 Morphological Characterization

Transmission light microscopy was used to analyze the CNT macrodispersion on thin sections of extruded strands prepared using the microtome RM2155 (Leica Microsystems GmbH, Germany). A BH2 light microscope and DP71 camera (both from Olympus Deutschland GmbH, Germany) were applied to record at least 10 images, with a total area of approximately 3 mm^2. The agglomerate area ratio A_A in the light microscopic images was determined as the ratio between the area of black-appearing agglomerates related to the image area (in %) using the software ImageJ version 1.43o in accordance to ISO 18553, whereby only agglomerates with an equivalent circular diameter larger than 5 µm were taken into account.

7.3 Results and Discussion

7.3.1 Characterization of the Dispersability of CNT Materials

The term "dispersability of CNT materials" describes the simplicity to individualize as-produced primary CNT agglomerates into their smallest units, namely single CNTs.

The properties of CNTs, *e.g.* number of walls, diameter, length, chemical purity, bulk density, waviness, degree of entanglement, and dispersability, are strongly influenced by the synthesis method and conditions. While some of these properties can be determined relatively easily by microscopic, thermal or physical methods, the dispersability is a parameter which is difficult to characterize and has been rarely considered in a quantitative way. The dispersability of MWCNTs is mainly influenced by the degree of entanglement in the primary CNT agglomerates, the interactions between the tubes, and, thus, the agglomerate strength. The degree of entanglement depends strongly on the structure of the catalyst and support materials used for the synthesis of CNTs, which, next to the synthesis conditions, determine the nanotubes' waviness caused by defects of individual CNTs, as well as the density of nanotube growth. In addition, the strong van der Waals forces between the CNTs result in a remarkable tendency to agglomeration. The purity and functional groups on the CNTs surface also influence the dispersability of CNTs in various media due to different wetting and infiltration characteristics of the media into the CNT agglomerates.

As the incorporation of CNTs into polymers, *e.g.* by melt mixing, with subsequent characterization of the achieved dispersion is a time-consuming process, a simpler method was aimed at developing an estimate of the dispersability of CNTs.[2,10] Here, the CNT material is dispersed in an aqueous surfactant solution using ultrasound under defined conditions. The isolated CNTs are stabilized due to the formation of surfactant micelles around these CNTs, which prevents reagglomeration.[11] Remaining non-dispersed CNT agglomerates are surrounded by the surfactant and such agglomerates have a higher mass and show a more pronounced sedimentation tendency. The stability of such dispersions is quantified by means of CSA using the analytical centrifuge LUMiSizer®. The results serve as a measure of the degree of dispersion achieved under the selected dispersion conditions. For the detection of the sedimentation process, the intensity of the incident parallel light (transmission) from a near-infrared (NIR) light source during centrifugation under defined conditions is time- and location-dependent and was measured along the sample cell.[12] Additional information concerning the dispersability can be gained if the sonication conditions during the preparation of the dispersions are varied, as done here for the ultrasonic treatment time.[1] Thus, it is possible to compare not only the state of CNT dispersion at a given energy input, but also to determine the energy necessary to disperse the CNT agglomerates completely and to achieve stable dispersions. In addition to the method applied here, the sedimentation behavior of CNT dispersions can be also studied using a disc centrifuge.[13]

Figure 7.1 shows the integral transmission of such dispersions after a centrifugation time of 45 min at 3000 rpm (filled symbols) plotted *versus* the sonication time and the energy input for four different types of commercial

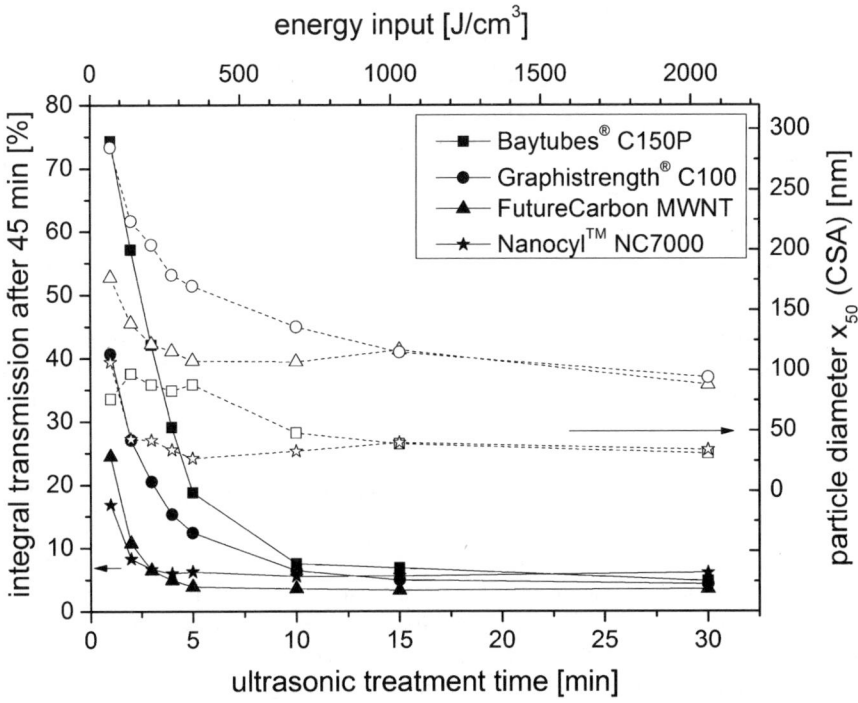

Figure 7.1 Dispersions of four different types of commercial MWCNTs in aqueous solution investigated using CSA in the LUMiSizer®. Integral transmission after a centrifugation time of 45 min at 3000 rpm and particle diameter x_{50} calculated from the CSA plotted *versus* sonication time and energy input.

MWCNTs. For FutureCarbon MWNT and Nanocyl™ NC7000, a 2 min sonication time at 40 W is sufficient to generate low and furthermore constant transmission values and thus stable dispersions. However, the dispersions containing Baytubes® C150P and Graphistrength® C100 require a 10 min sonication before the transmission values achieve constant low values characteristic for stable dispersions.

In addition to the assessment of the stability of the dispersions from CSA, the particle sizes in the dispersions were determined. In the interpretation of such particle sizes it is important to note that non-spherical particles are oriented in the direction of the centrifugal force.[14] The non-spherical particles settle with the speed of a sphere that corresponds to its smallest diameter. For a nanotube with a diameter of 10 nm and an aspect ratio of 50, an equivalent diameter of 21 nm can be calculated, as non-spherical particles in the centrifugal field are not oriented completely in the force direction.[15] Therefore, the particle size distributions and average particle diameters determined by CSA can give information about the individualization of the CNTs when compared with their diameters. Also, in Figure 7.1 the intensity-weighted average particle diameters are plotted (open symbols) as a function of the ultrasonic treatment time. Taking into account average CNT diameters of 10 nm for Nanocyl™ NC7000,[16] 10.5 nm for Baytubes® C150P,[16] 15 nm for FutureCarbon MWNT,[4] and 10–15 nm for Graphistrength® C100,[1] only Nanocyl™ NC7000 and Baytubes® C150P show particle sizes in their diameter range. Thus, it can be concluded that only for these CNT types a separation into individualized tubes was possible under the selected ultrasound dispersion conditions. For FutureCarbon MWNT and Graphistrength® C100, even after a 30 min ultrasonic treatment, mean particle diameters in the range of 90 nm were found, indicating the presence of remaining CNT bundles or agglomerates.

The results of CSA show that, depending on the CNT material, different energy inputs are needed to achieve stable dispersions. In addition, based on the calculated average particle diameter compared with the CNT diameter, differences in the ability to individualize the CNT agglomerates into single tubes can be derived.

In Table 7.1, the z-average hydrodynamic particle diameter $d_{h,z\ ave}$ for different kinds of CNTs from DLS measurements as a function of ultrasonic treatment time are summarized. Also, the DLS measurements showed a decrease in particle diameter with longer ultrasonic treatment time. The initial decrease (during the first 5 min of ultrasonic treatment) of the z-average particle size $d_{h,z\ ave}$ was faster for Nanocyl™ NC7000 and Graphistrength® C100 compared with Baytubes® C150P and FutureCarbon MWCNT. The differences in particle diameters between the dispersions of Nanocyl™ NC7000, Baytubes® C150P, and FutureCarbon MWCNT are diminished after 10 min of treatment. However, the values of $d_{h,z\ ave}$ of Graphistrength® C100 showed that these nanotubes and MWCNT agglomerates were the largest in comparison with the other kinds of nanotubes, which is also

Table 7.1 z-Average hydrodynamic particle diameter, $d_{h,z\ ave}$, calculated from the DLS measurement for different kinds of CNTs as a function of the ultrasonic treatment time.

Ultrasonic treatment time (min)	z-Average hydrodynamic particle diameter $d_{h,z\ ave}$ (DLS) (nm)			
	Nanocyl^{TM} NC7000	Baytubes® C150P	FutureCarbon MWNT	Graphistrength® C100
1	347.3	297.6	296.9	399.3
2	323.3	261.5	282.1	380.3
3	262.6	253.5	259.2	387.4
4	293.3	240.4	266.6	353.7
5	222.8	233.9	263.3	293.6
10	164.3	182.4	191.8	307.7
15	205.6	191.4	168.0	231.1
30	126.3	140.1	151.9	211.9

illustrated by the intensity-weighted particle size distributions shown previously.[1] The polydispersity index obtained by DLS was found to be 0.4 for Baytubes® C150P, Nanocyl^{TM} NC7000, and Graphistrength® C100, indicating very heterogeneous particles (single nanotubes as well as MWCNT agglomerates) existing in the dispersions. For FutureCarbon MWCNT, the agglomerates are more homogeneous in size, as illustrated by a polydispersity index of 0.2.

The comparison of the particle diameters calculated from CSA (Figure 7.1) and DLS (Table 7.1) shows that the particles in the dispersion are non-spherical, because of the higher particle diameters determined by DLS. Only if the results of DLS and CSA agree a spherical shape of the dispersed aggregates can be assumed.

Interestingly, the different dispersabilities of the CNT materials as assessed by CSA are also reflected in the morphology of melt-mixed composites, which was characterized on transmission light microscopy images. Figure 7.2 shows examples for melt-mixed composites with different polymers and MWCNTs together with the measured values of the agglomerate area ratios, A_A. It can be seen that the composites based on LDPE and PC with Baytubes C150P® show quantitatively more agglomerates and significantly higher values of A_A than the composites containing the Nanocyl^{TM} NC7000 material. These results correlate well with the worse dispersability of Baytubes® C150P in the aqueous surfactant solution. In another example of polyamide12 filled with 2 wt% CNTs, the lowest agglomerate area ratio A_A was found for the well-dispersible FutureCarbon MWCNTs. Although for Baytubes® C150P and Nanocyl^{TM} NC7000 identical agglomerate area ratios have been identified, fewer particles of Nanocyl^{TM} NC7000 per unit area were detected.[4] The highest agglomerate area ratio was determined for the composite with worse-dispersible Graphistrength® C100 material.

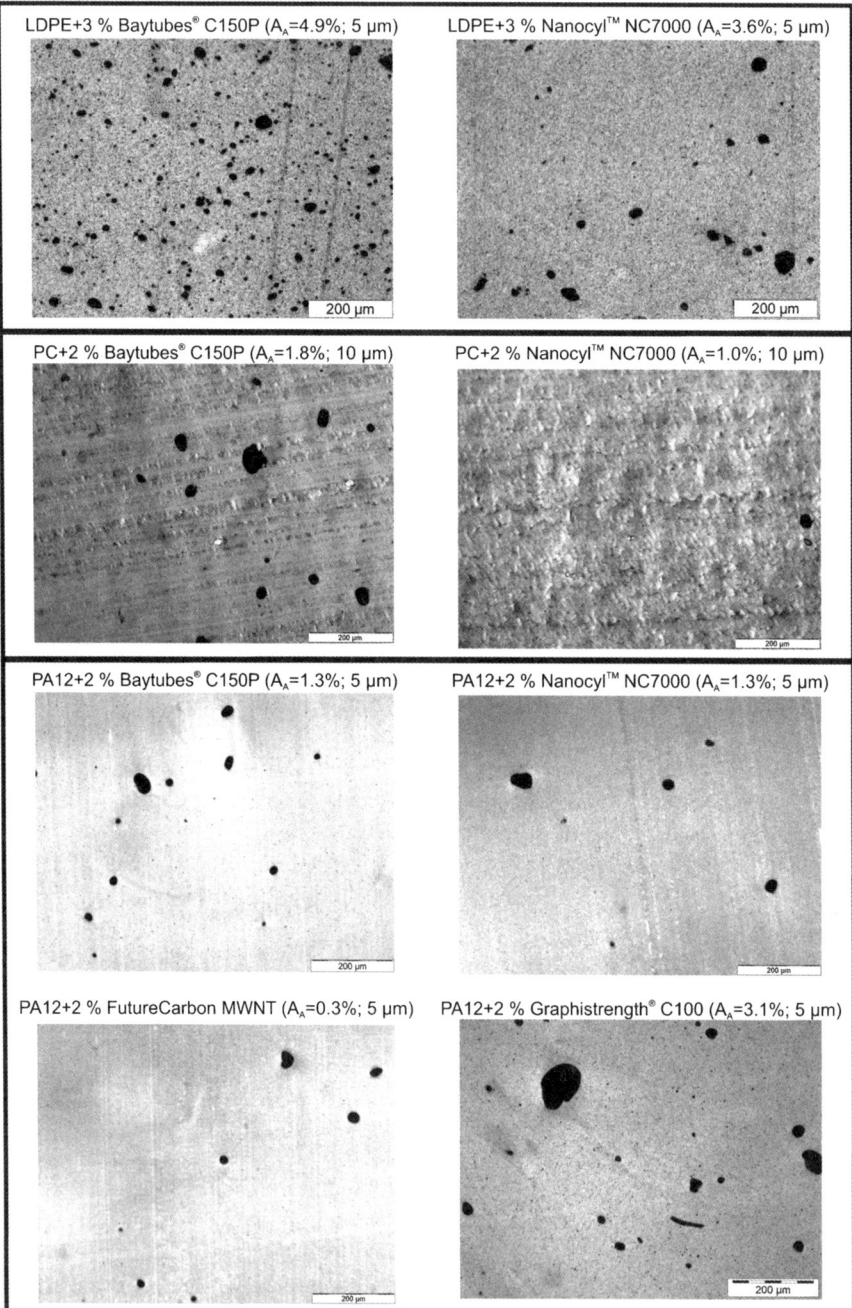

Figure 7.2 Melt-mixed composites based on different polymer matrices and types of MWCNTs. Light microscopy images of thin sections, including the composition, values of the agglomerate area ratio A_A, and the thickness of the thin sections.[4]

7.3.2 Length Analysis of CNTs Before and After Melt Processing

The electrical percolation threshold of CNTs in composites is strongly dependent on the aspect ratio of the CNTs and thus on their lengths.[16,17] However, the determination of the length distribution of CNTs is more difficult than measuring their diameters and is not often reported in literature. On one hand, CNTs are entangled and have to be separated non-destructively before their length can be measured. For this purpose, quite often ultrasonic treatment is used,[18–21] which was reported to lead to CNT shortening dependent on the applied energy.[22–24] On the other hand, the chosen method of length quantification has to be suitable to characterize the CNTs with their high aspect ratio, meaning that such magnifications have to be selected which show the full length of individualized CNTs. For this task, various imaging techniques, as atomic force microscopy (AFM),[21,25] scanning electron microscopy (SEM),[18–20] or transmission electron microscopy (TEM) are applicable.

In the literature, CNT length analyses have been described either for the as-produced CNT material or for CNTs dissolved from the composites. However, to give a quantitative statement on the aspect ratio as a function of processing approaches, it is necessary to determine the CNT length before and after processing using the same method, which was not described so far in literature. Therefore, a method was developed[26] for which the sample preparation scheme is shown in Figure 7.3. The CNTs or the composite are dispersed in a suitable solvent, whereby the solvent has to be a good dispersion medium for CNTs, as well as for the polymeric component of the composite, to extract the processed CNTs. A CNT concentration of less than 0.03 g L^{-1} in the dispersion was determined to be a suitable concentration in order to obtain individual nanotubes. For as-produced nanotube materials, chloroform, in particular,

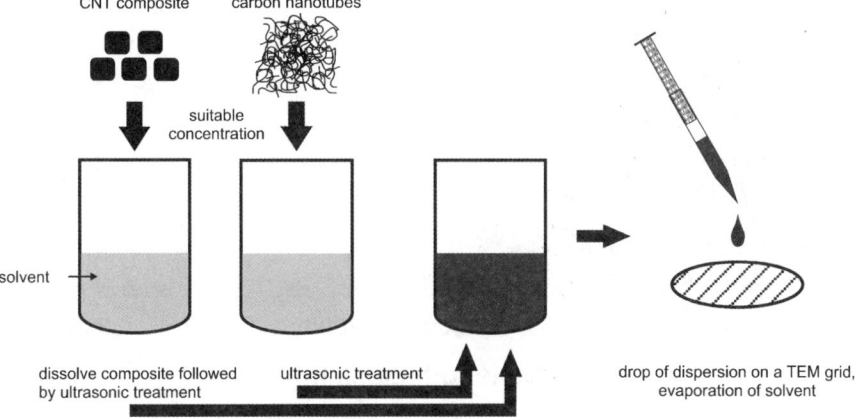

Figure 7.3 Scheme of the sample preparation used for the quantification of the length distribution of MWCNTs.

seems to be a very suitable solvent, as described previously.[26] After a brief treatment of the dispersions in an ultrasonic bath for 3 min at 80 W, from which no significant reduction in the CNT length is expected, a drop of dispersion is put on to a TEM grid which is covered with a perforated carbon film. After evaporation of the solvent, the sample is examined by TEM. The lengths of the CNTs were measured manually by means of the image analysis software SCANDIUM 5.1 (Olympus Soft Imaging Solutions GmbH) using the function of an open polyline.

As a result, the CNT length distributions can be illustrated in the form of histograms or a cumulative distributions of CNT lengths (see Figure 7.4). In addition, x_{10}, x_{50}, and x_{90} values can be determined, as shown in Figure 7.4. These values indicate that 10%, 50%, and 90% of the nanotubes are shorter than the named value.

As a first example, PC-based composites containing 3 wt% MWCNTs (either Nanocyl™ NC7000 or Baytubes® C150P) melt mixed using twin-screw extrusion were selected. Chloroform was used as a solvent medium and CNT suspensions with a concentration of 0.03 g L^{-1} were prepared. The as-received nanotube materials as well as the melt-mixed composites were dispersed for 3 min in an ultrasonic bath at 80 W. Figure 7.5 shows the example of a TEM image which was used to determine the lengths of CNTs extracted from a

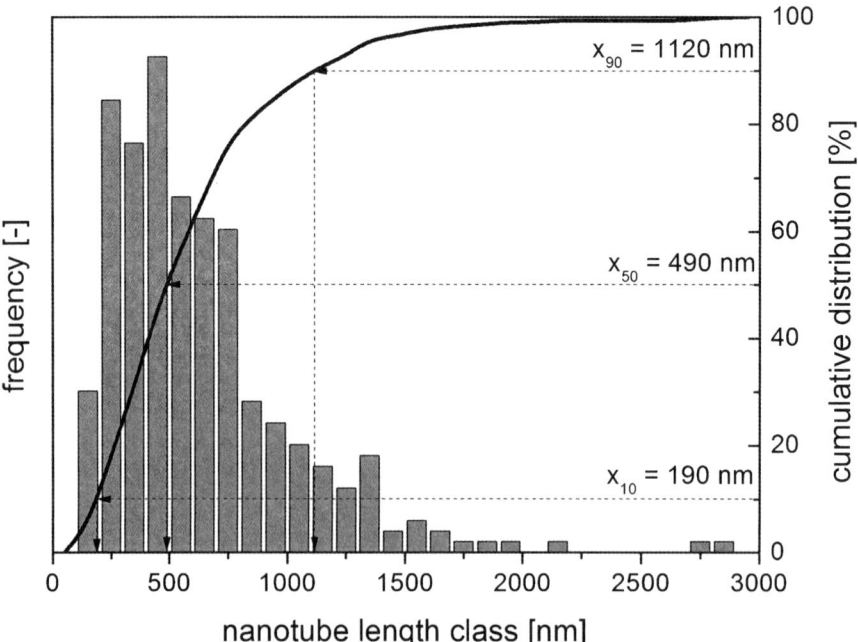

Figure 7.4 Example of nanotube length distribution as a histogram and cumulative distribution, including the way for determining the values for x_{10}, x_{50}, and x_{90}.

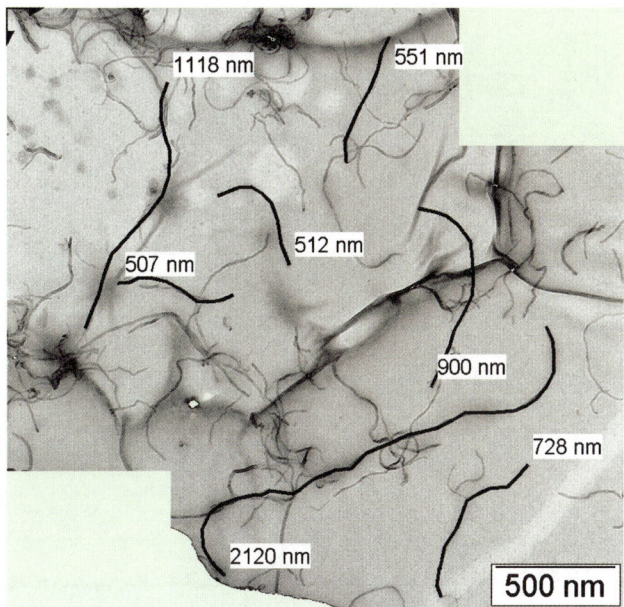

Figure 7.5 Stitched TEM images of Nanocyl™ NC7000 dissolved from PC composites containing 3 wt% MWCNT including lines to show the length of selected highlighted nanotubes.

composite. Note that after dropping the polymer–solvent–CNT dispersion on to the TEM grid, the polymer remaining after solvent evaporation (see Figure 7.5) does not interfere with the measurement of CNT lengths. As the true length of CNTs can only be evaluated for CNTs not touching the edges, in some cases very long CNTs were assessed by stitching together several TEM images, as shown in Figure 7.5 using the example of two TEM images.

In Figure 7.6, the CNT length distributions of both types of as-received MWCNT material are compared with those of CNTs dissolved from the PC–MWCNT composites. It can be concluded that under the selected melt compounding conditions for both types of MWCNTs, a significant length reduction took place. The initially longer Nanocyl™ NC7000 nanotubes were shortened to one-third and the initially shorter Baytubes® C150P to half the value of the original length (based on x_{50} values). Although the degree of length reduction is higher for the Nanocyl™ NC7000, the processed Nanocyl™ NC7000 are still longer than the processed Baytubes® C150P.

As a second example, the effect of treatment of MWCNTs using dry ball milling is shown. In a previous paper,[5] it was concluded that the grinding of CNTs results in a significant reduction in the CNT length as shown by nanotube length distributions in Figure 7.7. According to this Figure, Nanocyl™ NC7000 material was shortened after 5 and 10 h milling to 54% and 35% of its initial length, respectively (based on x_{50} values). In the subsequent melt processing step for all three types of MWCNTs, an additional

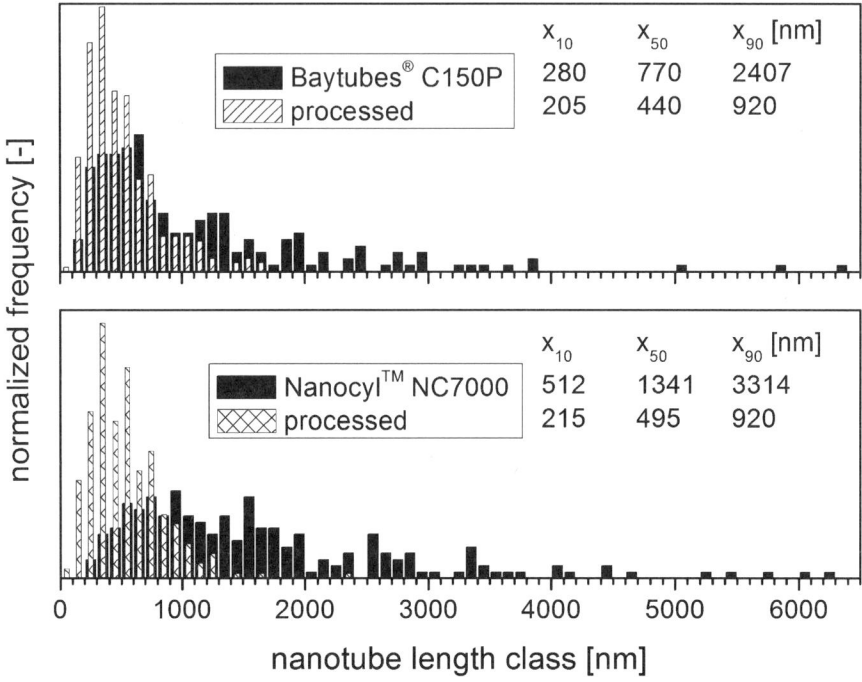

Figure 7.6 Length distribution of as-grown and processed MWCNTs dissolved from PC composites (3 wt% MWCNT, melt extruded at 500 rpm, 5 kg h^{-1}): Baytubes® C150P (top) and Nanocyl™ NC7000 (bottom).

length reduction could be observed, whereby the degree of reduction was the highest for the longest CNTs. When setting into relation the nanotube length and the achieved electrical percolation threshold in the composites based on the ball milled nanotubes,[5] the lowest threshold was found for composites with untreated Nanocyl™ NC7000, for which the aspect ratio of 42, even after processing, was the highest as compared with those of the CNTs already pre-shortened in the ball mill (37 and 31). In the composites containing the nanotubes with lower aspect ratios, higher percolation thresholds were found.

7.3.3 Relation Between SME, CNT Dispersion, CNT Length, and Electrical Properties of Melt Mixed Composites

In previous studies, macrodispersion was improved with increasing SME during the processing of melt mixed composites based on polyamide 6[6] and PC.[7,27] At the same time, the relation between electrical resistivity and SME showed a minimum which was found in both of the examples mentioned at approximately 1400–1800 J cm^{-3} (0.4–0.5 kW h kg^{-1}). Further increases in SME resulted in increasing electrical resistivity values, even if the macro-dispersion improves with SME or levels off at low values of A_A. Thus, a certain

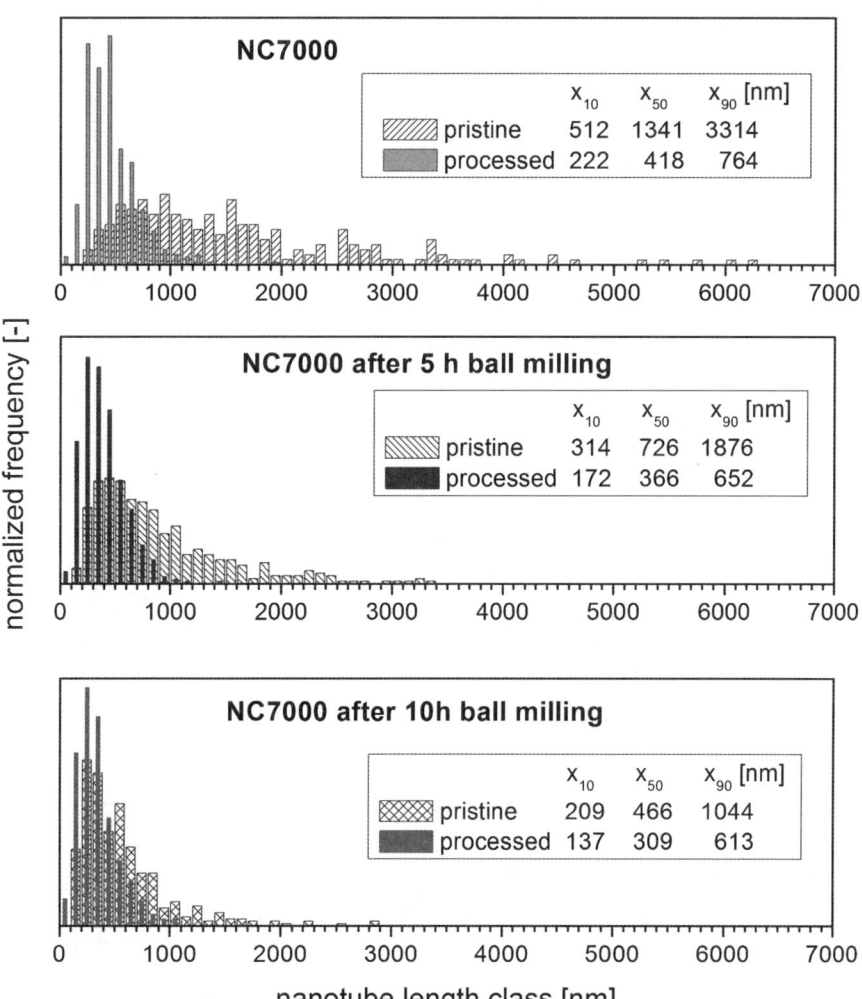

Figure 7.7 Length distributions of as-grown (pristine) ball milled (5 or 10 h) and processed Nanocyl™ NC7000 (dissolved from 2 wt% MWCNT/PC composites).

SME seems to be necessary in order to form a percolation network, but on the other hand too intense mixing resulted in an increase in the electrical resistivity. This effect is particularly pronounced for polyamide 6 composites.

The results of the electrical resistivity and the agglomerate area ratio A_A as a function of SME for polyamide 6^6 composites filled with 5 wt% Nanocyl™ NC7000 prepared under different melt mixing conditions are summarized in Figure 7.8. To vary the SME, different rotation speeds and mixing times were used. The lowest volume resistivity value was found at 2000 Ω cm for the composites mixed at 50 rpm and 15 min corresponding to a SME of 0.5 kW h kg^{-1}

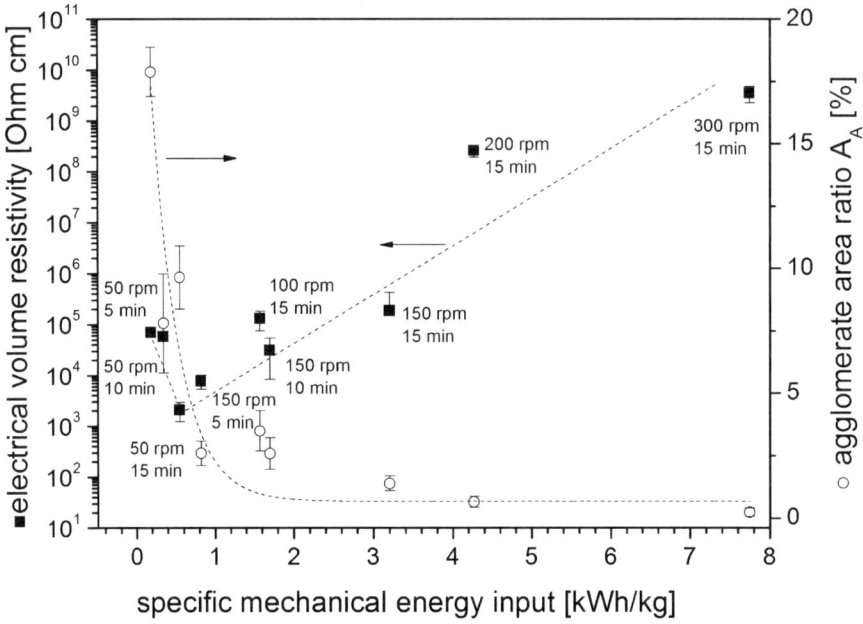

Figure 7.8 Electrical volume resistivity and agglomerate area ratio A_A as a function of the SME for polyamide 6 composites containing 5 wt% Nanocyl™ NC7000 (adapted from Krause et al.[6]).

(1800 J cm^{-3}). With further increasing SME values, the resistivity values significantly increase up to values in the order of 10^9 Ω cm. The macrodispersion increases indicated by exponentially decreasing values of the agglomerate area ratio A_A with the energy input. The reason for the increase in resistivity above a certain SME value despite a better dispersion was assumed to be CNT shortening.

In melt mixed PC-based composites filled with 1 wt% Baytubes® C150HP by using a DACA microcompounder, Kasaliwal et al.[7] studied the influence of SME on the electrical resistivity and the MWCNT macrodispersion by varying the mixing speed and the mixing time. It was also found that with increasing the SME, the agglomerate area ratio A_A decreased and leveled off at a plateau value (Figure 7.9), starting at approximately 2 kW h kg^{-1}. At SME values lower than 0.4 kW h kg^{-1}, high resistivity values were measured. After a minimum at 0.4 kW h kg^{-1}, further SME increase resulted in constant or only slightly increased resistivity values. However, the concentration of nanotubes used in the composites is above the electrical percolation threshold of approximately 0.6 wt%.[16] Therefore, the resistivity values are not extremely sensitive to changes in the CNT network, as well as the possible CNT shortening processes.

To investigate the relation between SME, CNT dispersion, CNT length, and electrical properties of melt mixed composites in a more detailed way, three studies on composites based on PCL and different PCs were performed.

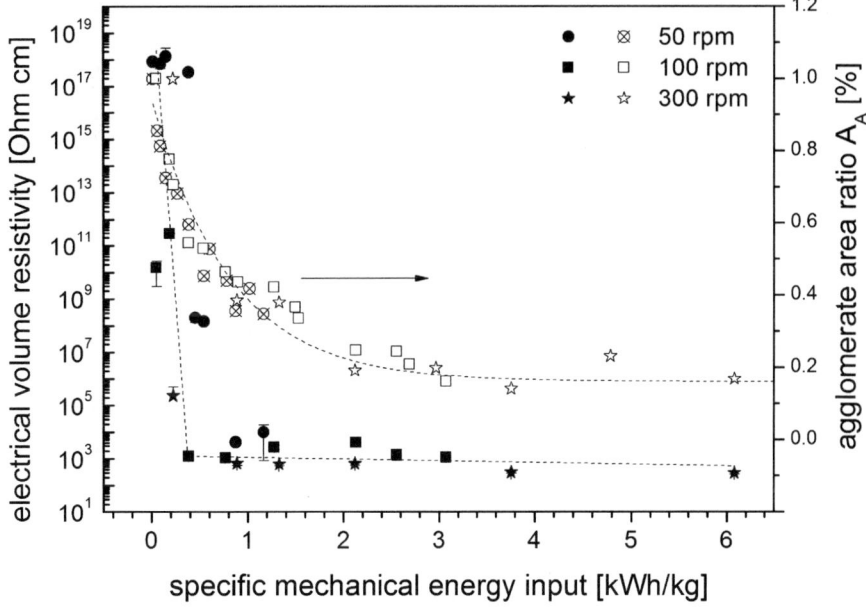

Figure 7.9 Electrical volume resistivity and agglomerate area ratio A_A as a function of SME for PC composites filled with 1 wt% Baytubes® C150HP prepared at different processing conditions (adapted from Kasaliwal et al.[7]).

The influence of the rotation speed during the melt mixing in small-scale and the related SME on the CNT macrodispersion, the CNT length, and the electrical surface resistivity of the composites was studied on PCL composites filled 0.5 wt% Nanocyl™ NC7000, which represents a concentration near the electrical percolation threshold[8] (Figure 7.10). This concentration was selected, as changes in CNT aspect ratio due to CNT shortening have a very pronounced influence on the electrical resistivity in the percolation concentration range. Even during mixing at the lowest mixing speed of 25 rpm for 2 min, a significant CNT shortening took place. The measured length value after processing was 41% of the initial value (based on the x_{50} value). The CNT length decreased up to 31% of the initial length at the highest SME of 0.7 kW h kg^{-1} achieved at 400 rpm. The lowest electrical surface resistivity of approximately 2×10^3 Ω sq^{-1} was found at a mixing speed of 75 rpm or a SME value of 0.5 kW h kg^{-1}. Above this minimum, with further increasing speed or SME, the electrical surface resistivity increases to values of approximately 1×10^{11} Ω sq^{-1}. The dependency of the agglomerate area ratio A_A on SME shows an exponential decrease up to approximately 100 rpm (0.7 kW h kg^{-1}) where it levels off. However, it can be also seen that only at very high screw speeds (400 rpm) can dispersions nearly free of agglomerates be achieved. The results indicate that, concerning the electrical resistivity, the effects of increasing macrodispersion and length shortening are counteracting,

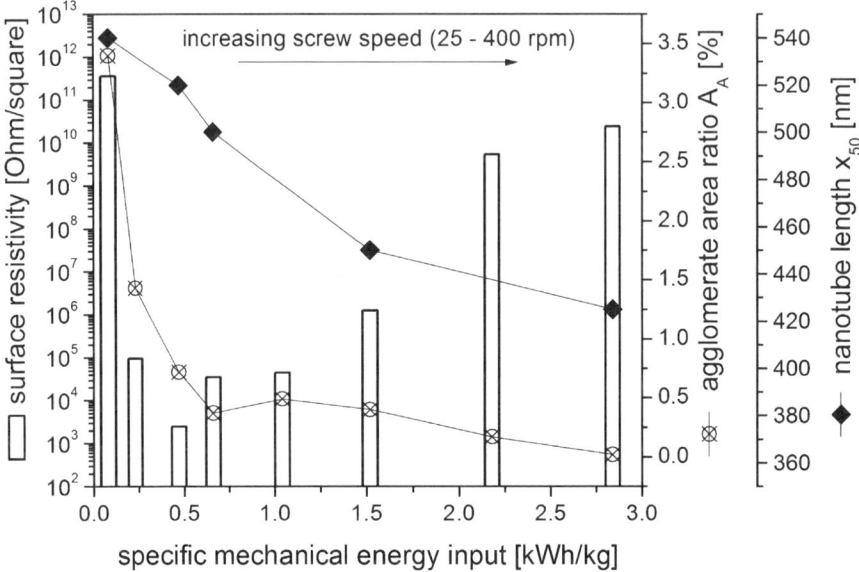

Figure 7.10 Surface resistivity, agglomerate area ratio A_A, and nanotube length x_{50} as a function of the SME for PCL composites containing 0.5 wt% Nanocyl™ NC7000.

leading to the lowest resistivity values at SME values of approximately 0.5 kW h kg^{-1}, which interestingly was also the range of a minimum or start of the plateau in the examples discussed before for polyamide 6 and PC.

On PC-based composites, the influence of matrix viscosity on the CNT shortening in a microcompounder was studied.[9] Due to the different matrix viscosities, different levels of shear stresses and SME were introduced during melt mixing under constant mixing conditions. Table 7.2 summarizes the viscosity of the matrix, the SME, and the composite properties concerning the nanotube length, the agglomerate area ratio A_A, and the electrical resistivity of compression moulded composites. The higher viscosity of the PC2 matrix resulted in a two-fold higher SME value during melt mixing compared with the low viscosity PC1. This difference leads to a better macrodispersion, as indicated by a lower agglomerate area ratio A_A but also to more significant CNT shortening. Based on the electrical volume resistivity values, the composite based on the low viscosity PC1 has longer nanotubes but worse CNT macrodispersion, whereas the composite based on PC2 has isolated nanotube structures. In particular, the x_{90} values of the CNT length distributions show strong differences between both composites, resulting in significant differences in the corresponding aspect ratios. After processing, the CNTs in the composite with low matrix viscosity show an x_{90} value of 1 μm, whereas in the composite with the high viscous matrix this value is only 570 nm. TEM studies showed that in the composite with high matrix viscosity a distinct individualization and separation of the CNTs occurred, whereas in the

Table 7.2 Characterization of Baytubes® C150P and Baytubes® C150P dissolved from PC composites.

	PC matrix viscosity at 100 rad s^{-1}, 280 °C (without CNTs) (Pa s)	SME input (kW h kg^{-1})	CNT length x value (nm)	Electrical volume resistivity (Ω cm)	Agglomerate area ratio A_A (%)
Baytubes® C150P[26]	–	–	x_{10}: 290; x_{50}: 770; x_{90}: 2407	–	–
PC1+0.75 wt% Baytubes® C150P[9]	224	1.38	x_{10}: 189; x_{50}: 418; x_{90}: 935	2.5×10^3	0.3 (only a few agglomerates)
PC2+0.75 wt% Baytubes® C150P[9]	931	2.80	x_{10}: 164; x_{50}: 350; x_{90}: 570	5.4×10^{16}	0.2 (nearly free of agglomerates)

Properties of composites of PC and Baytubes® C150P prepared using different matrix viscosity.

composite based on the low viscosity matrix a loosely packed network structure with secondary agglomerated CNTs was observed.[9] Thus, next to the effect of the matrix viscosity on CNT shortening and dispersion, also the secondary agglomeration seems to play an important role in the formation of the conductive network, which is more pronounced at lower matrix viscosity.

Finally, for a series of composites based on PC with 3 wt% Baytubes® C150P prepared using an extruder, the dependency of the CNT shortening and the CNT macrodispersion during melt mixing on SME varied by changing speed and throughput was investigated. In Figure 7.10, the agglomerate area ratio A_A and the x_{50} value of the CNT length distributions depending on SME are shown. With increasing SME, the agglomerate area ratio A_A,[28] as well as the CNT length, decrease and reach a plateau value starting at approximately 0.5 kW h kg^{-1}. This result illustrates that, with increasing SME, macrodispersion is improved, but at the same time the CNTs are more shortened. The electrical resistivity of these composites stay constant at a low resistivity value, as the CNT concentration of 3 wt% is well above the percolation range so that changes in CNT length are hardly detectable by means of electrical measurements. To investigate the influence of CNT shortening on the electrical properties, the composites with 3 wt% MWCNT extruded at different speeds and a throughput of 5 kg h^{-1} were diluted using a DSM Microcompounder under constant conditions of 280 °C, 50 rpm, and 5 min. The lowest electrical percolation threshold between 0.63 to 0.75 wt% was found for the composites for which the starting material with 3 wt% was melt mixed at 100 rpm and contained the longest CNTs compared with the other samples. With increasing speed and thus a higher SME during the production of the starting composites with 3 wt% MWCNT, the percolation threshold of

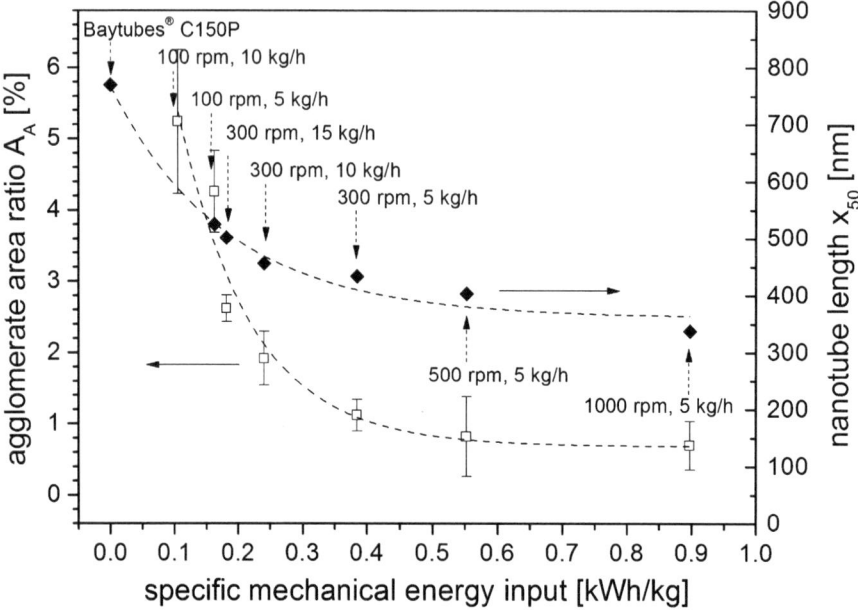

Figure 7.11 Agglomerate area ratio A_A and nanotube length x_{50} as a function of SME during melt extrusion of PC composites containing 3 wt% Baytubes® C150P.

the diluted PC composites increased to 1 wt%. Thus, the MWCNT shortening in the starting composites is clearly reflected in their dilutions which may have consequences on masterbatch dilution procedures.

Beside the effects of shortening on electrical percolation described here, a modified CNT aspect ratio also has consequences on the mechanical properties. Castillo *et al.*[16] studied PC composites containing CNTs of different lengths and found that the modulus increases with the aspect ratio of the MWCNTs. On the other hand, better macrodispersion leads to better mechanical properties, as remaining agglomerates act as imperfections and induce breaking at lower loads as well as higher scattering in the properties, *e.g.* in stress–strain experiments.

7.4. Summary and Conclusion

Two methods were developed and discussed with examples that allow the characterization of the dispersability of CNT materials and the CNT length distribution in polymer matrices. Applying the determined CNT properties using the methods described in this Chapter allows for a better understanding of the relations between energy input during the dispersion of CNT materials, the achievable macrodispersion, and electrical properties which are strongly dependent on the length of CNTs and the state of dispersion.

The results presented show that the energy input into aqueous surfactant dispersions or polymer melts plays an important role for the dispersion of CNTs. It was shown that the particle size of the remaining CNT agglomerates in aqueous surfactant solutions is reduced with increasing ultrasonic treatment time. The investigated industrial MWCNT materials differ significantly in their dispersability either in aqueous solutions or in polymer melts. In order to obtain stable dispersions of different MWCNTs in a surfactant solution, between 2 and 10 min (140–700 J cm^{-3}) ultrasound treatment are required under the selected conditions. The CNT dispersability shown in the dispersion experiments in aqueous solutions correlated well with the state of macrodispersion in melt mixed thermoplastic CNT composites. At constant mechanical energy inputs, less and smaller CNT agglomerates were found in the composites when using nanotubes with better dispersability.

On the other hand, higher SME resulted in all presented composites in a better CNT macrodispersion. At the same time, the CNTs were shortened significantly with increasing SME, leading to higher resistivity values or even crossing the percolation concentration. In composites with shorter nanotubes, an increase in electrical percolation threshold was shown.

The significant effect of shortening has been possibly underestimated so far in research on the properties of melt mixed CNT-based composites and should be taken into consideration in further studies on that topic. This was certainly due—in part—to the difficulties to measure the length after processing. However, the presented study clearly shows a way to measure the length as the importance to consider this quantity.

For practical applications, a balance of the counteracting processes of the better dispersion and more pronounced CNT length shortening with increasing SME has to be achieved. It has to be considered that for improved mechanical properties a good CNT dispersion is essential, as the remaining CNT agglomerates lead to defects in the composite. A good dispersion is achieved at high SME which on the other hand shortens the nanotubes more severely. However, for a high electrical conductivity of CNT composites the nanotubes should remain as long as possible which only can be achieved at low SME.

Thus, depending on the aspired property profile and in addition economic aspects, suitable energy inputs and mixing conditions have to be selected. In addition, it has to be considered that final shaping in most cases requires a second processing step, such as compression or injection molding. In this step, in particular, the process of secondary agglomeration of dispersed nanotubes influences the development of the percolation network and the electrical properties of the parts, which again depends on the shaping conditions.

Acknowledgements

We thank the German Federal Ministry of Education and Research (project grant ID 03X3006E and within the Innovation Alliance Inno.CNT, project

CarboDis, grant ID 03X0042) and the German Federal Ministry of Economics and Technology (project grant ID AiF 336 ZBG) for financial support.

References

1. B. Krause, M. Mende, P. Pötschke and G. Petzold, *Carbon*, 2010, **48**, 2746–2754.
2. B. Krause, G. Petzold, S. Pegel and P. Pötschke, *Carbon*, 2009, **47**, 602–612.
3. D. Lerche, *J. Dispers. Sci. Technol.*, 2002, **23**, 699–709.
4. R. Socher, B. Krause, R. Boldt, S. Hermasch, R. Wursche and P. Pötschke, *Compos. Sci. Technol.*, 2011, **71**, 306–314.
5. B. Krause, T. Villmow, R. Boldt, M. Mende, G. Petzold and P. Pötschke, *Compos. Sci. Technol.*, 2011, **71**, 1145–1153.
6. B. Krause, P. Pötschke and L. Häußler, *Compos. Sci. Technol.*, 2009, **69**, 1505–1515.
7. G. R. Kasaliwal, S. Pegel, A. Göldel, P. Pötschke and G. Heinrich, *Polymer*, 2010, **51**, 2708–2720.
8. P. Pötschke and T. Villmow, presented at ANTEC, Orlando, FL, USA, 2012, contribution 1259181.
9. R. Socher, B. Krause, M. T. Müller, R. Boldt and P. Pötschke, *Polymer*, 2012, **53**, 495–504.
10. S. T. Buschhorn, M. H. G. Wichmann, J. Sumfleth, K. Schulte, S. Pegel, G. R. Kasaliwal, T. Villmow, B. Krause, A. Göldel and P. Pötschke, *Chem.-Ing.-Tech.*, 2011, **83**, 767–781.
11. M. S. Strano, V. C. Moore, M. K. Miller, M. J. Allen, E. H. Haroz, C. Kittrell, R. H. Hauge and R. E. Smalley, *J. Nanosci. Nanotechnol.*, 2003, **3**, 81–86.
12. T. Sobisch and D. Lerche, *Chem.-Ing.-Tech.*, 2008, **80**, 393–397.
13. M. Nadler, T. Mahrholz, U. Riedel, C. Schilde and A. Kwade, *Carbon*, 2008, **46**, 1384–1392.
14. J. E. Butler and E. S. G. Shaqfeh, *J. Fluid Mech.*, 2002, **468**, 205–237.
15. A. R. Henn, *Part. Part. Syst. Charact.*, 1996, **13**, 249–253.
16. F. Y. Castillo, R. Socher, B. Krause, R. Headrick, B. P. Grady, R. Prada-Silvy and P. Pötschke, *Polymer*, 2011, **52**, 3835–3845.
17. D. Stauffer and A. Aharony, *Introduction in Percolation Theory*, Taylor and Francis, London, 1994.
18. R. K. Duncan, X. G. Chen, J. B. Bult, L. C. Brinson and L. S. Schadler, *Compos. Sci. Technol.*, 2010, **70**, 599–605.
19. J. Albuerne, A. Boschetti-de-Fierro and V. Abetz, *J. Polym. Sci., Part B: Polym. Phys.*, 2010, **48**, 1035–1046.
20. S.-Y. Fu, Z.-K. Chen, S. Hong and C. C. Han, *Carbon*, 2009, **47**, 3192–3200.
21. B. Lin, U. Sundararaj and P. Pötschke, *Macromol. Mater. Eng.*, 2006, **291**, 227–238.

22. S. Badaire, P. Poulin, M. Maugey and C. Zakri, *Langmuir*, 2004, **20**, 10367–10370.
23. K. L. Lu, R. M. Lago, Y. K. Chen, M. L. H. Green, P. J. F. Harris and S. C. Tsang, *Carbon*, 1996, **34**, 814–816.
24. Y. Wang, J. Wu and F. Wei, *Carbon*, 2003, **41**, 2939–2948.
25. M. F. Islam, E. Rojas, D. M. Bergey, A. T. Johnson and A. G. Yodh, *Nano Lett.*, 2003, **3**, 269–273.
26. B. Krause, R. Boldt and P. Pötschke, *Carbon*, 2011, **49**, 1243–1247.
27. G. R. Kasaliwal, A. Göldel and P. Pötschke, *J. Appl. Polym. Sci.*, 2009, **112**, 3494–3509.
28. I. Alig, P. Pötschke, D. Lellinger, T. Skipa, S. Pegel, G. R. Kasaliwal and T. Villmow, *Polymer*, 2012, **53**, 4–28.

CHAPTER 8

Methods for Improving the Integration of Functionalized Carbon Nanotubes in Polymers

L. VALENTINI*, D. PUGLIA AND J. M. KENNY

University of Perugia, Dipartimento di Ingegneria Civile e Ambientale, Strada di Pentima 4, Terni, 05100, Italy
*E-mail: mic@unipg.it

8.1 Introduction

Due to their exceptional mechanical, thermal and electrical properties, in addition to the low density with respect to the class of organic and inorganic tubes, carbon nanotubes (CNTs) are extremely promising for the development of high-performance nanostructured materials. Since their discovery in 1993, the research in this exciting field has been in continuous evolution, with most of the research focused on the assessment of the CNT properties and the development of advanced structural composites based on CNTs.[1,2] However, the incorporation of nanotubes is not a trivial task, particularly as a good dispersion for a chemical grafting to the polymer matrix is mandatory to maximize the advantage of nanotube reinforcement. In fact, the affinity to adhere to each other renders as-grown CNTs intractable and indispersable in common solvents.[3]

On the other hand, it has been demonstrated that CNTs can interact with different classes of compounds.[4–20] The formation of supramolecular complexes allows a better processing of CNTs for the fabrication of innovative nanodevices. In addition, CNTs can undergo chemical reactions that make them more soluble for their integration into organic systems. Two of the key challenges that are preventing the realization of composites made out of CNTs are securing a reliable control over their surface chemistry through either covalent or non-covalent modification and achieving dispersion.

In this Chapter, we report on some examples of nanocomposites with CNTs, highlighting a meshwork of interactions between the mechanical, electrical and optical properties of CNTs and the interface with the polymer matrix. CNTs are considered ideal materials for reinforcing fibers due to their exceptional mechanical properties. Functionalization of CNTs seems to be the most effective way to incorporate these nanofibers into the polymer matrix. It is generally accepted that the fabrication of high-performance nanotube–polymer composite depends on the efficient load transfer from the host matrix to the tubes. If the percentage of nano-reinforcements is very low or if it is well-dispersed, there are more strong interfaces that slow down the progress of a crack.[21,22]

To address these issues, several strategies for the preparation of such composites have been reported. Here we report some of these strategies involving physical mixing in solution, infiltration of monomers in the presence of nanotube sheets and chemical functionalization of CNTs by plasma treatment.

8.2 *In Situ* Polymerization Methods

8.2.1 Poly(methyl methacrylate) (PMMA)-based Nanocomposites

It is known how the selective localization of nanoparticles [*i.e.* carbon black (CB)] at the interface of a polymer blend[23,24] represents an alternative route to obtain conductive materials. Following this concept, it was recently proposed[25] that the localization of the nanotubes at an interface instead of a homogeneous dispersion within the whole composite volume enhances the conductivity of the material with a low nanotube content. A compromise between the nanotube dispersion and their localization at the interface can be obtained by preparing thin films of CNTs containing polymers. In this regard, more recently, a novel approach was proposed in which pre-aligned arrays of multi-walled carbon nanotubes (MWNTs) were grown on a substrate by chemical vapour deposition,[26] then a monomer (*i.e.* methyl methacrylate) was infiltrated and polymerized into these arrays. The resulting composite films showed good dispersion of nanotubes in the polymer matrix with an enhanced thermal stability. Similar synthesis approaches have been used to develop composite architectures consisting of intercalated networks of nanotubes and polymers.[27–29]

Zhang *et al.*[30] obtained a CNT–polymer composite by infiltration of a monomer liquid into aligned CNT aerogel fibers with subsequent *in situ* polymerization. PMMA/MWNT composite fibers showed that the PMMA filled the spaces of the nanotube fibers and bound the nanotubes together. PMMA in the composite fibers exhibited local order. The resultant composite fibers with 15 wt% nanotube loading exhibited a 16-fold and a 49-fold increase in tensile strength and Young's modulus, respectively, compared with the control PMMA.

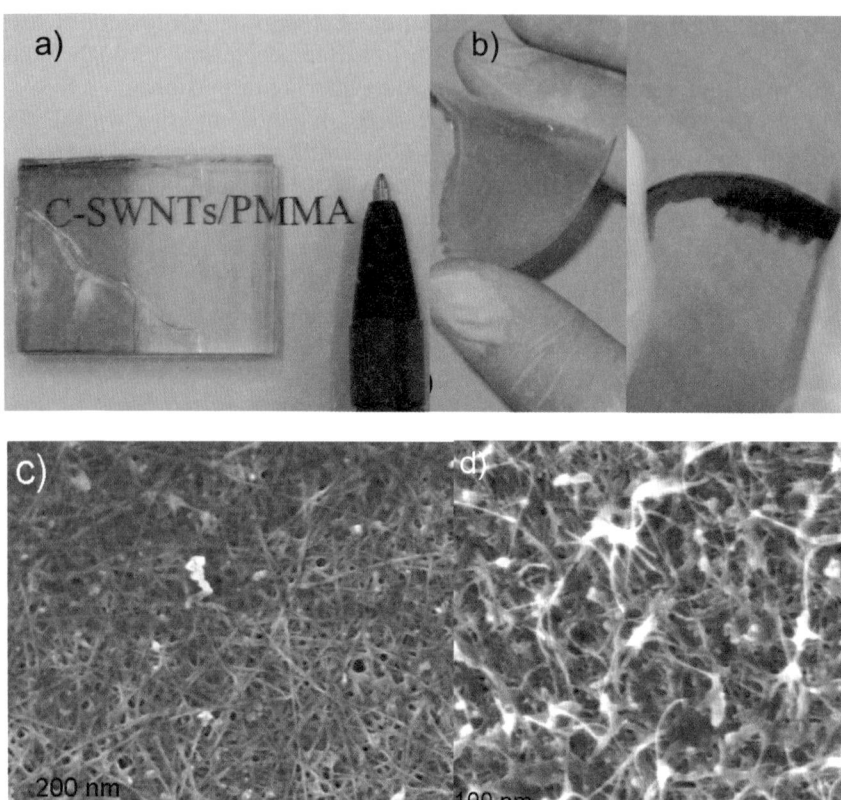

Figure 8.1 (a) Image showing semi-transparent carboxy-SWNT after methyl methacrylate (MMA) infiltration and polymerization. (b) Image of the PMMA (left image) and carboxy-SWNT/PMMA (right image) samples after the polymerization and subsequent peeling off from the fluorine tin oxide surface. (c) Field emission SEM (FESEM) top view image of nanotube after PMMA polymerization. (d) A back-side view of the nanotubes/PMMA composite after the peeling off from the substrate. Reproduced with permission from Valentini et al.[33]

Carboxylated single-walled carbon nanotubes (SWNTs) suspended in acetonitrile were elecrophoretically deposited between two electrodes under the influence of a DC electric field.[31,32] This process enabled the anchoring of the tubes in the form of bundles on the positive electrode, thus providing an easy route to obtain functional nanostructures by infiltrating and polymerizing a methyl methacrylate-based monomer solution. The nanotube/PMMA system obtained with this procedure demonstrated outstanding mechanical properties functioning as semi-transparent conductive thin films.[33]

Optical images (Figure 1a) of the composite film prepared by methyl-methacrylate infiltration and subsequent polymerization, show that the films detached from the substrate are flexible and can be deformed easily (Figure 1b). The sample retains the nanotube original shape and size inside the resulting composite matrix (Figure 1c), even after the polymerization process and the subsequent peeling off from the substrate (Figure 1d). The resulting PMMA is found to coat the CNTs as observed from a high-resolution scanning electron microscopy (SEM) image of composites shown in Figure 1(d).

8.2.2 Hybrid Conducting Polymers

The development of transparent conductive electrodes based on SWNT thin films represented an outstanding scientific breakthrough for applications in the area of optoelectronics.[34] However, to build integrated CNT–polymer-based systems it is necessary to engineer the interfaces between the two constituents through organized nanotube architectures.

More generally for the manufacturing sector 'flexible' is synonymous of 'rollable', thus the development of deposition techniques fully compatible with plastic substrates, low temperature processes and solution processable materials, and suitable for flexible substrates for roll-to-roll manufacturing technologies is mandatory.

One-step electropolymerization deposition processes are envisaged as the most interesting deposition methods in view of potential technological applications. For this reason, there is a growing interest in the integration of materials obtained in such way to obtain multifunctional nanostructured composites and hybrid materials. The selective localization through electrophoretic deposition of nanotubes at the interface of polymers to obtain flexible, transparent and conductive materials could represent a novel approach.

Fluorene polymers and copolymers (either obtained by electrochemical[35,36] or chemical processes[37–39]) are proven to be among the most promising materials in the field of organic light-emitting devices (OLEDs).

The properties of supramolecular structures of polyfluorene–CNT hybrids have been recently reported.[40–42]

The use of SWNT films as electrodes for electropolymerization of the fluorene units has been investigated. The unique structural and electronic properties of CNTs render them an ideal candidate to function as a model carbon nanoelectrode for electrodeposition. This represents a direct and effective method for fabricating CNT–polymer composites by incorporating CNTs into a polymer matrix without considering organized nanotube architectures in selected polymer matrixes as well as engineering the interfaces between the two constituents. The enhanced conductivity of fluorene polymer chains[43] can be attributed to the entrapped nanotubes and nanotube bridging (Figure 8.2). The one-dimensional structure of CNTs may also induce and promote oriented polymerization, hence yielding an enhanced supramolecular order and higher conductivity.[43]

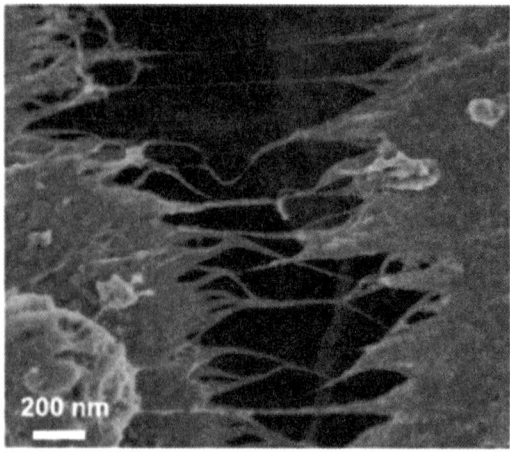

Figure 8.2 FESEM image of the polyfluorene deposited on the carboxy-SWNTs/fluorine–tin-oxide electrode. Reproduced with permission from Valentini et al.[43]

8.3 Melt Blending and Solvent Dispersion

8.3.1 Thermoplastic Polymers

In general, the mechanical properties and glass transition temperature are reported to be enhanced in the presence of CNTs. A practical method of producing CNT–polymer composites is by melt mixing. For this purpose, the goal of this study was to prepare SWNT–polymer composites by melt mixing using a shear mixer.

Here an isotactic polypropylene matrix (iPP) was selected as an example because thermoplastics are materials with a higher consumption due to their well-balanced physical and mechanical properties and their easy processability at a relatively low cost that makes them a versatile material.[44]

The objectives of this study included: (1) to study the non-isothermal crystallization behaviour of SWNT-reinforced iPP in detail and to compare the results to those of neat iPP, and (2) to describe the effect of SWNTs on the mechanical properties of neat iPP. The results were then compared with those obtained for nanocomposites containing CB as a filler.

The thermal parameters of the crystallization temperature (T_c), melting temperature (T_m), crystallization enthalpy (ΔH_c), heat of fusion (ΔH_f), supercooling temperature ($\Delta T = T_m - T_c$) and percentage of crystallinity (X_c) were obtained and reported in Table 8.1.

The supercooling temperatures reported in Table 8.1 are significantly shorter for reinforced iPP. This result indicates that the induction time for PP crystallization is reduced by the addition of low percentage SWNT fractions. In addition, the obtained results show that the crystallization peak temperature, T_c, increases when both nanotubes and CB fillers are

Table 8.1 Thermal parameters of the prepared samples. Reproduced with permission from Lòpez Manchado et al.[44]

Material	T_c (°C)	ΔH_c (J g^{-1})	χ_c (%)	T_m (°C)	ΔT (°C)
PP	115.22	−97.63	51.38	162.06	46.84
PP–SWNTs (0.25%)	117.62	−95.33	50.17	161.47	43.85
PP–SWNTs (0.5%)	118.39	−94.89	49.94	160.37	41.98
PP–SWNTs (0.75%)	119.85	−96.12	50.59	161.12	41.27
PP–SWNTs (1%)	120.38	−96.39	50.73	162.16	41.78
PP–CB (0.25%)	115.80	−95.06	50.03	162.21	46.41
PP–CB (0.5%)	116.46	−93.68	49.30	161.06	44.60
PP–CB (0.75%)	116.57	−95.22	50.11	162.97	46.40
PP–CB (1%)	117.14	−97.65	51.39	161.28	44.10

incorporated in the polymer matrix, this effect being more evident in the presence of SWNTs. As the filler concentration was increased, the T_c continued to increase. It is assumed that an increase in T_c is associated with an increased number of heterogeneous nuclei for crystallization. These results lead us to the conclusion that SWNTs behave as effective nucleating agents for PP crystals in the composite, even at very low percentages.

Analogous results have been recently obtained by other research groups.[45–48] The effect of -COOH- and phenol-functionalized CNTs on mechanical, dynamic mechanical and thermal properties of polypropylene (PP) nanocomposites were considered, and the results confirmed in both cases that the percentage crystallinity was found to increase on phenol and carboxylic functionalization of MWNTs.

8.3.2 Confinement of CNTs in Block Copolymer Matrix

The fabrication of a solution of CNTs with good dispersion and stability has been of great interest because most of the potential applications of SWNTs are based on solution processes such as spin coating and dip coating.[49–51] The physical adsorption on to the nanotube surface of surfactants and/or macromolecules has been shown as a possible way to stabilize SWNTs in both aqueous and organic media.[52–57]

Block copolymers have attracted significant attention as nanostructured materials, as they self-assemble to form well-defined, ordered, periodic nanoscale morphologies.[58–63] The versatility of this unique class of polymers offers tremendous potential for their use as templates and scaffolds for applications in microelectronics, separation devices, optics and optoelectronics.[64–67]

More generally to separate and solubilize as-synthesized nanotubes, they have to be functionalized. It was demonstrated that the nanotube treatment with octadecylamine (ODA) and other amines, such as nonylamine, dodecylamine, pentacosylamine, tetracontylamine, pentacontylamine and alkylaryl amine as well as aromatic amines, aid the tube solubilization in several solvents (*e.g.* toluene, chlorobenzene, dichlorobenzene, dimethylformamide, tetrahydrofuran, hexamethylphosphoramide and dimethylsulfoxide).[68]

Along with the utilization of block copolymers as CNT dispersants, the role of an organic surfactant for the confinement of the nanotubes inside the self-assembled block copolymer matrix was reported. More generally a surfactant consists of two parts, one hydrophilic and the other one hydrophobic, and due to this architecture it is used as a bridge between nanofillers and polymeric matrices.[69]

In particular, when a block copolymer matrix is used, an adequate surfactant can be used to selectively disperse nanofillers in one of the blocks of the block copolymer matrix. This selectivity is important in designing the properties of block copolymer-based composites.

Starting from the approach reported elsewhere for metallic nanoparticles we focused our attention on a hybrid system consisting of a poly(styrene-*b*-isoprene-*b*-styrene) (SIS) block copolymer and soluble nanotubes.[70] The nanotubes have been modified with a surfactant that interacts with only one block of the self-assembling block copolymer matrix. In particular, we have investigated the possibility of modifying ODA-functionalized CNTs (ODA-SWNTs) with dodecanethiol as a surfactant and to use this mixture as nanofiller for the production of composites based on SIS block copolymer. When the surfactant is used, the nanotubes are aligned along the block copolymer domains and appear to be embedded in the styrene phase, as shown in Figure 8.3 in which the interlamellar periodicity of approximately 20 nm is regular along the sample.

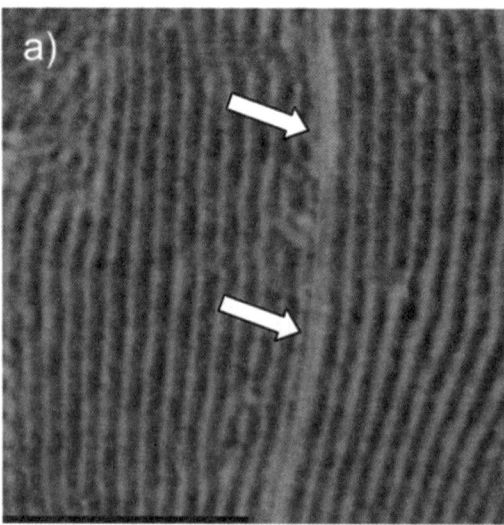

Figure 8.3 Tapping mode atomic force microscopy (TM-AFM) height image of dodecanethiol (DT)–ODA-SWNTs/SIS composite film. The arrow serves to indicate the nanotube location. Scale bar, 500 nm. Reproduced with permission from Peponi *et al.*[70]

8.3.3 Solvent Dispersion

Production processes for CNTs often produce mixtures of solid morphologies that are mechanically entangled or that self-associate into aggregates. Entangled or aggregated nanoparticles often need to be dispersed into fluid suspensions in order to develop materials that have unique mechanical characteristics or transport properties.

While mechanical dispersion methods, such as ultrasonication, separate nanotubes from each other, but can also fragment the nanotubes, decreasing their aspect ratio during processing, chemical methods use surfactants or functionalization to change the surface energy of the nanotubes, improving their wetting or adhesion characteristics and reducing their tendency to agglomerate in the continuous phase solvent.

The solubility of CNTs may be necessary for their chemical and physical examination, as it allows easy characterization and facilitates their manipulation. Exhaustive research was addressed to overcome the poor solubility of CNTs in either water or organic solvents. With the aid of surfactants,[71,72] organic molecules[73,74] or small biomolecules,[75,76] SWNTs can be dispersed in aqueous solution. Conjugated polymers and small molecules can also be employed in organic solvents.[77,78] Covalent sidewall functionalization is another effective method to improve the solubility and stability of SWNTs in solution.[79] However, the introduction of a third component and the modification of the sidewall would affect the pristine properties of the tubes which should be avoided. Over the years, constant effort has been devoted to finding appropriate media to solubilize pristine nanotubes. Various solvents have been investigated in order to solubilize and disperse SWNT aggregates. Non-hydrogen-bonding Lewis bases, such as dimethylformamide (DMF) and N-methylpyrrolidone (NMP), have demonstrated the ability to readily form stable dispersions of SWNTs produced by different techniques.[80–82]

Availability of stable CNT suspensions is a prerequisite for production of polymer composites using a liquid phase compounding strategy. In a recent work, nanocomposite films based on SWNTs and poly(DL-lactide-co-glycolide) copolymer (50:50 PLGA) were processed and analyzed.[83] Specifically, using a solvent casting process, dispersion of pristine and COOH-functionalized CNTs were considered, with the aim of investigating the effect of different functionalization systems on the physical stability and morphology of the obtained PLGA films. Both covalent and non-covalent functionalization of CNTs were considered in order to control the interactions between PLGA and SWNTs and to understand the role of the filler in the biodegradation properties. It has been observed that both the system composition and SWNT functionalization may play a crucial role on the autocatalytic effect of the degradation process. These studies suggest that the degradation kinetics of the films can be engineered by varying the CNT content and functionalization. Analogous conclusions were reported in a study by Armentano et al.,[84] in which poly(L-lactide) (PLLA)/SWNTs nanocomposite films were produced using the solvent casting method, and morphological, thermal and mechanical properties were investigated. The role of SWNT incorporation and functionalization on PLLA bio-polymers was investigated. Pristine SWNTs and carboxy-SWNTs were considered in order to control the interaction between PLLA and the nanotubes. Biological investigations showed osteoblasts cultured on PLLA/carboxy-SWNTs nanocomposites had higher cell adhesion and proliferation than osteoblasts cultured on PLLA and PLLA/SWNTs nanocomposites. Recent results on the introduction of functionalized CNTs in biodegradable matrices using solvent dispersion can be found in the studies by Yang et al.[85] and Lin et al.,[86] in which electrospinning of the biodegradable polylactide (PLA) and its composites containing CNTs was studied in terms of solution concentrations and solvents effects, as well as CNT loadings. The results revealed that the PLA fibers obtained from the solutions using the mixed solvents of chloroform/assistant solvent showed better morphologies than those from the solutions using chloroform as the single solvent. This is due to the synergistic effect of the improved conductivity and altered viscosity with addition of the assistant solvent.

The combination of SWNTs and silver (Ag) nanoparticles with a biodegradable polymer matrix was also considered.[87] Different SWNT amounts were mixed with Ag nanoparticles and introduced in the poly(ε-caprolactone) (PCL) polymer matrix by a solvent casting process. Results showed a good dispersion of nanostructures in the PCL matrix and an increase of the Young's modulus with Ag content in the binary systems. The PCL/Ag composites exhibited poor electrical properties, whereas in the PCL/Ag/SWNTs ternary films higher values of conductivity were measured compared with both of the binary composites (Figure 8.4). Results obtained in this research indicate that the addition of a small percentage of SWNTs significantly promoted the electrical properties of PCL/Ag nanohybrid films. Biocompatibility of binary and ternary composites, evaluated by bone marrow-derived human mesenchymal stem cells (hBM-MSCs), suggested that the combination of Ag nanoparticles and SWNTs with a biodegradable polymer opens new perspectives for biomedical applications.

8.4 Chemical and Physical Methods of CNT Dispersion

8.4.1 Epoxy Nanocomposites

The direct fluorination of SWNTs and their subsequent derivatization provide a versatile tool for the preparation and manipulation of nanotubes with variable sidewall functionalities.[88–93] Recent studies have shown that covalently attached fluorine moieties in SWNTs can be efficiently displaced by alkylamino functionalities.[93] The nucleophilic substitution offers an opportunity for SWNTs to be integrated into the structure of the epoxy systems through the sidewall amino functional groups.

Methods for Improving the Integration of Functionalized Carbon Nanotubes in Polymers 243

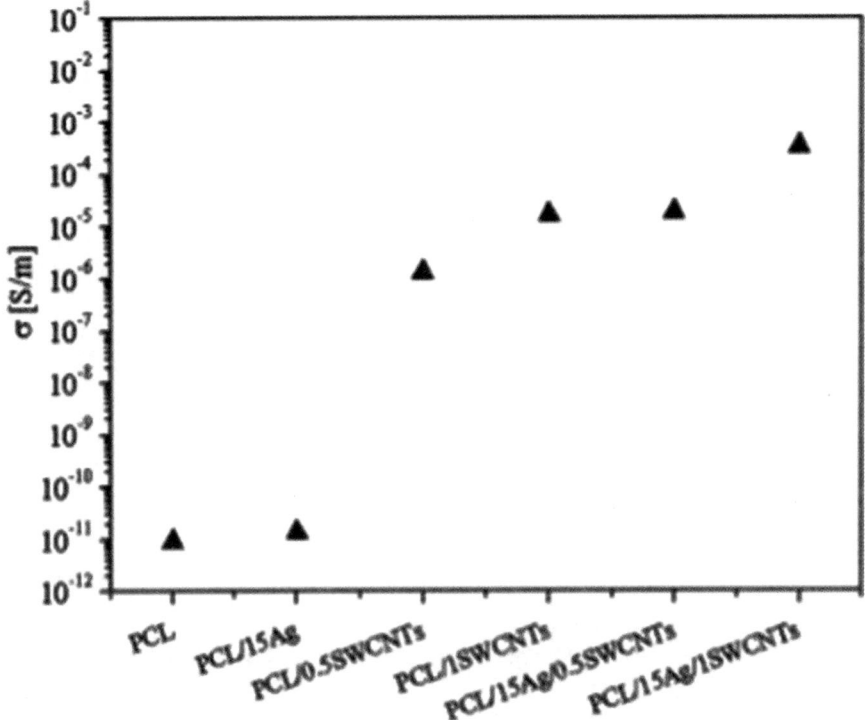

Figure 8.4 DC electrical resistivity of binary PCL/SWNTs, PCL/Ag and ternary PCL/Ag/SWNTs composites. Reproduced with permission from Fortunati et al.[87] PCL (neat matrix); PCL/15Ag (PCL + 15% wt. Ag nanoparticles); PCL/0.5SWCNTs (PCL + 0.5% wt. SWNTs); PCL/1SWCNTs (PCL + 1% wt. SWNTs); PCL/15Ag/0.5SWCNTs (PCL + 15% wt. Ag nanoparticles + 0.5% wt. SWNTs); PCL/15Ag/1SWCNTs (PCL + 15% wt. Ag nanoparticles + 1% wt. SWNTs).

The functionalization of CNTs through plasma treatment represents a novel and easy approach to scale up towards industrial applications. In more recent works, there were many attempts to fluorinate CNT sidewalls in such manner.[94–97] The CF_4 plasma treatment of SWNT sidewalls was demonstrated to enhance the reactivity of tubes with aliphatic amines.[96] The cure reaction of diglycidyl ether of bisphenol A-based epoxy resin (DGEBA), when reacted with butylamine molecules (BAMs) anchored on to the plasma treated fluorinated SWNTs, was reported. The advantage of this method was that the functionalization could be achieved through a simple approach, which is widely used in thin film technologies. As covalently modified CNTs with fluorine groups offer the opportunity for chemical interactions with the amine systems, it was recently demonstrated that this

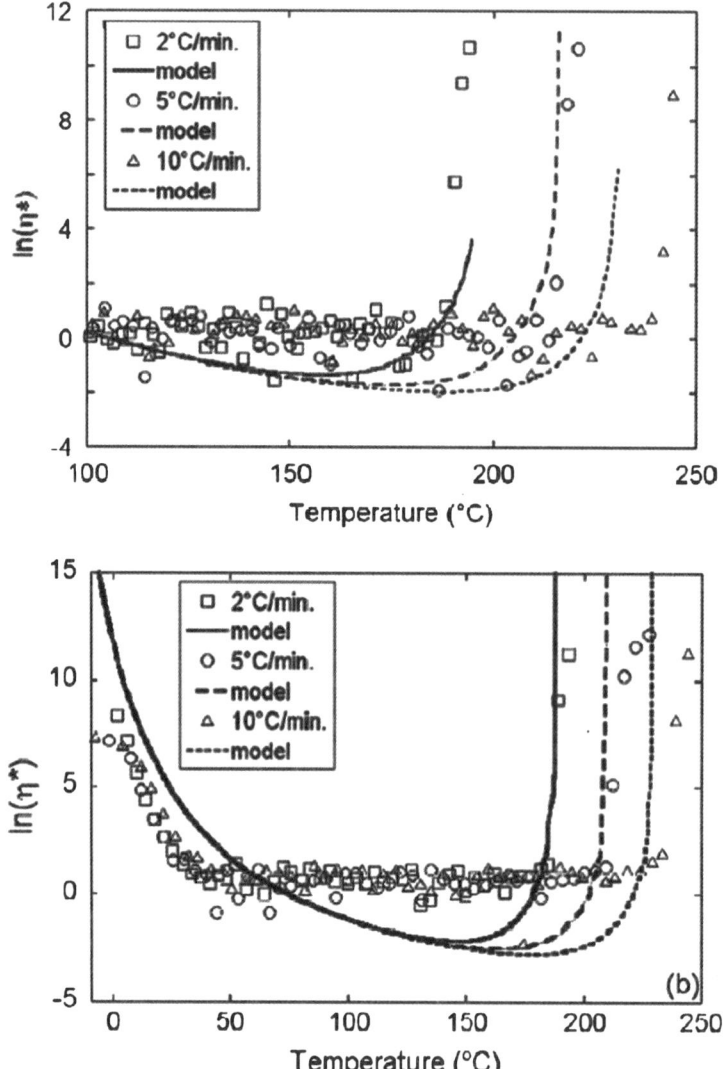

Figure 8.5 Comparison between experimental data and the chemorheological model for (top) DGEBA/MDEA and (bottom) DGEBA/MDEA/C-DWNTs systems, respectively. Reproduced with permission from Terenzi et al.[98]

reaction proceeds through the intermolecular elimination of HF and the formation of C–N bond.[93–95]

Another example of how the functionalization of CNTs represents an open issue for the preparation and manipulation of CNT-based nanocomposites with multifunctional properties was reported by Terenzi et al.[98] It was shown that CNT dispersion affected the rheological properties of double-walled CNT (DWNT)-based polymer nanocomposites appreciably. Grafted *N*-methyldiethanolamine (MDEA) on carboxy-functionalized DWNTs were used and the chemorheology of the amino-functionalized DWNTs during the cure reaction of an epoxy system was studied. Calorimetric and rheological tests (Figure 8.5) revealed how the presence of MDEA functionality on DWNTs has a strong influence on the maximum degree of cure and on the gel time of the epoxy system. The use of kinetic and chemorheological models showed an excellent agreement with the experimental data. Recent results from our group on the modeling of SWNT alignment in an epoxy liquid monomer by the application of a DC electric field was also reported;[99] experimental tests, performed to verify the effectiveness of the model, are based on the application of an electric field to a liquid epoxy–SWNT colloid (0.025 wt%), while measuring the current and observing the system by optical microscopy. A good agreement was found between the model results and the experimental measurements. According to these results, a few minutes were sufficient to obtain a highly oriented SWNT suspension in the epoxy resin.

8.4.2 Assembly of CNTs

The main available approaches for the alignment of CNTs can be grouped into two main categories: (a) the post synthesis assembly approaches, which involve dispersing CNTs in solutions, followed by aligning them using spin-coating, Langmuir–Blodgett assembly,[100] mechanical shearing[101] or blown bubble[102] films technique, and then fixed the aligned CNT structures/patterns by solvent evaporation or resin solidification; and (b) the *in situ* growth approaches by direct growing aligned CNTs by controlled chemical vapor deposition (CVD) and arc discharge techniques. Several reviews have summarized the studies focused on CNT assemble and characterization,[102] as well as direct growth of aligned CNTs by the CVD method.[103,104] One important issue remains the selective positioning with predetermined orientations at a large scale on a substrate. Several attempts have been based on the chemical patterning of the substrate,[105,106] electrophoretic deposition[107] or dielectrophoresis.[108] More recently, Shimoda *et al.*[109] reported that CNTs were aligned on hydrophilic glass parallel to solution surface, whereas Russell *et al.*[110] demonstrated that different alignments were obtained by the evaporation method and dip-coating on to indium-tin-oxide-coated glass. The flow-directed alignment has been also reported by Lieber and co-workers.[111] Another convenient method to obtain an organized assembly of CNTs consists on controlling the wettability of the conducting surfaces and their orientation by changing the pulling speed.

In suggested mechanisms of nanotube orientation (Figure 8.6a), during the drying process of the liquid film on the vertical substrate under the gravitational force, the downwardly de-wetting liquid film exerts a hydrodynamic drag force

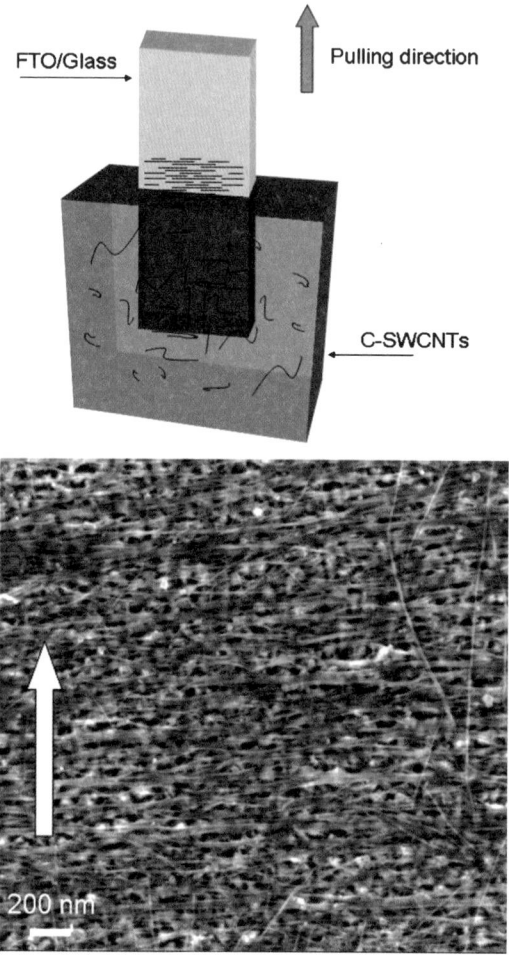

Figure 8.6 The dip-coating deposition of C-SWCNT solutions on fluorine doped tin oxide: (a) A schematic outline of the dip-coating procedure that was used to assemble carboxy-SWNTs. (b) FESEM image of carboxy-SWNTs deposited on to neat fluorine-tin-oxide sample from acetonitrile solution by vertically dip-coating the neat fluorine-tin-oxide in carboxy-SWNT dispersion with a pulling speed of 0.05 mm min^{-1} (the arrow indicates the pulling direction). Reproduced with permission from Valentini et al.[113]

defined as $F_h = \eta l V$, where η is the liquid viscosity, l is the nanotube length and V is the velocity of the hydrodynamic flow estimated as the rate of de-wetting, $V = \gamma \theta 3/6\eta L$ where θ is the contact angle, γ is the surface tension of the liquid and L is a constant of order 10.[112] Accordingly to the data reported by Russell et al.,[110] the high evaporation rate of the acetonitrile with respect to the dip-coating rate and the low hydrodynamic drag force exerted by the substrate with lower wettability induce the nanotube orientation parallel to the solution surface.[113]

Combing the substrate oxygen plasma treatment and the nanotube dispersion in acetonitrile, it was observed that the tubes orient perpendicular to the flow direction, favoring their assembly in networks perpendicular to pulling direction (Figure 8.6b). The control of the wettability of a surface and as well as the dip-coating speed makes possible the realization of highly organized CNT-based architectures.

8.5 Conclusions

The primary objective of this report is a "mix and match" approach towards innovative classes of multifunctional nanostructured composites and hybrid materials. Such materials are expected to stimulate evolutionary advances and revolutionary breakthroughs in emerging key technology areas.

Polymers hold great promises for a widely applicable design and fabrication of structural and functional material. A new paradigm of ordering polymeric materials at the mesoscale is required. Importantly, a novel approach must assist in bridging the gap between intrinsic properties of organic/inorganic materials and the final hybrid material. In this report we propose a few concepts for interfacing polymers and nanotubes—which possess highly specific and widely variable functions, yielding nanostructured composites. Hereby, the surface of the nanofillers will serve as a template for assembling, patterning and ordering the organic materials. One of the major benefits of this approach is that it strictly involves the high aspect ratio of the nanofillers with organic compounds, which is foreseen to enhance the incorporation of organic and inorganic constituents markedly. The approach is also modular, as all components are fully exchangeable (*i.e.* "mix and match") to meet specific requirements. In terms of perspective, the scientific focal points will be based on:

1. Demonstrate "proof of concept", that is, to achieve interfacing electronically active nanofillers with polymers. More generally for the manufacturing sector, polymer is synonymous of flexible, thus the development of deposition techniques fully compatible with plastic substrates, low temperature processes and solution processable materials, and suitable for flexible substrates for roll-to-roll manufacturing technologies is mandatory.
2. Demonstrate the fabrication of nanostructured surfaces and the mutual adhesion of components through either covalent chemical or non-covalent supramolecular means. The challenge in this field is to reduce the manufacturing steps for the assembly of nanocomposites trying to preserve, where possible, the efficiency with respect to the conventional architectures and to improve their stability.
3. Demonstrate that the surfaces of these hybrid nano-objects are "responsive" to external stimuli, such as light or their chemical environment, by grafting polymeric structures on to surfaces.

References

1. K. Kratschmer, L. D. Lamb, K. Fostiropoulos and R. D. Huffman, *Nature*, 1990, **347**, 354.
2. S. Iijima and T. Ichihashi, *Nature* 1993, **363**, 603.
3. R. H. Baughman, A. A. Zakhidov and W. A. de Heer, *Science*, 2002, **297**, 787.
4. E. T. Thostenson, Z. Ren and T. W. Chou, *Compos. Sci. Technol.*, 2001, **61**, 1899.
5. L. Dai and A. W. H. Mau, *Adv. Mater.*, 2001, **13**, 899.
6. A. Hirsch, *Angew. Chem., Int. Ed.* 2002, **41**, 1853.
7. J. L. Bahr and J. M. Tour, *J. Mater. Chem.*, 2002, **12**, 1952.
8. S. B. Sinnott, *J. Nanosci. Nanotechnol.*, 2002, **2**, 113.
9. R. Andrews, D. Jacques, D. Qian and T. Rantell, *Acc. Chem. Res.*, 2002, **35**, 1008.
10. J. E. Fischer, *Acc. Chem. Res.*, 2002, **35**, 1079.
11. C. N. R. Rao, B. Satishkumar, A. Govindaraj and M. Nath, *ChemPhysChem*, 2001, **2**, 78.
12. Y.-P. Sun, K. Fu, Y. Lin and W. Huang, *Acc. Chem. Res.*, 2002, **35**, 1096.
13. S. Niyogi, M. A. Hamon, H. Hu, B. Zhao, P. Bhowmik, R. Sen, M. E. Itkis and R. C. Haddon, *Acc. Chem. Res.*, 2002, **35**, 1105.
14. S. Banerjee, T. Hemraj-Benny and S. S. Wong, *Adv. Mater.*, 2005, **17**, 17.
15. G. de la Torre, W. Blau and T. Torres, *Nanotechnology*, 2003, **14**, 765.
16. J. J. Davis, K. S. Coleman, B. R. Azamian, C. B. Bagshaw and M. L. H. Green, *Chem.–Eur. J.* 2003, **9**, 3732.
17. D. Tasis, N. Tagmatarchis, V. Georgakilas and M. Prato, *Chem.–Eur. J.*, 2003, **9**, 4000.
18. M. Terrones, *Annu. Rev. Mater. Res.*, 2003, **33**, 419.
19. E. Katz and I. Wilner, *ChemPhysChem*, 2004, **5**, 1084.
20. X. Lu and Z. Chen, *Chem. Rev.*, 2005, **105**, 3643.
21. P. D. Calvert, *Nature*, 1999, **399**, 210.
22. R. Andrews and M. C. Weisenberger, *Curr. Opin. Solid State Mater. Sci.*, 2004, **8**, 31.
23. F. Gubbels, R. Jerome, E. Vanlathem, R. Deltour, S. Blacher and F. Brouers, *Chem. Mater.*, 1998, **10**(5), 1227.
24. F. Gubbels, R. Jerome, E. Vanlathem, R. Deltour, S. Blacher and F. Brouers, *Macromolecules*, 1995, **28**(5), 1559.
25. V. Bocharova, A. Kiriy, U. Oertel, M. Stamm, F. Stoffelbach and R. Jerome, *J. Phys. Chem. B*, 2006, **110**, 14640.
26. N. R. Raravikar, L. S. Schadler and A. Vijayaraghavan, *Chem. Mater.*, 2005, **17**, 974.
27. J. Li, L. Hu, L. Wang, Y. Zhou, G. Gruner and T. J. Marks, *Nano Lett.*, 2006, **6**, 2472.
28. B. Vigolo, A. Penicaud, C. Coulon, C. Sauder, R. Pailler and C. Journet, *Science*, 2000, **290**, 1331.

29. J. N. Coleman, W. J Blau, A. B. Dalton, E. Munoz, S. Collins and B. G. Kim, *Appl. Phys. Lett.*, 2003, **82**, 1682.
30. S. J. Zhang, L. B. Zhu, C. P. Wong and S. Kumar, *Macromol. Rapid Commun.*, 2009, **30**, 1936–1939.
31. S. D. Seo, I. S. Hwang, S. H. Lee, H. W. Shim and D. W. Kim, *Ceram. Int.*, 2012, **38**, 3017–3021.
32. J. Y. Lin, C. H. Lien and S. W. Chou, *J. Solid State Electrochem.*, 2012, **16**, 1415–1421.
33. L. Valentini, S. Bittolo Bon and J. M. Kenny, *Carbon*, 2007, **45**, 2685.
34. M. Zhang, S. Fang, A. A. Zakhidov, S. B. Lee, A. E. Aliev, C. D. Williams, K. R. Atkinson and R. H. Baughman, *Science*, 2005, **309**, 1215.
35. J. Rault Berthelot and J. Simonet, *J. Electroanal. Chem. Interfacial Electrochem.*, 1985, **182**, 187.
36. G. Schiavon, G. Zotti and G. Bontempelli, *J. Electroanal. Chem. Interfacial Electrochem.*, 1985, **186**, 191.
37. M. Fukuda, K. Sawada and K. Yoshino, *J. Polym. Sci., Part A: Polym. Chem.*, 1993, **31**, 2465.
38. Q. Pei and Y. Yang, *J. Am. Chem. Soc.*, 1996, **118**, 7416.
39. M. Ranger, D. Rondeau and M. Leclerc, *Macromolecules*, 1997, **30**, 7686.
40. J. M. A. E. J. F. de Carvalho and M. C. dos Santos, *ACS Nano*, 2011, **5**, 3993.
41. N. Berton, F. Lemasson, J. Tittmann, N. Sturzl, F. Hennrich, M. M. Kappes and M. Mayor, *Chem. Mater.*, 2011, **23**, 2237.
42. J. Gao, M. Kwak, J. Wildeman, A. Hermann and M. A. Loi, *Carbon*, 2011, **49**, 333.
43. L. Valentini, F. Mengoni, L. Mattiello and J. M. Kenny, *Nanotechnology*, 2007, **18**, 115702.
44. M. A. López Manchado, L. Valentini, J. Biagiotti and J. M. Kenny, *Carbon*, 2005, **43**, 1499.
45. P. B. Tambe, A. R. Bhattacharyya, S. S. Kamath, A. R. Kulkarni, T. V. Sreekumar, A. Srivastav, K. U. B. Rao, Y. D. Liu and S. Kumar, *Polym. Eng. Sci.*, 2012, **52**(6), 1183.
46. S. A. Girei, S. P. Thomas, M. A. Atieh, K. Mezghani, S. K. De, S. Bandyopadhyay and A. Al-Juhani, *J. Thermoplast. Compos. Mater.*, 2012, **25**(3), 333.
47. P. S. Thomas, S. A. Girei, A. A. Al-Juhani, K. Mezghani, S. K. De and M. A. Atieh, *Polym. Eng. Sci.*, 2012, **52**(3), 525.
48. P. Biswajit, A. R. Bhattacharyya and A. R. Kulkarni, *Polym. Eng. Sci.*, 2011, **51**(8), 1550.
49. M. A. Meitl, Y. X. Zhou, A. Gaur, S. Jeon, M. L. Usrey and M. S. Strano, *Nano Lett.*, 2004, **4**(9), 1643.
50. N. P. Armitage, J. C. P. Gabriel and G. Gruner, *J. Appl. Phys.*, 2004, **95**, 3228.
51. H. E. Unalan, G. Fanchini, A. Kanwal, A. Du Pasquier and M. Chhowalla, *Nano Lett.*, 2006, **6**, 677.
52. M. Moniruzzaman and K. I. Winey, *Macromolecules*, 2006, **39**, 5194.

53. M. Zheng, A. Jagota, E. D. Semke, B. A. Diner, R. S. Mclean and S. R. Lustig, *Nat. Mater.*, 2003, **2**, 338.
54. G. R. Dieckmann, A. B. Dalton, P. A. Johnson, J. Razal, J. Chen and G. M. Giordano, *J. Am. Chem. Soc.*, 2003, **125**, 1770.
55. G. W. Lee and S. Kumar, *J. Phys. Chem. B*, 2005, **109**, 17128.
56. M. F. Islam, E. Rojas, D. M. Bergey, A. T. Johnson and A. G. Yodh, *Nano Lett.*, 2003, **3**, 269.
57. W. Wenseleers, I. I. Vlasov, E. Goovaerts, E. D. Obraztsova, A. S. Lobach and A. Bouwen, *Adv. Funct. Mater.*, 2004, **14**, 1105.
58. I. W. Hamley, *Nanotechnology*, 2003, **14**, R39.
59. M. W. Matsen and F. S. Bates, *J. Polym. Sci. Part B: Polym. Phys.*, 1997, **35**, 945.
60. S. C. Warren, L. C. Messina, L. S. Slaughter, M. Kamperman, Q. Zhou, S. M. Gruner, F. J. Disalvo and U. Wiesner, *Science*, 2008, **320**, 1748.
61. C. Park, J. Yoon and E. L.Thomas, *Polymer*, 2003, **44**, 6725.
62. S. B. Darling, *Prog. Polym. Sci.*, 2007, **32**, 1152.
63. M. Li, C. A. Coenjarts and C. K. Ober, *Adv. Polym. Sci.*, 2005, **190**, 183.
64. M. Lazzari and M. A. López-Quintela, *Adv. Mater.*, 2003, **15**, 1583.
65. R. E. Segalman, *Mater. Sci. Eng. Rev.*, 2005, **48**, 191.
66. W. Caseri, *Macromol. Rapid Commun.*, 2000, **21**, 705.
67. K. A. Mauritz, R. F. Storey, D. A. Mountz and D. A. Reuschle, *Polymer*, 2002, **43**, 4315.
68. D. Chattopadhyay, I. Galeska and F. Papadimitrakopoulos, *J. Am. Chem. Soc.*, 2003, **125**, 3370.
69. S. Niu and R. F. Saraf, *Nanotechnology*, 2007, **18**, 125607.
70. L. Peponi, A. Tercjak, J. Gutierrez, M. Cardinali, I. Mondragon, L. Valentini and J. M. Kenny, *Carbon*, 2010, **48**, 2590.
71. M. F. Islam, E. Rojas, D. M. Bergey, A. T. Johnson and A. G. Yodh, *Nano Lett.*, 2003, **3**, 269.
72. V. C. Moore, M. S. Strano, E. H. Haroz, R. H. Hauge, R. E. Smalley, J. Schmidt and Y. Talmon, *Nano Lett.*, 2003, **3**, 1379.
73. C. G. Hu, Z. L. Chen, A. G. Shen, X. C. Shen, H. Li and S. S. Hu, *Carbon*, 2006, **44**, 428.
74. M. J. O'Connell, P. Boul, L. M. Ericson, C. Huffman, Y. H. Wang, E. Haroz, C. Kuper, J. Tour, K. D. Ausman and R. E. Smalley, *Chem. Phys. Lett.*, 2001, **342**, 265.
75. N. Nakashima, S. Okuzono, H. Murakami, T. Nakai and K. Yoshikawa, *Chem. Lett.*, 2003, **32**, 456.
76. M. Zheng, A. Jagota, E. D. Semke, B. A. Diner, R. S. Mclean, S. R. Lustig, R. E. Richardson and N. G. Tassi, *Nat. Mater.*, 2003, **2**, 338.
77. S. A. Curran, P. M. Ajayan, W. J. Blau, D. L. Carroll, J. N. Coleman, A. B. Dalton, A. P. Davey, A. Drury, B. McCarthy, S. Maier and A. Strevens, *Adv. Mater.*, 1998, **10**, 1091.
78. T. G. Hedderman, S. M. Keogh, G. Chambers and H. J. Byrne, *J. Phys. Chem. B*, 2004, **108**, 18860.

79. D. Tasis, N. Tagmatarchis, V. Georgakilas and M. Prato, *Chem.–Eur. J.*, 2003, **9**, 4001.
80. K. D. Ausman, R. Piner, O. Lourie, R. S. Ruoff and M. Korobov, *J. Phys. Chem. B*, 2000, **104**, 8911.
81. C. A. Furtado, U. J. Kim, H. R. Gutierrez, L. Pan, E. C. Dickey and P. C. Eklund, *J. Am. Chem. Soc.*, 2004, **126**, 6095.
82. J. L. Bahr, E. T. Mickelson, M. J. Bronikowski, R. E. Smalley and J. M. Tour, *Chem. Commun.*, 2001, 193.
83. I. Armentano, M. S. Dottori, D. Puglia and J. M. Kenny, *J. Mater. Sci.: Mater. Med.*, 2008, **19**(6), 2377.
84. I. Armentano, L. Marinucci, M. S. Dottori, S. Balloni, E. Fortunati, M. Pennacchi, E. Becchetti, P. Locci and J. M. Kenny, *J. Biomater. Sci. Polym. Ed.*, 2011, **22**(4–6), 541.
85. T. Yang, D. F. Wu, L. L. Lu, W. D. Zhou and M. Zhang, *Polym. Compos.*, 2011, **32**(8), 1280.
86. C. L. Lin, Y. F. Wang, Y. Q. Lai, W. Yang, F. Jiao, H. G. Zhang, S. F. Ye and Q. Q. Zhang, *Colloids Surf. B.*, 2011, **83**(2), 367.
87. E. Fortunati, F. D'Angelo, S. Martino, A. Orlacchio, J. M. Kenny and I. Armentano, *Carbon*, 2011, **49**(7), 2370.
88. N. O. V. Plank, G. A. Forrest, R. Cheung and A. J. Alexander, *J. Phys. Chem. B.*, 2005, **109**, 22096.
89. Y. Q. Wang and P. M. A. Sherwood, *Chem. Mater.*, 2004, **16**, 5427.
90. L. Zhang, V. U. Kiny, H. Peng, J. Zhu, R. F. M. Lobo, J. L. Margrave and V. N. Khabashesku, *Chem. Mater.*, 2004, **16**, 2055.
91. P. E. Pehrsson, W. Zhao, J. W. Baldwin, C. Song, J. Liu, S. Kooi and B. Zheng, *J. Phys. Chem. B*, 2003, **107**, 5690.
92. J. L. Stevens, A. Y. Huang, H. Peng, I. W. Chiang, K. V. N. Habashesku and J. L. Margrave, *Nano Lett.*, 2003, **3**, 331.
93. V. N. Khabashesku, W. E. Billups and J. L. Margrave, *Acc. Chem. Res.*, 2002, **35**, 1087.
94. B. N. Khare, P. Wilhite and M. Meyyappan, *Nanotechnology*, 2004, **15**, 1650.
95. A. Felten, C. Bittencourt, J. J. Pireaux, G. Van Lier and J. C. Charlier, *J. Appl. Phys.*, 2005, **98**, 074308.
96. L. Valentini, D. Puglia, I. Armentano and J. M. Kenny, *Chem. Phys. Lett.*, 2005, **403**, 385.
97. L. Valentini, I. Armentano, F. Mengoni, D. Puglia, G. Pennelli and J. M. Kenny, *J. Appl. Phys.*, 2005, **97**, 114320.
98. A. Terenzi, C. Vedova, G. Lelli, J. Mijovic, L. Torre, L. Valentini and J. M. Kenny, *Compos. Sci. Technol.*, 2009, **68**(7–8), 1862.
99. M. Monti, M. Natali, L. Torre and J. M. Kenny, *Carbon*, 2012, **50**(7), 2453.
100. Y. F. Ma, B. Wang, Y. P. Wu, Y. Huang and Y. S. Chen, *Carbon*, 2011, **49**, 4098.
101. A. Sulong and J. Park, *J. Compos. Mater.*, 2011, **45**, 931.

102. G. W. Hsieh, J. J. Wang, K. Ogata, J. Robertson, S. Hofmann and W. I. Milne, *J. Phys. Chem. C*, 2012, **116**, 7118.
103. Y. H. Yan, M. Chan-Park and Q. Zhang, *Small*, 2007, **3**(1), 24.
104. Z. F. Liu, L. Y. Jiao, Y. G. Yao, X. J. Xian and J. Zhang, *Adv. Mater.*, 2010, **22**(21), 2285.
105. L. M. Huang, Z. Jia and S. J. O'Brien, *J. Mater. Chem.*, 2007, **17**, 3863.
106. S. Auvray, V. Derycke, M. Goffman, A. Filoramo, O. Jost and J. P. Bourgoin, *Nano Lett.*, 2005, **5**(3), 451.
107. D. E. Johnston, M. F. Islam, A. G. Yodh and A. T. Johnson, *Nat. Mater.*, 2005, **4**(8), 589.
108. A. R. Boccaccini, J. Cho, J. A. Roether, B. J. C. Thomas, E. J. Minay and M. S. P. Shaffer, *Carbon*, 2006, **44**(15), 3149.
109. H. Shimoda, S. J. Oh, H. Z. Geng, R. J. Walker, X. B. Zhang, L. E. McNeil and O. Zhou, *Adv. Mater*, 2002, **14**(12), 899.
110. J. M. Russell, S. J. Oh, I. LaRue, O. Zhou and E. T. Samulski, *Thin Solid Films*, 2006, **509**(1–2), 53.
111. Y. Huang, X. F. Duan, Q. Q. Wie and C. M. Lieber, *Science*, 2001, **291**, 630.
112. G. Reiter and A. Sharma, *Phys. Rev. Lett.*, 2001, **87**(16), 166103.
113. L. Valentini, D. Bagnis and J. M. Kenny, *Carbon*, 2008, **46**, 365.

CHAPTER 9
Raman Spectroscopy of Carbon Nanotube–Polymer Hybrid Materials

KONSTANTINOS PAPAGELIS*

Department of Materials Science, University of Patras, 26504 Rion Patras, Greece, and FORTH/ICE-HT, Stadiou Str., 265 04 Rion Patras, Greece
*E-mail: kpapag@upatras.gr

9.1 Introduction

Carbon nanotubes (CNTs) have received much attention as a class of next-generation nanomaterials due to their unique physical and chemical properties. The potential utility of CNTs in a variety of technologically important applications, such as polymer composites, supercapacitors, lithium rechargeable batteries, sensors, photovoltaic and solar cells, high-resolution printable conductors and so on, is now well established.[1,2] However, due to their poor solubility and strong aggregation, enhancing the processability of CNTs while preserving their properties has been a main challenge before CNTs could be integrated into functional hybrids for the fabrication of advanced devices.

The hybridization of various types of CNTs, such as single-walled CNTs (SWCNTs), double-walled CNTs (DWCNTs) or multi-walled CNTs (MWCNTs), with insulating or conducting polymers at the molecular level is an efficient strategy to: (i) obtain the emerging properties through the synergistic effects of the two components while overcoming the intrinsic limitations of the individual materials, (ii) reinforce the macromolecular compound and (iii) introduce novel electronic properties based on morphological modifications or interactions between the two constituents.[3] Moreover, from the materials engineering perspective, the characteristics of the CNT–polymers interface should be rationally designed to make CNTs chemically compatible for polymeric matrices with enhanced mechanical/electrical properties for the final composite.[4]

In the past decade, the development of Raman spectroscopy (RS) in the field of CNTs and, recently, of graphene is truly impressive. The level of information and details that have become available has allowed the development of nanotechnology from a more fundamental perspective. In this process, Raman spectroscopy is one of the most useful and versatile tool to characterize and investigate CNT samples. It is a fast and non-invasive probing technique with high resolution giving the maximum structural, vibrational and electronic information. Raman spectroscopy has been widely used to investigate the structure, physical and chemical properties of CNTs. The unique optical properties observed in SWCNTs are largely due to the one-dimensional confinement of electronic and phonon states, resulting in the so-called van Hove singularities (vHs) in the nanotube electronic density of states (DOS) (see below in the text). In combination with the intriguing electronic structure, the resonantly enhanced Raman scattering intensity allows one to obtain detailed information about the vibrational properties of nanotubes, even at the isolated individual SWCNT level.[5]

Raman scattering has been employed successfully to determine the diameter, chirality and curvature of nanotubes, their structural integrity, the metallic *vs.* semiconducting behaviour, physical adsorption and/or wrapping of polymers to CNTs surface, defects and other crystal disorder, as well as strain or stress that the nanotubes experience under external loads or environmental effects. Also, it is extremely sensitive for detecting local changes in the hybridization state of carbon atoms from sp^2 to sp^3 and/or modifications of the electronic and phonon structure caused by the functionalization of the graphitic network. Charge transfer effects between the CNTs and the polymer molecules and the concomitant alterations of the electronic structure can be effectively traced using Raman spectroscopy.

This Chapter provides an overview of micro-Raman spectroscopy of CNTs and its application in studying CNT reinforced polymer composites. By choosing and presenting various important and characteristic paradigms, we intend to elucidate the usefulness of Raman scattering as a valuable tool to characterize different types of composite materials and probe the interactions between CNTs and polymer molecules. More detailed information is available in topical reviews.[6-8]

9.2 Chemical Modification of CNTs with Polymers

As already mentioned, the usage of CNTs as a starting material in different applications has been largely limited due to their poor processability, insolubility and infusibility. *To bypass this disadvantage, an efficient key is the functionalization of CNTs with polymers, leading to functional composite materials.* The modification strategies of CNTs by polymers can be divided into two categories based on the type of bonds between CNTs and polymers, namely non-covalent or covalent bonding.

Non-covalent CNT modification concerns the physical adsorption and/or wrapping of polymers to the surface of the CNTs. The graphitic sidewalls of CNTs provide the possibility for π-stacking interactions with conjugated polymers, as well as organic polymers containing heteroatoms with free electron pairs. An advantage of non-covalent functionalization is that it does not destroy the conjugated system of the CNT sidewalls and therefore it does not affect the final structural properties of the material.[3]

The second modification is covalent chemical bonding (grafting) of polymer chains to CNTs, in which strong chemical bonds between CNTs and polymers are created. There are two main methodologies for the grafting of CNTs depending on the building of polymer chains. Firstly, the "grafting to" approach involves the synthesis of a polymer with a specific molecular weight terminated with reactive groups or radical precursors. In a subsequent reaction, this polymer chain is attached to the surface of nanotubes by addition reactions.[9] A disadvantage of this method is that the grafted polymer content is limited because of the relatively low reactivity and high steric hindrance of macromolecules. In comparison, the "grafting from" approach involves growing polymers from CNT surfaces *via in situ* polymerization of monomers initiated by chemical species immobilized on the CNT sidewalls and tips. The advantage of this method is that the high reactivity of monomers makes efficient, controllable and tailored grafting feasible.[3]

9.3 Background of Raman Spectroscopy of CNTs

9.3.1 Electronic Structure of CNTs

Owing to their one-dimensional nature, the π-derived electronic DOS of a SWCNT (the prototype material of CNT family) forms sharp singularities, the so-called van Hove singularities. The sharp vHs define narrow energy ranges for which the electronic DOS intensity becomes quite large (Figure 9.1a–9.1c). Therefore, a SWCNT exhibits a "molecular-like" behaviour, with well-defined electronic energy levels at each singularity. The vHs closer to the Fermi level originate from cutting lines closer to the K point in the 2D Brillouin zone.[7] To a first approximation, the vHs are energetically symmetrical with respect to the Fermi level of the individual SWCNT. Dipole-allowed optical transitions (designated as E_{ii}) occur between the i-th valence band vHs and the i-th conduction band vHs. Each individual (n, m) SWCNT exhibits a different set of valence and conduction band vHs and a different set of optical transition energies E_{ii}. For this reason, optical experiments can be used for the structural determination of a given (n, m) CNT.[7]

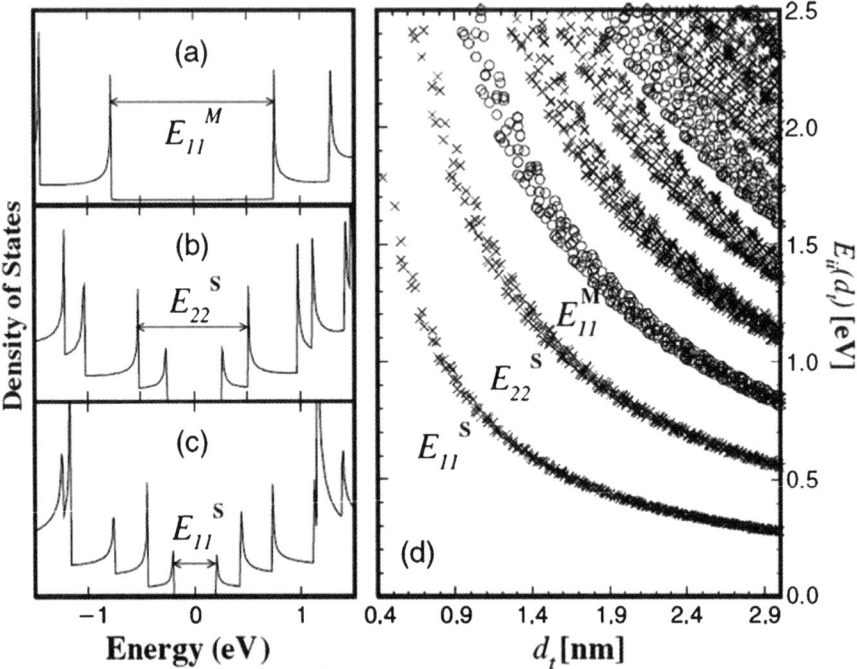

Figure 9.1 (a–c) Electronic DOS of different metallic (M) and semiconducting (S) CNTs, and (d) Kataura plot derived from a simple first neighbour tight-binding approximation. Adapted from Jorio et al.[5]

It is quite useful in CNT characterization by Raman spectroscopy to consider the plots of E_{ii} vs. d_t (nanotube diameter). Figure 9.1(d) presents $E_{ii}(d_t)$ values for all the (n, m) SWCNTs with diameters between 0.5 and 3.0 nm, calculated using the tight-binding approximation.[10] As can be clearly seen, the electronic transition energies vary, exhibiting an almost $1/d_t$ dependence. Each point in this plot corresponds to one optically allowed E_{ii} from a given (n, m) semiconducting (Figure 9.1d, crosses) or metallic (Figure 9.1d, circles) SWCNT. The so-called Kataura plot[11] has been widely used to interpret the optical spectra from CNTs.

More specifically, an observable Raman signal from a CNT can be obtained when the laser excitation energy (E_{laser}) is equal to the energy separation between two vHs in the valence and conduction bands. Because Raman scattering is a resonance process in CNTs, Raman spectra even at the single nanotube level allow us to study the electronic and phonon structure of SWCNTs in great detail. Since the observable Raman spectra come predominantly from tubes in resonance with E_{laser}, a Kataura plot specifies which nanotubes will be detected for a particular laser line. When Raman spectra of SWCNT bundle samples are taken, only those SWCNTs with E_{ii} in resonance with the E_{laser} will contribute strongly to the detected Raman signal. It should be stressed that the transition energies that apply to an individual suspended SWCNT, e.g. wrapped by a surfactant, do not necessarily hold for

SWCNT in a bundle where the transition energies for bundled SWCNTs are downshifted and the resonance widths are found to be broadened.[5,12]

9.3.2 Raman Spectrum of CNTs

Figure 9.2 shows the main spectral features of the Raman spectrum of a SWCNT, namely the radial breathing modes (RBMs) (150–300 cm^{-1}), the D band at 1250–1450 cm^{-1}, the G-band at 1580 cm^{-1} and the 2D band at 2500–2750 cm^{-1}. Each feature corresponds to different vibration modes associated with the structure of SWCNTs and will be discussed briefly below.

9.3.2.1 The Radial Breathing Modes

The RBM bands correspond to the coherent vibration of the carbon atoms where all the tube atoms vibrate radially in phase. These features are unique in CNTs and occur with frequencies between 120 and 350 cm^{-1} for SWCNTs with diameters in the range 0.7 nm $< d_t <$ 2 nm. The RBM frequency (ω_{RBM}) varies as $1/d_t$ through the relation: $\omega_{RBM} = A/d_t + B$, where the parameters A and B are determined experimentally. Different values of the constants A and B have been reported in the literature,[7,10] whereas the variations in the A and B parameters are often attributed to environmental effects, namely whether the SWCNTs are present as isolated, supported or in the form of bundles.[13] Therefore, from the ω_{RBM} measurement of an individual SWCNT, it is possible to obtain its diameter value. Also, by recording Raman spectra using

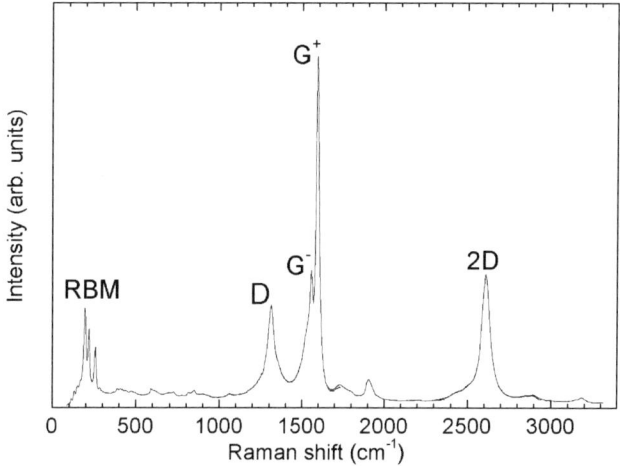

Figure 9.2 Typical Raman spectrum of SWCNT excited at 514.5 nm; the more intense Raman features are marked.

many laser lines, the nanotubes diameter distribution in a sample can be extracted. In this line, the RBM spectrum of SWCNT bundles contains RBM peaks corresponding to different SWCNTs in resonance with the E_{laser}.

The RBM is a highly resonant feature and very sensitive to electronic and structural changes as well as environmental, e.g. polymer wrapping on to CNT sidewalls. The Raman intensity is enhanced by some orders of magnitude if the energy of the incident or scattered light matches the energy of an electronic transition E_{ii}. These resonance conditions are called incoming and outgoing resonance, respectively. A resonance profile for a given RBM mode is rendered by recording the Raman intensity as a function of excitation energy using tunable laser systems. From the resonance profiles it is possible to obtain the transition energies E_{ii} and assign the chiral index (n, m).

The Raman resonance profile for a single resonance is a superposition of an incoming and an outgoing resonance and can be described by eqn (1):[14]

$$I(E_{laser}) = \left(\frac{Mc}{\hbar\omega_{RBM}}\right)^2 \left|\frac{1}{(E_{laser} - E_{ii} - i\gamma/2)} - \frac{1}{(E_{laser} - \hbar\omega_{RBM} - E_{ii} - i\gamma/2)}\right|^2 \quad (1)$$

where E_{laser} is the laser energy, E_{ii} is the energy of the allowed optical transition and γ is the lifetime broadening of the intermediate electronic states. The M contains all of the matrix elements and c summarizes the remaining factors. An incoming resonance occurs when $E_{laser} = E_{ii}$ and an outgoing resonance when $E_l = E_{ii} + \hbar\omega_{RBM}$. If the incoming and outgoing resonances are not resolved in the resonance profile, the recorded spectral profile exhibits a Lorentzian-like shape with a resonance maximum at approximately $E_{ii} + 0.5\hbar\omega_{RBM}$. It should be stressed that in practice the energy distance between the maxima of the incoming and outgoing resonances is in the range of 20 to 40 meV, which is too small to appear as two distinct maxima.

The above-mentioned RBM properties have been extensively employed to characterize chemically modified CNTs. CNT–polymer interactions that are sensitive to the diameter as well as the electronic structure (metallic or semiconducting) of the SWCNTs can be easily followed. On the other hand, recording resonance profiles in the case of functionalized material only makes sense by comparing both reference and functionalized nanotubes, as the absolute values of the optical transitions are often modified during preparation procedures and due to environmental changes. Covalent modification disrupts the electronic band structure, thus reducing the resonance enhancement of the Raman process significantly affecting the resonant profiles. In many cases the optical transitions are broadened due to functionalization which causes longer lifetimes and thus larger values for the γ parameter.

9.3.2.2 The D-band

The D-band is a disorder-induced feature arising from double resonance Raman scattering process from a non-zero-centre phonon mode.[15] It is also an energy dispersive feature, namely the peak frequency increases with increasing laser excitation energy. It is well-documented that sidewall functionalization breaks the translational symmetry along the tube axis causing this mode to become Raman active. Therefore, an increase in the D-band intensity comprises a fingerprint for successful sidewall functionalization. In general, the ratio of intensities of the D- to the G-band (see below) is widely used to evaluate the degree of CNTs graphitization and the functionalization degree of chemically modified nanotubes. Alternatively, Maultzsch et al.[16] argue that the G-band itself can be defect-induced and suggest that the intensity of the D-band should be normalized to the intensity of the second order mode 2D (overtone of the D-mode) as a measure for the defect concentration in SWCNTs. It should be mentioned that the presence of amorphous or disordered carbon in a pristine material also contributes in the recorded Raman intensity of the D-peak.

Finally, the D' band, which is a weak shoulder of the G-band at $\sim$1615 cm^{-1} (with 514.5 nm excitation), is also a double resonance energy dispersive feature induced by disorder and defects and provides a complementary peak to assess the degree of modification in carbon-based nanomaterials.

9.3.2.3 The G-band

The high energy region in the Raman spectrum of SWCNTs contains two main first-order components, resulting from the in-plane C–C carbon displacements parallel and perpendicular to the tube axis (tangential G-band), usually labelled as G$^+$ (1590 cm^{-1}) and G$^-$ (1570 cm^{-1}).[7] In metallic CNTs, the G$^-$ component exhibits a broad asymmetric Breit–Wigner–Fano (BWF) line-shape, resulting from the phonon coupling to an electronic continuum, which is assumed to be proportional to the DOS at the Fermi level and quantified by the asymmetry parameter $1/q$.[17] This BWF coupling results in the broadening and softening of the G$^-$ peak, as well as in a more pronounced inverse tube diameter (d_t) frequency dependence than that in semiconducting tubes.[18]

As pointed out by Dresselhaus et al.,[7] G$^+$ is sensitive to charge transfer arising from dopant additions to SWCNTs. The G$^+$ mode up-shifts for acceptors and downshifts for donors as in graphite intercalation compounds.[7] Additionally, charge transfer to SWCNTs can lead to an intensity increase or decrease in the BWF feature. Also, the ω(G$^+$) is essentially independent of d_t or the chiral angle, whereas ω(G$^-$) is dependent on d_t (not on chiral angle) and whether the SWCNT is metallic or semiconducting.[7] The G$^+$ – G$^-$ splitting is relatively small for large diameter MWCNTs and smeared out because of the diameter distribution in a certain sample. As a result, in MWCNTs the G feature predominantly exhibits a weakly asymmetric characteristic line-shape, with a peak appearing at the graphite frequency $\sim$1580 cm^{-1}.[5]

9.4 Raman Characterization of CNT–Polymer Hybrid Materials

The transition from fundamental research to materials engineering and applications involves good knowledge of the physico-chemical properties of the investigated materials, and their electronic and vibrational features provide primary information. In the following, *via* the presentation of some specific cases, we demonstrate the ability of Raman scattering to characterize, in detail, chemically modified CNT–polymers hybrid materials.

Tasis *et al.*[19] successfully introduced polyacrylamide chains on to SWCNT and MWCNT surfaces *via* ceric ion-induced redox radical polymerization (Figure 9.3). The resulting polymerized nanotubes were carefully analyzed by UV–Vis spectroscopy, X-ray photoelectron spectroscopy (XPS), Raman spectroscopy and scanning electron microscopy (SEM) techniques. Raman spectra showed strong evidence for the significant alteration of the conjugated network of the graphitic cylinders. The D- to G-band intensity ratios, $I(D)/I(G)$ for starting and polyacrylamide-modified SWCNTs are 0.07 and 0.16, respectively. Thus, the enhancement of the $I(D)/I(G)$ ratio indicates a relative increase in the defect sites due to covalent attachment of chemical functionalities, resulting in a conversion of a significant amount of sp^2- to sp^3-hybridized carbon.

In the same line, high-Pressure carbon monoxide (HiPco) SWCNTs were functionalized along their sidewalls with hydroxyalkyl groups using a radical addition scheme.[20] These moieties were found to be active in the polymerization of acrylic acid from the surface of the nanotubes by a redox radical mechanism. In Figure 9.4 (left-hand panel) the Raman response of pristine, ethanol-treated and polymer-modified samples is illustrated. Compared with

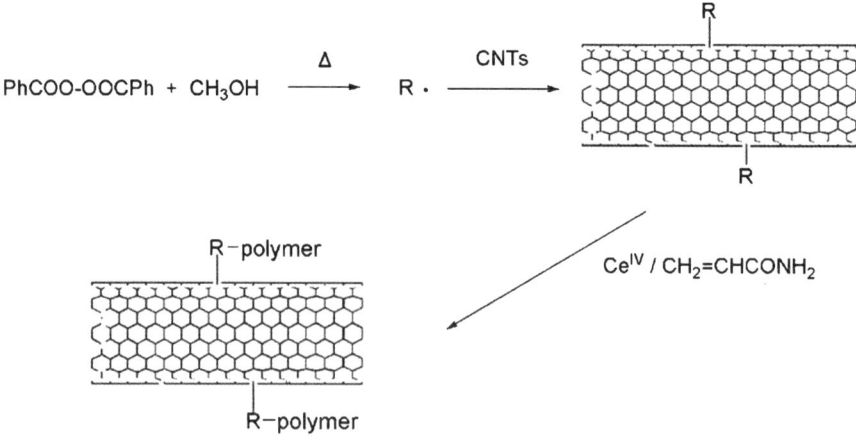

Figure 9.3 Functionalization strategy. The first step is a radical addition to the CNTs sidewalls, whereas the second one is a redox radical grafting of polyacrylamide chains. Adapted from Tasis *et al.*[19]

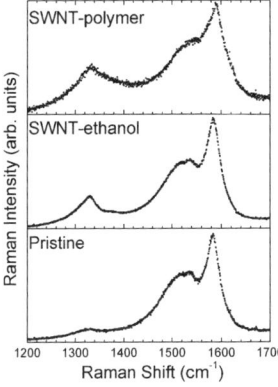
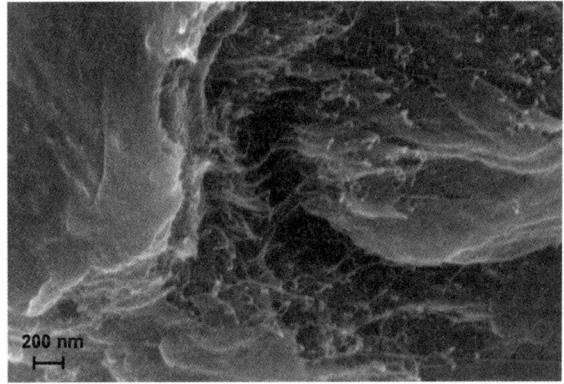

Figure 9.4 Raman spectra in the D- and G-band regions, ethanol-treated and polymer-modified CNTs excited with the 514.5 nm laser line (left-hand panel); and SEM image of polymer-modified SWCNTs (right-hand panel). Adapted from K. Papagelis et al.[20]

the pristine, the modified materials exhibit drastic differences in the relative intensity of the D-band with respect to the main G^+ peak assigned to C–C vibrations along the nanotube axis. Evaluation of the relative intensities $I(G^+)/I(D)$ for the studied samples shows that the ratio is 5.4 for pristine, 1.62 for SWCNT–ethanol and 0.76 for SWCNT–polymer, indicating a significant alteration of the graphitic network. It is worth mentioning that Raman scattering originated from the polymer chains themselves is expected to contribute in the spectral intensity of polymer-modified CNTs. Also, the absence of the strong peaks at 1450 cm^{-1} and 1700 cm^{-1} belonging to the Raman spectrum of neat poly(acrylic acid)[21] implies a minor contribution of the polymer in the observed Raman scattering intensity. Therefore, it is suggested that the polymer grafting not only takes place from the initiating groups but from some polymer chains that are attached directly to the sidewalls. The absorption spectroscopy of CNTs is a complementary experimental technique which provides direct information for the vHs in the visible region and their changes upon functionalization. In the absorption spectrum of pristine and ethanol-treated tubes, the characteristic vHs are clearly observed.[20] On the contrary, the spectrum of the polymer-modified material shows a complete loss of the van Hove transitions. This is clear evidence of covalent attachment of chemical species on to the π-system.

Müller et al.[22] presented a resonant Raman scattering study on SWCNTs decorated with side chain dendritic terpyridine–Ru(II)–terpyridine [tpy–Ru(II)–tpy] complexes in order to understand how the latter substances will affect the processability and, most importantly, the electronic properties of CNTs. The resonance profiles for the nanotube RBMs were recorded using a tunable excitation laser (Ti:sapphire) and a dye laser. The scattering signal of CaF$_2$ was used to normalize the signal intensity at different excitation energies. Following the procedure of Maultzsch et al.[14] from the transition energies of

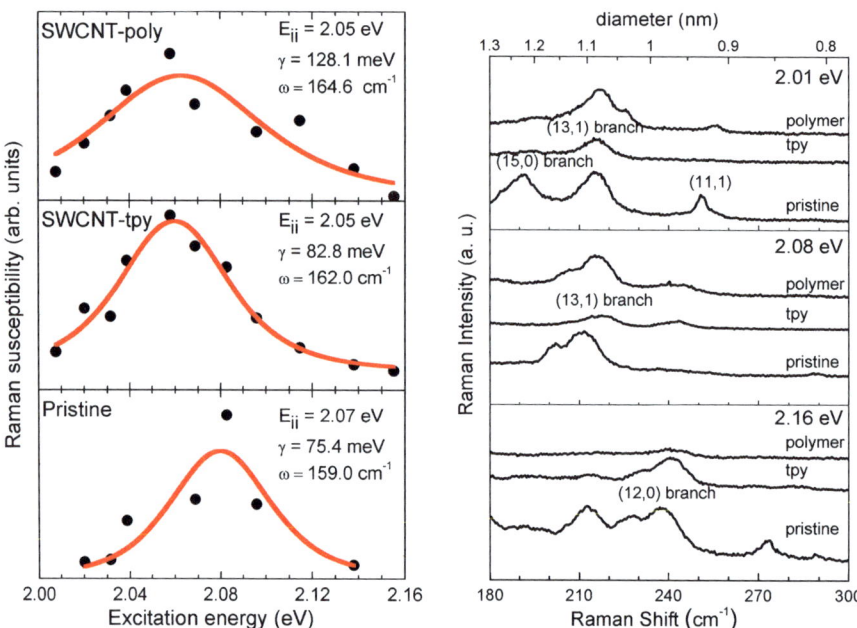

Figure 9.5 Left-hand panel: resonance profiles of the decorated and non-decorated reference sample with E_{ii} (transition energy), ω (RBM frequency) and γ (fitting parameter related to width) [eqn (1)]; (right-hand panel) RBM spectra at various excitation energies. Adapted from M. M. Müller et al.[22]

the E_{22}^S (semiconducting tubes) and E_{11}^M (metallic tubes), the signal is assigned to certain tube chiral index (n, m). Comparing the pristine tubes with the tpy- and poly-tpy-decorated samples, an additional downshift of the transition energies is observed with functionalization, as it can be seen in Figure 9.5 (left-hand panel). This trend holds for all the recorded profiles. Also a broadening of the resonance profile is obvious for the polymer-decorated tubes (Figure 9.5). The full width at half maximum is increased by approximately 50%. For the monomeric sample, this effect is less pronounced.

Concerning the RBM frequencies, a hardening with functionalization can be observed. Shifts of approximately 2 and 5 cm^{-1} for monomeric or polymeric decoration are observed, respectively (see Figure 9.5, right-hand panel). The pronounced shift in the polymeric sample is most probably due to wrapping of the polymer chain around the tube, as it has been observed for certain polymers.[23]

The frequency shifts are accompanied by drastic effects on the intensity of the RBM signal. In general, the RBM intensity decreases with functionalization. For excitation energies below 1.8 eV, Raman signal from both monomeric and polymeric samples could hardly be recorded, because the signal almost vanishes with functionalization. Interestingly, the decrease of the intensity and therefore the chemical reaction seems to be driven by two effects:

small diameters are highly affected, as it can be observed, *e.g.*, for the (11, 1) tube, but also for tubes with relatively larger diameters like the entire (15, 0) branch (Figure 9.5, right-hand panel). This effect holds for different tube chiralities observed at different excitation energies (Figure 9.5). The preferred reaction of small diameter tubes is expected due to higher strain relaxation and has been reported for various moieties.[24,25] Also, it may possibly be due to defect functionalization, because, even if the defect density is the same in all starting materials, larger tube diameters lead to a higher amount of defects in absolute numbers. Steric effects may also come into play, with less curved material offering more space for the reactants to attack the CNT sidewalls.

Further interesting works on CNTs functionalized with polymeric chains and extensively characterized by means of Raman spectroscopy have been published previously.[26,27]

In another study, Ma and co-workers[28] prepared thermotropic liquid crystalline polyester (TLCP) nanocomposites reinforced with a small quantity of MWCNTs (up to 5 wt%) by an *in situ* polymerization method. The MWCNTs were well dispersed in the host matrix due to aromatic interactions. The interactions between the TLCP molecules and CNTs were tuned through Raman spectroscopy using an excitation wavelength of 532 nm. RBMs are particularly sensitive probes to study interactions between CNTs and "foreign" molecules absorbed on to the CNT surface, as they correspond to vibrations of the carbon atoms perpendicular to the nanotube axis. The authors detected a distinct Raman shift of the RBMs, suggesting that CNTs interacted with the surrounding liquid crystal molecules, most likely through aromatic interactions (π–π stacking). Also, in a blank experiment in which nanotubes were embedded in a different liquid crystal host, formed by molecules having a similar molecular structure but without phenyl rings, no shift was observed in the RBM band.[28] This further justifies the observation that the RBM shift is directly associated with π–π stacking interactions at the CNT surface, involving the aromatic core structure of the TLCP molecules.

Hybrid materials consisting of SWCNTs and a conductive block copolymer, perchlorate-doped poly(3,4-ethylenedioxythiophene)-block-poly(ethyleneoxide) (P-PEDOT-*b*-PEO), were successfully prepared.[29] Using an excitation wavelength of 532 nm, the authors observed several SWCNT RBM bands in the range of 180–250 cm^{-1} corresponding to a diameter range from 0.9 to 1.3 nm. According to the authors,[29] the noticeable weakness of the RBM bands in SWCNT/P-PEDOT-b-PEO relative to the bundled SWCNTs for frequencies lower than 200 cm^{-1} indicates that P-PEDOT-b-PEO effectively coated, isolated and dispersed the small-diameter nanotubes because of the weak intertube interactions of small SWCNTs.[30] Also, the band broadening and shift in the low-frequency range of 150–200 cm^{-1} was attributed to effective exfoliation of SWCNTs. The negligible changes in the $I(D)/I(G)$ ratio of SWCNTs indicate that the non-covalent functionalization by P-PEDOT-*b*-PEO prevents the destruction of intrinsic electronic structures of SWCNTs.

Based on the observations that the UV–Vis and Fourier transform infrared (FT-IR) spectra of P-PEDOT-b-PEO were changed after the hybridization with SWCNTs, the authors assumed that charge transfer may occur in the supramolecular assembly of hybrids. The G-band of SWCNTs is quite sensitive to the charge transfer.[31] As a result, the authors compared the G-band of SWCNTs with the one of SWCNT/P-PEDOT-b-PEO nanohybrids to verify the presence of electronic interactions. As a matter of fact, the G-band of the pristine material (approximately 1578 cm^{-1}) was blue-shifted by 8 cm^{-1} for SWCNT/P-PEDOT-b-PEO hybrids. The up-shift of the G-band was attributed to either p-doping (such as bromine as an electron acceptor) or oxidation (similar to that by HNO_3 or H_2SO_4).[32] The carbon bonds are stiffened due to a reduction in the delocalized electron density on the SWCNTs, resulting in hole injection from P-PEDOT-b-PEO to SWCNT. Consequently, the spectroscopic analysis of SWCNT/P-PEDOT-b-PEO materials indicates that the charge transfer in the CNT–polymer nanohybrids was facilitated by the electronic interactions between SWCNTs and P-PEDOT blocks.[29]

Conducting polythiophene (PTh)–SWCNTs composites were synthesized by the *in situ* chemical oxidative polymerization method.[33] Using an excitation wavelength of 514 nm, the authors studied the pristine and the derived materials. The Raman spectrum for the SWCNT–PTh composites is clearly an addition of the corresponding spectra of PTh and SWCNTs, demonstrating that SWCNTs served as templates in the formation of a co-axial nanostructure for the composites.[33]

In the work of Yi *et al.*,[34] the authors prepared a composite material by grafting a carbonizable polymer, poly(furfuryl alcohol) (PFA), to arylsulfonic acid (SA)-modified SWCNTs (Figure 9.6). The Raman spectrum of the PFA-SWCNT has similar features to those of the SA–SWCNT. The intensity ratio of the D-band to G-band was almost the same as that of the SA–SWCNT, suggesting that the PFA wrapping did not alter the hybridization of the carbon atoms within the SWCNT framework. They also observed that the RBMs of the PFA–SWCNT composite became even weaker than that of the SA–SWCNT. This would be expected if the attachment of the macromolecule would restrict the radial breathing of SWCNTs more than the SA group alone.

On the other hand, the RBM peaks of the carbonized, at 600 °C, nanocomposite—yielding a nanoporous carbon–CNT hybrid (NPC/SWCNT)—slightly shifted to higher wavelength numbers. As the RBMs of the NPC/SWCNT hybrid material show intensities comparable with those of the purified material, the authors concluded that there is no chemical bonding or charge transfer between the SWCNT and NPC. They attribute these shifts to molecular forces being exerted by the NPC on the SWCNT. It has been well established that the Raman features exhibit a blue shift under the influence of hydrostatic pressure.[35] Therefore, it can be calculated that the molecular pressure of the NPC on the SWCNT is hundreds of MPa based on RBM shifts. Although there is an obvious frequency change in RBMs, tangential modes remain at almost the same frequency. This experimental finding is attributed to

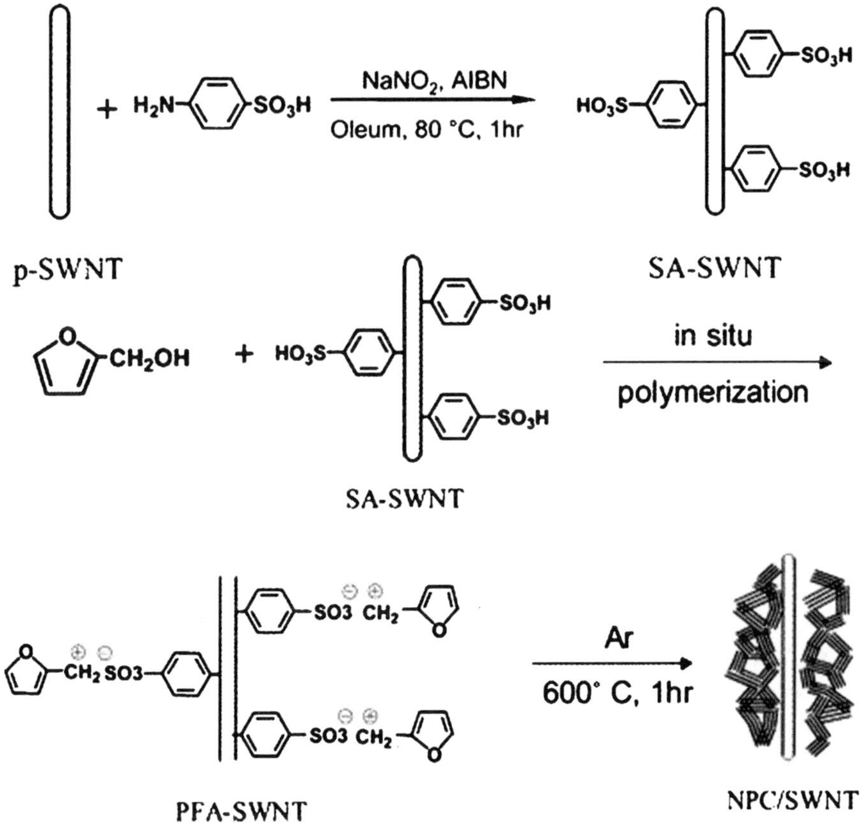

Figure 9.6 Preparation of purified SWCNTs (p-SWNT) functionalized with arylsulfonic acid (SA-SWNT). A nanocomposite carbon material (PFA-SWNT) was prepared by grafting PFA to a SWCNT. Pyrolysis of the PFA-SWCNT produced a nanoporous carbon nanocomposite (NPC/SWNT). Adapted from Yi et al.[34]

the anisotropic behaviour of the NPC on the SWCNTs. When PFA was pyrolyzed to form NPC, there is considerable volume shrinkage in all dimensions, which may induce compressive stress on the SWCNT. The compressive stress more strongly affects the radially modes than the tangentially ones.

In another work, Nogueira et al.[36] presented a covalent modification approach with thiophene groups located at the edges and defects of SWCNTs in order to modify the interaction with the polymer matrix with the aim of its application in solar cells. Raman spectra of the pristine SWCNTs, purified material (SWCNT-COOH) and the modified material (SWCNT-THIOP) were obtained. For the non-purified SWCNT, at least four distinct tube radii were observed in the Raman spectrum excited with a wavelength of 632.8 nm. The main peak for the RBMs occurs at ~ 162 cm^{-1} (1.4 nm diameter), whereas other less intense peaks appear at ~144, 176 and 186 cm^{-1}. According to the authors,[36] the introduction of carboxylic and thiophene groups is expected to change the force constant of the RBM mode, and as a consequence this band was reduced in intensity and shifted to higher energy values. This upshift is difficult to quantify, given the low frequency and the weak nature of the RBM after the chemical treatment. Only two weak peaks were observed at ~163 and 187 cm^{-1} after both the purification (SWCNT-COOH) and functionalization (SWCNT-THIOP) processes. The G-band profile for the pristine SWCNT shows a higher-frequency component (approximately 1573 cm^{-1}) and a lower-frequency component (approximately 1540 cm^{-1}) with a BWF line-shape. These features are attributable to the presence of metallic SWCNT in the sample used. Also, after purification significant upshifts by approximately 4, 10 and 21 cm^{-1} for the D-, G- and 2D-bands were observed, respectively. As mentioned above, the removal of electrons from SWCNTs (*i.e.*, p-doping or oxidizing conditions) results in an upshift of the G-peak,[37] whereas the concomitant upshifts of the D- and 2D-bands are another signal of charge transfer. After the chemical modification with thiophene-containing groups, G- and 2D-bands upshifted again by 4 and 3 cm^{-1}, respectively.

In a recent work, Imin *et al.*[38] have successfully synthesized a new class of highly soluble alternating copolymers of fluorene and dithieno[3,2-*b*:2',3'-*d*]pyrrole. Sample preparation for Raman measurements involved drop casting dilute polymer–SWCNT solutions in tetrahydrofuran (THF) on to a glass microscope slide and air-drying prior to measurement. The authors collected Raman data using the 785 nm excitation wavelength because the RBM spectra at this excitation wavelength can be used to evaluate the extent of aggregation occurring in a sample.[39,40] RBM spectra of the as-received HiPco sample, sodium dodecylbenzenesulfonate (SDBS)-wrapped SWCNTs as well as the soluble polymer–SWCNT hybrids were reported. All of the peaks of the polymer–SWCNT complexes and SWCNT–SDBS show a characteristic red shift of 3–6 cm^{-1} relative to the same peaks in the spectrum of the starting material. The signal at 284 cm^{-1} is much more dominant in the spectra of pristine bundled SWCNTs, and this feature nearly disappeared in the polymer-functionalized samples as well as the SDBS (surfactant)-dispersed sample. This result indicates that the nanotubes are individually dispersed by the polymeric chains in solution, and there is no evidence of aggregation when they are drop-casted on to the glass substrate. Interestingly, both polymer–SWCNT complexes exhibit a strong signal at 247 cm^{-1} (d_t = 0.96 nm), indicating that both polymers selectively bring a specific nanotube species into resonance when excited at 785 nm.

9.5 Conclusions

Modification of CNTs with polymers has been widely investigated with the primary purpose to control their solubility and processability, either through covalent or non-covalent bonding. The non-covalent attachment does not alter the structure of the SWCNTs, As the interactions between the wrapping polymeric chains and the nanotube are, in principle, weak. Besides, much stronger interactions between SWCNTs and polymer molecules might be achieved by covalent modification using a variety of functionalization strategies. Raman spectroscopy has been successfully utilized to determine the influence of chemical functionalization on the vibrational and electronic properties of CNTs. The dependence of the resonant Raman spectrum of CNT–polymer hybrids on the laser excitation energy can be used to understand the type of bonding between the two constituents and the effect of individual polymer molecules on the structure and the overall symmetry of the tubes. Following the resonance of particular tube chirality, the different reactivity of nanotubes can be traced. Metallic or semiconducting nanotubes exhibit different reactivity, quantified by the different DOS at the Fermi level, and may be affected differently by polymer reactions. Also, the reactivity depends on the steric demands of the functionalities. Moreover, the charge transfer between the CNTs and the addends creates alterations in the electronic structure of the tubes, causing red shifts in the optical transitions, accompanied by a broadening of the resonance conditions. Finally, covalent sidewall functionalization can be proven due to the intensity enhancement of the defect-induced D-mode in combination with attenuated RBMs.

Acknowledgements

The author would like to thank Costas Galiotis, Joannis Kallitsis, Matthias Müller, Janina Maultzsch and Christian Thomsen for their contribution to parts of the presented work.

References

1. D. P. Hashim, N. T. Narayanan, J. M. Romo-Herrera, D. A. Cullen, M. G. Hahm, P. Lezzi, J. R. Suttle, D. Kelkhoff, E. Muñoz-Sandoval, S. Ganguli, A. K. Roy, D. J. Smith, R. Vajtai, B. G. Sumpter, V. Meunier, H. Terrones, M. Terrones and P. M. Ajayan, *Sci. Rep. (Nature)*, 2012, **2**, 1–7.
2. V. Ahir and E. M. Terentjev, *Nat. Mater.*, 2005, **4**, 491–495.
3. Z. Spitalsky, D. Tasis, K. Papagelis and C. Galiotis, *Prog. Polym. Sci.*, 2010, **35**, 357–401.
4. S. Lefrant, M. Baibarac and I. Baltog, *J. Mater. Chem.*, 2009, **19**, 5690–5704.
5. A. Jorio, M. A. Pimenta, A. G. Souza Filho, R. Saito, G. Dresselhaus and M. S. Dresselhaus, *New J. Phys.*, 2003, **5**, 139.1–139.17.
6. M. S. Dresselhaus, G. Dresselhaus, A. Jorio, A. G. Souza Filho and R. Saito, *Carbon*, 2002, **40**, 2043–2061.
7. M. S. Dresselhaus, G. Dresselhaus, R. Saito and A. Jorio, *Phys. Rep.*, 2005, **409**, 47–99.
8. R. Graupner, *J. Raman Spectrosc.*, 2007, **38**, 673–683.

9. D. Tasis, N. Tagmatarchis, A. Bianco and M. Prato, *Chem. Rev.*, 2006, **106**, 1105–1136.
10. R. Saito, G. Dresselhaus and M. S. Dresselhaus, *Physical Properties of Carbon Nanotubes*, Imperial College Press, London, 1998.
11. H. Kataura, Y. Kumazawa, Y. Maniwa, I. Uemezu, S. Suzuki, Y. Ohtsuka and Y. Achiba, *Synth. Met.*, 1999, **103**, 2555-2558.
12. A. Jorio, M. S. Dresselhaus, R. Saito, G. Dresselhaus, *Raman Spectroscopy in Graphene Related Systems*, Wiley-VCH, Berlin, 2011.
13. M. S. Dresselhaus, A. Jorio, M. Hofmann, G. Dresselhaus and R. Saito, *Nano Lett.*, 2010, **10**, 751–758.
14. J. Maultzsch, H. Telg, S. Reich, and C. Thomsen, *Phys. Rev. B*, 2005, **72**, 205438.1–205438.16.
15. C. Thomsen and S. Reich, *Phys. Rev. Lett.*, 2000, **85**, 5214–5217.
16. J. Maultzsch, S. Reich, C. Thomsen, S. Webster, R. Czerw, D. L. Carroll, S. M. C. Vieira, P. R. Birkett, and C. A. Regoet, *Appl. Phys. Lett.*, 2002, **81**, 2647–2649.
17. S. D. M. Brown, A. Jorio, P. Corio, M. S. Dresselhaus, G. Dresselhaus, R. Saito and K. Kneipp, *Phys. Rev. B*, 2001, **63**, 1554143.1–1554143.8.
18. A. Jorio, A. G. Souza Filho, G. Dresselhaus, M. S. Dresselhaus, A. K. Swan, M. S. Unlu, B. B. Goldberg, M. A. Pimenta, J. H. Hafner, C. M. Lieber and R. Saito, *Phys. Rev. B*, 2002, **65**, 155412.1–155412.9.
19. D. Tasis, K. Papagelis, M. Prato, I. Kallitsis and C. Galiotis, *Macromol. Rapid Commun.*, 2007, **28**, 1553–1558.
20. K. Papagelis, M. Kalyva, D. Tasis, J. Parthenios, A. Siokou and C. Galiotis, *Phys. Status Solidi B*, 2007, **244**, 4046–4050.
21. S. Koda, H. Nomura and M. Nagasawa, *Biophys. Chem.*, 1983, **18**, 361–367.
22. M. Müller, K. Papagelis, J. Maultzsch, A. A. Stefopoulos, E. K. Pefkianakis, A. K. Andreopoulou, J. K. Kallitsis and C. Thomsen, *Phys. Status Solidi B*, 2009, **246**, 2721–2723.
23. Y. K. Kang, O.-S. Lee, P. Deria, S. H. Kim, T.-H. Park, D. A. Bonnell, J. G. Saven and M. J. Therien, *Nano Lett.*, 2009, **9**, 1414–1418.
24. A. Hirsch and O. Vostrowsky, *Top. Curr. Chem.*, 2005, **245**, 193–237.
25. M. Müller, J. Maultzsch, D. Wunderlich, A. Hirsch and C. Thomsen, *Phys. Status Solidi B*, 2008, **245**, 1957–1960.
26. A. A. Stefopoulos, C. L. Chochos, M. Prato, G. Pistolis, K. Papagelis, F. Petraki, S. Kennou and J. K. Kallitsis, *Chem.–Eur. J.*, 2008, **14**, 8715–8724.
27. A. A. Stefopoulos, E. K. Pefkianakis, K. Papagelis, A. K. Andreopoulou and J. K. Kallitsis, *J. Polym. Sci. Part A: Polym. Chem.*, 2009, **47**, 2551–2559.
28. X. Wang, J. Wang, W. Zhao, L. Zhang, X. Zhong, R. Li and J. Ma, *Appl. Surf. Sci.*, 2010, **256**, 1739–1743.
29. H. S. Park, B. G. Choi, W. H. Hong and S.-Y. Jang, *J. Phys. Chem. C*, 2012, **116**, 7962–7967.

30. Y. Liu, L. Gao and J. J. Sun, *J. Phys. Chem. C*, 2007, **111**, 1223−1229.
31. C. Engtrakul, M. F. Davis, T. Gennett, A. C. Dillon, K. M. Jones and M. J. Heben, *J. Am. Chem. Soc.*, 2005, **127**, 17548−17555.
32. Q. H. Yang, P. X. Hou, M. Unno, S. Yamauchi, R. Saito and T. Kyotani, *Nano Lett.*, 2005, **5**, 2465−2469.
33. M. R. Karim, C. J. Lee and M. S. Lee, *J. Polym. Sci. Part A: Polym. Chem.*, 2006, **44**, 5283–5290.
34. B. Yi, R. Rajagopalan, H. C. Foley, U. J. Kim, X. Liu and P. C. Eklund, *J. Am. Chem. Soc.*, 2006, **128**, 11307–11313.
35. J. Arvanitidis, D. Christofilos, K. Papagelis, T. Takenobu, Y. Iwasa, H. Kataura, S. Ves and G. A. Kourouklis, *Phys. Rev. B*, 2005, **72**, 193411.1–193411.4.
36. A. F. Nogueira, B. S. Lomba, M. A. Soto-Oviedo, C. R. D. Correia, P. Corio, C. A. Furtado and I. A. Hummelgen, *J. Phys. Chem. C*, 2007, **111**, 18431−18438.
37. A. M. Rao, P. C. Eklund, S. Bandow, A. Thess, R. E. Smalley, *Nature*, 1997, **388**, 257–259.
38. P. Imin, M. Imit, and A. Adronov, *Macromolecules*, 2011, **44**, 9138–9145.
39. M. J. O'Connell, S. Sivaram, S. K. Doorn, *Phys. Rev. B*, 2004, **69**, 235411–235415.
40. D. A. Heller, P. W. Barone, J. P. Swanson, R. M. Mayrhofer, M. S. Strano, *J. Phys. Chem. B*, 2004, **108**, 6905–6909.

Subject Index

Page number in *italics* refer to entries in tables or figures.

acrylamides *124,* 127, 140–2, 260
acrylates 124, 127–37, *138, 139*
actuation
　electromechanical 36, *39*
　electrothermal 40, *41, 42*
　photo-induced *33,* 34–6, *37*
actuators 32–8
agarose gels 185–6
agglomerate area ratio A_A 219, 225–9
agglomerates 216, 218
air spraying 197, 198, 202
alumina filters *199,* 200
amines 239–40
　see also individual amines
anionic polymerization 157–8, *159*
anthracene 191
arylsulfonic acid 264, *265*
aspect ratio 213, 221, 224, 230
atom transfer radical polymerization (ATRP) *75,* 80, 120–42
　binary-grafting *163, 164*
　CNT-based macroinitiators 121–4, *125–6*
　SCVP 165, *166*
　in situ polymerization of acrylamides 140–2
　in situ polymerization of methacrylates 124, 127–37, *138, 139*
　in situ polymerization of styrenics 137–40
atomic force microscopy (AFM) 63, *64, 99, 240*

2,2′-azobisisobutyronitrile (AIBN) 87, 144, *145*

band-to-band transition 53–6
batteries 12–14, 184
benzoxazines 83
binary-grafting 160–2, *163, 164, 165*
biodegradation 241–2
biosteel *96*
bipolaron formation *6*
block copolymers (BCs)
　amphiphilic *76, 77*
　ATRP approach 135–7, *138*
　CNT confinement in 239–40
　RAFT approach 144
Breit-Wigner-Fano (BWF) feature 189, 259
bromine CNT macroinitiators 122–7, *128*
Buchner filtration 91, 92
buckypaper 91–2, 198
　and mechanical properties 106–7
butyl lithium 98
butylamine molecules (BAMs) 243

calcium carbonate 133–4
ε-caprolactam 88, 148, 151
carbon black (CB) 31, 238, *239*
carbon nanotube (CNT)–polymers 6–9, 22–3
　applications 9–17, 212
　electrically conductive 191–7

Subject Index

hybrid conducting polymers 237–8
IR sensors 42–4
main fabrication techniques 85–93
mechanical properties 93–100
Raman characterization 260–6
shape-changing 31–42
shape-memory 23–31
in situ polymerization 235–7
via "grafting from" approach *see* "grafting from" approach
see also multi-walled CNTs (MWCNTs); single-walled CNTs (SWCNTs)
carbon nanotubes (CNTs)
 applications 22, 253
 assembly of 245–7
 in block copolymers 239–40
 chemical functionalization *see* chemical functionalization
 chemical modification with polymers 254–5
 chirality and conductivity *201*
 dispersability *see* dispersability
 electronic structure 255–7
 mechanical properties 72
 properties 7–8, 22, 212, 216
 Raman spectroscopy 255–9
carbon–carbon double bonds, coupling 122
carboxy/carboxylic acid groups 121–2, 133–4, 236, *238*, 241–2, *246*, 266
centrifugal separation analysis (CSA) 214–15, 217, 218, 219
chain transfer agents (CTAs)
 macroCTAs 143–4, *145*, 146
 processing 143
charge–discharge cycling 12, *13*
charge transfer 58, 68
chemical functionalization 72–3, 109–10
 covalent 77–83
 covalent and non-covalent comparisons *86*
 covalent with non-covalent 83–5

and mechanical properties 93–9
non-covalent 73–7
SWCNTs 57, 59
chemical polymerization 2, 3, *4, 7*, 12
chemical vapor deposition (CVD) *61*, 68, *95*, 97, 101–2, 245, *246*
chemorheology *244*, 245
chiral index 182–3, 185
and electrical conductivity *201*
chlorinated polypropylene (CPP) 82
chloroform 221, 242
coagulation spinning 90–1
and mechanical properties 105–6
coatings
 conductive 184
 transparent conductive 197–203
coefficient of thermal expansion (CTE) 38, 39
column-based gel chromatography 185–6
CoMoCAT-SWNTs *43*, 44
compatibilizer (P2) 75
conducting polymers 1–2
 as CNT composite matrices 6–9
 conductivity and doping 3–6, *7*
 hybrid 237–8
 synthesis 2–3
conductors 4
confocal fluorescence *162, 171*
conjugated double bonds 2
conversion efficiency 9
coordination polymerization 158
covalent functionalization 57, 59, 77–83, 255
 compared to non-covalent *86*
 with non-covalent 83–5
 sidewall *8*
 see also "grafting from" approach
crystallinity 94, 100–1
 percentage of 238, *239*
crystallization enthalpy 238, *239*
crystallization temperature 238, *239*
cyanate ester (CE) 84
cyano poly(*p*-phenylene vinylene) (CNPPV) 65, *66*

D-Band 257, 258–9, *261,* 266
defect-group functionalization *8*
degree of polymerization (DB) 167
dendritic CNT–polymers 162–73
density gradient ultracentrifugation (DGU) 185, 202
density of states (DOS) 52, *53,* 254, 255, *256*
N,N-dicyclohexylcarbodiimide (DCC) 168, *169*
differential scanning calorimetry (DSC) 100, 141, *142,* 154
diglycidyl ether of bisphenol A (DGEBA) 243, *244*
dimethylformamide (DMF) 91, 241
dip-coating deposition *246*
1,3-dipolar cyclo-addition *77,* 78
dispersability 213
    characterization 216–20
    epoxy nanocomposites 242–5
    experimental methods 213–16
    results and discussion 216–30
    see also solvent dispersion
dodecanethiol 240
'domino pushing effect' 91
doping 2, 3–6
    interfacial bonding and *7, 8*
    methods *7*
double layer capacitance *10,* 12
double-walled CNTs (DWCNTs) 253
dry ball milling 223–4, *225*
dY/dV_f (rate of increase)of the Young's modulus 93, 94, *95–6,* 100–2, 108
dye-sensitized solar cells (DSSCs) 203
dynamic light scatterring (DLS) 214, *219*

electrical conductivity 3–6, *7*
    and CNT chirality *201*
    CNT–SMP composites 29
    SWCNTs 183, 191–5
    transparent coatings and films 197–203
electrical current density 195

electrical resistivity 224–9, *243*
electrically conductive nanocomposites
    non-enriched SWCNTs 191–5
    separated metallic SWCNTs 195–7
electroactive shape-changing CNT–polymers 36, 38–42
electrochemical polymerization 2–3, *4, 7*
    supercapacitors 11–12
electromechanical actuation 36, *39*
electron–hole pair discrimination 64–8
electron–hole pair generation 53, 56, 58
electronic structure, CNTs 255–7
electrophoretic deposition 237
electropolymerization 237
electrospinning 90–1
    and mechanical properties 105–6
electrostatic force microscopy (EFM) 63–4, *65*
electrothermal actuation 40, *41, 42*
endohedral functionalization *8*
epoxidation *84,* 88–9
epoxy composites *30, 96,* 101
    dispersability 242–5
epoxy resin thermosets 89–90, 103
excitons 14, 15, 58

F8BT 58, 66, *67, 68*
Fermi energy levels 52, 53
ferritin 98
fibre processing 87
field emission SEM (FESEM) *236, 238, 246*
films, transparent conductive 197–203
fluorene polymers 57–8, 69, 237, *238,* 242–3, 266
fluorine-tin-oxide *246*
F8T2 58, 63–7, *68*
fuel cells 184

G-Band 257, 259, *261,* 264, 266
gate voltage (V_G) 55, *60, 61,* 62–3, *66,* 184

Subject Index

gel permeation chromatography (GPC) 156, *170*
gemini-grafting 160, *162*
"grafting from" approach 79–81, 255
  dendritic CNT–polymer composites 162–73
  linear CNT–polymer composites 120–62
"grafting to" approach *79*, 81–3, 255
  binary-grafting 160–2, *163, 164, 165*
graphene 52, 182–3
Grubbs' catalysts 155, 156, 157

Halpin–Tsai equation 101, 108
heat of fusion 238
high-pressure carbon monoxide (HiPCO) *43*, 44, 184, 185, 186, 260
high-resolution transmission electron microscopy (HRTEM) 127, *141, 153*
hybrid conducting polymers 237–8
hydroxyl-functionalized CNTs 97, 166–8, *170*
hydroxylated poly(ether ether ketone) (HPEEK) 83
hyperbranched poly(ether ketones) (HPEKs) 172
hyperbranched polyglycerol (HPG) 167–8, *169, 170, 171*
hyperbranched polymers (HPs) 162–3, 165–6
hyperbranched poly(urea urethane) (HPU) 169, 171, *172*

in situ growth assembly 245
in situ polymerization 87–9
  acrylamides 140–2
  hybrid conducting polymers 237–8
  and mechanical properties 104–5
  methacrylates 124, 127–37, *138, 139*
  PMMA 235–7
  styrenics 137–40

indium tin oxide (ITO) 184, 197, 201, *202,* 203
inductive heating 28, *29*
infrared (IR) emission 53, *54*
infrared (IR) heating 25–8
infrared (IR) irradiation *33, 34*–6, *37*
infrared (IR) laser-excitation *55*
infrared (IR) sensors 42–4
inimer 165, 166
injection moulding 102
insulators 3, *4*
interfacial bonding/doping 7, 8
ion exchange chromatography 185
isotactic polypropylene (iPP) 238, *239*

Kataura Plot 53, *256*
Kevlar 77, *80*, 81, 93, *96*, 101

LaRC-EAP composites 36, *39*
laser ablation 186
laser excitation energy (E_{laser}) 256, 258
layer-by-layer (LbL) technique 92–3, 110, 130
  click grafting 158, 160, *161*
  and mechanical properties 107–8
  SSLbL 192
  SWCNT films 200
length
  CNTs 215, 221–4, *225*
  melt-mixed composites 224–30
light-driven shape-changing CNT–polymers 32–6, *37, 38*
linear CNT–polymer composites 120–62
liquid crystalline elastomers (LCEs) 32–7, *38*
lithiation *82*
lithium-ion batteries 12–14
load transfer 73, 74, 103
low-density polyethylene (LDPE) 215, 219, *220*
LUMiSizer® 214, 217

macroCTAs 143–4, *145,* 146

macroinitiators
  ATRP 121–4, *125–6*
  bromine 122–7, *128*
matrix viscosity 228, *229*
mechanical properties
  and buckypaper 106–7
  and chemical functionalization 93–9
  CNTs 72
  and LBL technique 107–8
  and melt processing 101–2
  and *in situ* polymerization 104–5
  and solution processing 100–1
  thermosets 103
melt-mixed composites 219, *220,* 222, 224–30
melt processing 86–7, 213, 215
  and length analysis 221–4, *225*
  and mechanical properties 101–2
  problems with 109
  thermoplastic polymers 238–9
melting temperature 238, *239*
metallic SWCNTs 183–4
  electrically conductive composites 195–7
  electronic DOS *256*
  harvesting 184–91
  transparent conductive coatings and films 197–203
metals, conductivity *4*
methacrylates 124, 127–37, *138, 139*
  see also poly(methyl methacrylate) (PMMA)
N-methyldiethanolamine (MDEA) *244,* 245
methylmethacrylate (MMA) *124, 125, 126,* 127, *236*
  see also poly(methyl methacrylate) (PMMA)
molecular dynamic simulations *74*
molecular tweezers approach *188,* 191
morphological characterization 216
multi-hydroxyl poly(GMA-OH)-grafted CNTs 133, *134*

multi-walled CNTs (MWCNTs) 23, 253
  ATRP approach to 127–34
  PDMS 32, *33,* 38–40, *41*
  TPU composites 25, *26,* 27, 29–30

N-methylpyrrolidone (NMP) 91, 93, 241
n-type doping 5
Nafion 192–3
nanoporous carbon (NPC) 264–5
nanotube oxidation 78
nitrene chemistry 122, *123*
nitroxide-mediated radical polymerization (NMRP) 146–8
non-covalent exohedral functionalization *8*
non-covalent functionalization 57, 59, 73–7, 255
  compared to covalent *86*
  with covalent 83–5
non-linear optical (NLO) response 6
nylon 6 *see* polyamide 6 (PA6)

octadecylamine (ODA) 186, *187,* 239, 240
optical absorption spectra *189*
optical transition energy (E_{ii}) 255–6, 258, *262*
optical transmittance 197–8, 200, 202
optoelectronic devices 59–63, 184, 237
organic light-emitting diodes (OLEDs) 16–17, 52, 56, 237
organic photovoltaic cells (OPVs) 52, 56
ozonolysis 78

P-PEDOT-*b*-PEO 263–4
p-type doping 5
palladium oxide *133*
particle diameter *217,* 218, 219, 256
particle size distribution 214, 218, 219
percentage of crystallinity 238, *239*

Subject Index

percolation theory/threshold 191, 192, 193, 201, 221, 224
photo-induced actuation 33, 34–6, 37
photo-induced charge transfer 14, 15
photodoping 6
photoelectrical responses 51–2, 58–69
photoluminescence excitation (PLE) 57, 58
photosensitive polymers 58–68
photovoltaic devices 9, 14–16
  organic 52, 56
plasma treatment 243, 247
polaron formation 6
poly(3-hexylthiophene) (P3HT) 195, 196
poly(3-octylthiophene) (P3OT) 15, 62
polyacetylene (PA) 1, 2, 3, 5
polyacrylamide 260
poly(acrylic acid) (PAA) 131–2, 133, 137, 139, 260–1
polyacrylonitrile (PAN) 90, 106, 192
poly(allylamine hydrochloride) (PAH) 92
polyamide 6 (PA6) 78, 88, 95, 97, 101–2, 148, 151
  dispersability 215, 225, 226
  HPU composites 171
polyamide 12 (PA12) 215, 219, 220
poly(amido amine) (PAMAM) 171–2
polyamido-amine generation-0 (PAMAM-0) 82, 96, 101
polyaniline (PANI) 2, 3, 9
  electrically conductive 193
  lithium-ion batteries 14
  mechanical properties 107
  supercapacitors 10, 11
polyBrAzPMA 160
poly(butyl methacrylate) (PBMA) 105
poly(ε-caprolactone) (PCL) 151, 152, 153, 154, 165
  dispersability 215, 226–30
  solvent dispersion 242, 243
polycarbonate (PC) 43, 44, 95, 102

dispersability 215, 219, 220, 222–3, 224, 226, 227
  matrix viscosity 228, 229
polycondensation 169, 171–3
poly(dicyclopentadiene) (polyDPCD) 155
poly[2-(diethylamino)ethyl methacrylate] (PDEAEMA) 130, 131
polydimethylsiloxane (PDMS) 32, 33, 38–40, 41, 77
  electrically conducting 193, 203
polydispersity index (PDI) 135, 219
polyelectrolyte-grafted CNTs 130–1
polyethylene (PE) 93, 102, 158
  LDPE 215, 219, 220
poly(ethylene glycol) (PEG) 155, 164
poly(ethylene terephthalate) (PET) 197, 201, 202
poly(3,4-ethylenedioxythiophene) (PEDOT) 193, 195, 196
  lithium-ion batteries 14
  P-PEDOT-b-PEO 263–4
  supercapacitors 10
poly(ethyleneimine) (PEI) 107–8
polyethylenoxide(PEO) 76, 154
  P-PEDOT-b-PEO 263–4
polyfluorene 57–8, 69, 237, 238, 242–3
poly(9,9-dioctylfluoreny-2,7-diyl) (PFO) 57
poly(furfuryl alcohol) (PFA) 264, 265
polyhydroxyethyl methacrylate (PHEMA) 135, 138, 146
polyimide (CP2) composite 40, 42
polyimide-graft-bisphenol diglyceryl acrylate (PIOH-BDA) 83, 84
polyimide-graft-bisphenoldiglyceryl acrylate (PI-BDA) 99
polyimide (PI) 98–9
poly(L-lactide) (PLLA) 151, 153, 241–2
poly(lactic acid) (PLA) 90
poly(DL-lactide-co-glycolide) (PLGA) 241

poly(*m*-phenylenevinylene-*co*-2,5-dioctyloxy-*p*-phenylenevinylene) (PmPV) 59, *60, 61, 74*
polyMAIG 135, *136, 137*
polymer brushes *124,* 127, 160, *164*
polymer wrapping 56–8, 74, *76*
polymerization-filling technique (PFT) 73
poly[2-methoxy-5-(2'-ethylhexyloxy)-1,4-phenylene vinylene] (MEHPPV) 65, *66*
poly(methyl methacrylate) (PMMA) 79–80, 84, 87, 94
  anionic polymerization approach 157–8
  ATRP approach 127–30, 134–5, *138*
  mechanical properties *95, 97,* 101, 104–5
  in situ polymerization 235–7
poly(*n*-butyl methacrylate) (pnBMA) *164*
poly(*N*-isopropylacrylamide) (PNIPAAm) 140–1, *142*
polynorbornene 155, *156,* 157
poly(*p*-phenylene ethynylene) (PPE) 35, *37*
poly(*p*-phenylene vinylene) (PPV) *3*
poly[(*p*-phenylenevinylene)-*co*-(2,5-dicotoxy-*m*-phenylenevinylene)] (PmPV) 16–17
poly-*p*-phenylene *3*
poly(*p*-phenylene benzoisoxazole) (PBO) 89, 104
poly(2,6-pyridinylenevinylene-*co*-2,5-dioctyloxy-*p*-phenylenevinylene) (PPyPV) 59, *74*
polypyrrole (PPy) *3,* 9, *74*
  electrically conductive 193
  polaron and bipolaron formation *6*
  supercapacitors 10, 11, 12
poly(sodium 4-styrenesulfonate) (PSS) 139, 193, 195, *196*
poly(styrene-*b*-isoprene-*b*-styrene) (SIS) 240

polystyrene (PS) 79, *95, 96,* 98
  anionic polymerization approach 157, 158, *159*
  ATRP approach 137, 138–9, *140*
  binary-grafting *165*
  in electrically conductive SWCNTs 192
  NMRP approach 146, 147, *148*
  RIGP approach 158–9
  TLIRP approach 160
poly(tert-butyl acrylate) P*t*BA 137, *139,* 157
polytetramethylene ether (PTME) 148
polythiophene (PT) *3, 74,* 264
polyurethane (PU) *95,* 98
  thermoplastic 25, *26,* 27–30
poly(vinyl acetate) 192
poly(vinyl alcohol) (PVA) 30, 78, 90, *95, 97,* 98
  in electrically conductive SWCNTs 191, 192–3
  mechanical properties 100, 107, 108
poly(vinyl carbazole) (PVK) 13, 100, 158
poly(vinyl pyrrolidone) (PVP) 107
polyvinylchloride (PVC) *95,* 101
poly(vinylidene fluoride) (PVDF) 14, 75, 83, 192–3
porphyrin 186, *187,* 188, 190
post-production separation 186, *187,* 188, 204
post synthesis assembly 245
pseudocapacitance *10,* 12
Pseudomonas lipase (PS lipase) 151, *152, 153*
pyrene 186, *187,* 188, 189–91
pyrene-main chain (PyMC) 36, *38*

quartz substrates 67–8

radial breathing modes (RBMs) 257–8, 261–2, 263, 264, 266

Subject Index

radiation-induced graft polymerization (RIGP) 158–9
radical addition 260
radical polymerization
  nitroxide-mediated 146–8
  redox 260
  thiol-lactam-initiated 158, 160
  see also atom transfer radical polymerization (ATRP)
radiofrequency (RF) actuations 28, 29
Raman spectra 189, *190, 257, 261*
Raman spectroscopy 254–9
  CNT–polymer composites 260–6
  electronic structure of CNTs 255–7
rechargeable lithium-ion batteries 12–14
recovery stress 23, *26*, 27, 30
redox radical polymerization 260
resistive heating 28–31
resonance profile 258
reversible-addition fragmentation chain-transfer (RAFT) polymerization 142–6
rhodamine 6B 168, *169, 171*
ring-opening metathesis polymerization (ROMP) 155–7
ring-opening polymerization (ROP) 148–55
  binary-grafting *164*
  dendritic polymers 166–9, *170, 171*
rod/wire coating 200
rotation (mixing) speed 227, *228*
rule of mixtures 104, 108, 109
ruthenium alkylidene 155, *156*
ruthenium oxides 10

S_{11} optical transition 42–3
scanning electron microscopy (SEM)
  ATRP approach *129, 130, 140*
  CNT–polymers *261*
  dendritic polymer-CNTs *168, 171*
  electrically conductive nanocomposites *194*
  FESEM 236, 238, 246

PmPV *60*
quartz *68*
ROP *152*
supercapacitors 11
sedimentation 217
selected area electron diffraction (SAED) *132*
self-condensing vinyl (co)polymerization (SCVP) 165–6
semiconducting SWCNTs 23, *24*, 53–6, 61, 183
  electronic DOS *256*
  separation 184, 186, 188–90
semiconductors 4, 5
sensors 183
shape-changing CNT–polymers 31–42
  electroactive 36
  light-driven 32–6, *37, 38*
shape-memory CNT–polymers (SMPs) 23–31
  applications and limitations 23
  inductive heating 28, *29*
  infrared heating 25–8
  multifunctional properties 24
  resistive heating 28–31
shape-memory cycle *25*
shape recovery 23, *25*, 27, 28, *29*, *30*
shear strength 94
sheet resistance 192, 193, 195, 197, 200–2
silica colloids 194–5
silver nanoparticles 242, *243*
  ATRP approach *132*
  dendritic PAMAM-grafted 172
single-stranded DNAs (ssDNAs) 185
single-walled CNT-FETs (SWCNT-FETs) *54, 55*, 59, *60*, 61–6
single-walled CNTs (SWCNTs) 23, 253
  applications 183, 191
  band structure and chirality dependence 52–3
  band-to-band transition 53–6
  composites 9

electrically conductive 183, 191–5
energy transfer from photosensitive polymers 58
epoxidation *84,* 88–9
formation 182–3
functionalization mechanisms *8*
hole and electron discrimination 64–8
IR sensors 42–4
optical properties 254
photoelectrical responses 51–2, 58–69
photovoltaics 15, 16–17
Raman spectra 189, *190, 257, 261*
shape-changing composites *37*
UV-Vis-near-IR spectra 23, *24*
wrapping with polymers 56–8
see also metallic SWCNTs; semiconducting SWCNTs
SMPs *see* shape-memory CNT–polymers (SMPs)
sodium cholate 185
sodium dodecyl sulfate (SDS) 91, 185, 193, 197–8
sodium dodecyl sulfonate (SDDBS) 214
sodium dodecylbenzene sulfonate (SDBS) *57,* 91, 200, 260
solubilization, SWCNTs 56–7
solution processing 86, 98
and mechanical properties 100–1
solvent dispersion 241–2, *243*
block copolymers 239–40
thermoplastic polymers 238–9
see also dispersability
source–drain current (I_{SD}) *60, 61,* 66
sp^2 hybridized bonds 7
specific capacitance 10, 11, 12
specific mechanical energy (SME) 213, 224–30
spin-spray layer-by-layer (SSLbL) assembly 192
spray coating 197–8, 202, 203
styrenes *124, 125, 126,* 127, *148*
in situ polymerization 137–40

see also polystyrene (PS)
supercapacitors 9–12, 184
supercooling temperature 238, *239*
surface-initiated anionic polymerization 158, *159*
surface resistivity *196,* 198, *199,* 200
surfactants 73–4, 91, 185–6, 240
see also headings beginning with sodium dodecyl
swelling under ultrasonication 93

tapping mode atomic force microscopy (TM-AFM) *240*
tensile strain 99
tensile strength 84, 98, 235
tensile stress 99
terpyridine–Ru(II)–terpyridine [tpy–Ru(II)–tpy] 261–2
2,2,6,6-tetramethylpiperidinyl-1-oxyl (TEMPO)-CNTs 146, *147*
thermal conductivity 23, 25
thermo-responsive LCEs *34*
thermo-responsive SMPs *25,* 27
thermogravimetric analysis (TGA) 127, *137,* 154, *170,* 189
thermoplastic polymers 86, 238–9
thermoplastic polyurethane (TPU) 25, *26,* 27–30
thermosets 89–90, 103
thermotropic liquid crystalline polyester (TLCP) 263
thiocarbonylthio CNTs 143, *144*
thiol-lactam-initiated radical polymerization (TLIRP) 158, 160
thiophene groups 265–6
polythiophene *3, 74,* 264
toughness 84, 98, 106
transistors 183
transmission electron microscopy (TEM)
anionic polymerization *159*
ATRP approach 130, 131, 132, 136, 138, 139, 140
binary-grafting *162, 164, 165*
CNT length *221, 222, 223*

dendritic polymer-CNTs *168, 171*
HRTEM 127, *141, 153*
matrix viscosity 228
nitrene *123*
PMMA *97*
transmission light microscopy 216, *220*
transmission values *217,* 218
transparent conductive coatings and films 197–203
triethylene tetramine (TETA) 89–90, 98
Triton X-100 200

ultrasonic treatment time *217,* 218, *219*
ultrasonication 86, 93, *221,* 222

vacuum filtration 198–9

van Hove singularity 52, *53,* 254, 255, 261
vaterite 133–4
vinyl polymerization 165–6
viscosity, matrix 228, *229*

wet-processing methods 198–201

X-ray photoelectron spectroscopy (XPS) 135

Young's modulus 72, 84, 93–8
  and melt processing 102
  and *in situ* polymerization 105, 235
  and solution processing 100
  and thermosets 103

z-average particle size $d_{h,z\ ave}$ 218, *219*
Ziegler–Natta polymerization 80